DYING PLANET

DYING

DUKE UNIVERSITY PRESS

DURHAM & LONDON

2005

DYING PLANET

Mars in Science and the Imagination

ROBERT MARKLEY

Printed in the United States of America on acid-free paper ∞
Designed by Rebecca Giménez
Typeset in Minion by Keystone Typesetting, Inc.
Library of Congress Cataloging-in-Publication Data appear
on the last printed page of this book.

FOR THE MARTIANS:

Michelle, Helen, Harrison, Jeanne, Dan, & Jeannette

"THERE IS NO IF . . ."

CONTENTS

ACKNOWLEDGMENTS

Writing any book is a long and involved process, and writing a book that encompasses more than one discipline extends and complicates that process. Simple thanks are often insufficient. That said, I owe sincere debts of gratitude to a number of friends and colleagues.

My initial interest in Mars was triggered by conversations with Molly Rothenberg, and the earliest versions of some of the key ideas in this book were developed in dialogue with her. A multimedia version of aspects of my argument in this book can be found on the DVD-ROM *Red Planet: Scientific and Cultural Encounters with Mars,* published in 2001 by the University of Pennsylvania Press. My coauthors in that four-year collaboration really do deserve more thanks than I can give them here: Harrison Higgs, Michelle Kendrick, Helen Burgess, Jeanne Hamming, Dan Tripp, and Jeannette Okinczyc. *Red Planet* includes excerpts of video interviews with planetary scientists, cultural critics, and science-fiction authors. For their insights and their patient responses to various questions about Mars, I am deeply indebted to Richard Zare, Jeff Moore, Kim Stanley Robinson, Chris McKay, Molly Rothenberg (again), Katherine Hayles, Philip James, Robert Zubrin, Carol Stoker, Frederick Turner, Henry Giclas, and Martyn Fogg. I owe thanks as well to other scientists with whom I have discussed Mars over the years, particularly John Barrow, Matt Golombek, Robert Craddock, Michael Meyer, Kevin Zahnle, Marc Buie, and especially Earl Scime. All are absolved from any responsibility for the interpretations advanced in this study.

At the Lowell Observatory, I benefited from the expertise, helpfulness, and genial good humor of Antoinette Beiser and Marty Hecht. Much of the research for this book was carried out while I held the Jackson Chair in British Literature at West Virginia University, and I would like to thank particularly Duane Nellis, then dean of the Eberly College of Arts and Science, for his support. Thanks are owed as well to others in Morgantown: Rudy Almasy, Pat Conner, David Stewart, Michele Marshall, Barbara Patterson, Bonnie Anderson, Rachel Ramsey, Elizabeth Johnston, Cynthia Klekar, Matt Packer, Neal Bukeavich, and Catherine Gouge. Many other colleagues and friends have supported, encouraged, and corrected me in various ways while I was writing this book. I would like to thank Laurie Finke, Stephen Mainville, Laurie Stabenow, Ed Prichard, Jeanne Gordon, John Orr, Lisa Orr, Elizabeth Soliday, Frances Loughrey, Paul Remley, Lisa Zunshine, Ronald Schleifer, Rajani Sudan, Tom DiPiero, Nick Smith, Bruce Clarke, Hugh Crawford, Richard Grusin, Kenneth Knoespel, Eric Halpern, Arkady Plotnitsky, Anne Balsamo, and David Gross. And as always, my children, Stephen and Hannah, who have now launched their own writing careers, have kept me reasonably sane and always entertained.

An earlier version of chapter 9 appeared in *Modern Fiction Studies*, and I am grateful for permission to reprint parts of that article here. The images from the Lowell Observatory archives and from Earl C. Slipher's *The Photographic Story of Mars* are reprinted with the kind permission of the observatory.

INTRODUCTION

Mars is essentially in the same orbit [as Earth] . . .
Mars is somewhat the same distance from the Sun, which
is very important. We have seen pictures where there are
canals, we believe, and water. If there is water, that means
there is oxygen. If there is oxygen, that means we can
breathe.—VICE PRESIDENT DAN QUAYLE,
August 11, 1989 (quoted in David
Grinspoon, *Venus Revealed*)

WHY MARS?

For well over a century, Mars has been at the center of scientific and philosophical debates about humankind's place in the cosmos. Since Giovanni Schiaparelli announced in 1878 that he had observed *canali* (channels or canals) crisscrossing its surface, the planet has been the subject of thousands of scientific articles, dozens of full-length studies, and the setting for hundreds of science-fiction novels, stories, and movies. A century before Schiaparelli, astronomers had begun to explain the seasonal changes they observed on Mars in terms of what they knew about the climate and biology of Earth. Although specific analogies between the two planets have changed dramatically since then, scientists still frequently resort to terrestrial analogies to describe Mars. Over time, such analogies have reflected changing conceptions of both worlds: while Mars has been perceived through the lenses of terrestrial sciences, the study of the red planet has shaped, and continues to shape, humankind's understanding of Earth. Since the height of the canal controversy a century ago, lessons extrapolated from, or imposed on, Mars as a "dying planet" have been invoked to support competing, even antithetical, views of the fate of our world and its inhabitants: a glorious future of techno-scientific progress or an irrevocable fall into environmental devastation, social chaos, and eventual extinction. Simply reading the titles of some of the scientific studies and science-fiction novels published during the last century—*Mars as the Abode of Life, Mars Beckons, Destination Mars, Mapping Mars, Outpost Mars, Red Mars, Green Mars,* and *Blue Mars*—seems to invite us to explore the persistence of the red planet in our cultural imagination. Even as orbiters and landers continue to explore its surface and atmosphere, Mars still dominates our daydreams of interplanetary colonization and haunts our nightmares about invaders from outer space.

Dying Planet explores the ways in which Mars has served as a screen on which we have projected our hopes for the future and our fears of ecological devastation on Earth. It draws on work in planetary astronomy, the history and cultural study of science, science fiction, literary and cultural criticism, ecology, and astrobiology to offer a cross-disciplinary investigation of changing perceptions of Mars as both a scientific object and a cultural artifact.[1] Even in 2004, as scientists pour over gigabytes of data from ongoing NASA and European Space Agency (ESA) missions, the red planet seems as dynamic, complex, and intriguing a world as it did a century ago when the *New York Times* and other newspapers routinely ran stories about its canal-building inhabitants. In tracing the history and analyzing the significance of our cultural fascination with Mars, I have three primary goals: to describe and contextualize important scientific debates about the planet; to explore the reasons for its tenacious hold on the scientific and popular imagination; and to analyze the dynamic interactions among planetary science, science fiction, and other disciplines, notably ecology, that have kept Mars on the front pages since the 1800s. The paths to these goals frequently intersect, and one of my purposes is to suggest that in order to make sense of current scientific thinking about Mars, we need to understand a complex sociocultural history that has cast the planet as a harbinger of the ecological fate of the Earth.

My title, "dying planet," was used for half a century to characterize Mars and to sum up its relationship to Earth. This label dates from the work of nineteenth-century astronomers, though it was popularized between 1912 and 1948 by Edgar Rice Burroughs in the eleven novels that take place on the planet he calls Barsoom. Yet long before Burroughs began writing, Mars occupied a unique place in the Western imagination, and many studies of the planet begin with a simple but important question: Why Mars? Why has the planet haunted our collective imagination and why has it been, far and away, the favorite site for interplanetary science fiction since 1880? (Markley et al. 2001; Godwin 2000: 210–11; Hartmann 2003, 4–6).[2] One answer is that Mars has proved to be the most interesting object in the night sky for earthbound observers since the invention of the telescope in the seventeenth century. The twenty-six-month cycle that brings the red planet within thirty-four million miles of Earth (as it did in summer 2003) has provided tantalizing glimpses of a world that bears some striking similarities to our own. Although Venus is closer to Earth and now seems, to some planetologists, more earth-like than Mars in its atmospheric and geological cycles, it is covered by dense, acidic clouds; its atmosphere only recently penetrated by space probes and radar to map its hellfire, 850 degree surface (Grinspoon 1997, 143–48). Mars, on the other hand, displays features that have been seen and sketched since the 1600s, notably the polar caps that wax and wane and the wave of darkening (the result of seasonal albedo changes) that spreads across the planet from the poles toward the equator during its spring and summer months.[3] Although the surface of Mars is difficult to see clearly through the dense and turbulent atmosphere of Earth, its dark patches and polar caps historically have suggested powerful, seemingly self-evident, analogies between the two planets. The very limitations of earth-based observation have provoked both painstaking study of the planet's surface and rampant speculation based on analogies to Earth's seasonal and hydrological cycles (Sheehan 1996; Hartmann 2003, 66).

During the last forty years, however, the data and photographs returned by spacecraft have revealed a world that poses complex questions about its history and composition. The oldest terrain on Mars preserves landforms three to four billion years old that provide a window into a geological past that has long since disappeared from Earth. The surface of Mars has been photographed with increasing precision and sophistication since Mariner 4 returned nineteen wide-angle black-and-white images in 1965, and each successive array has provoked a rethinking of fundamental assumptions about the planet as well as aesthetic and even (in the broadest sense) religious responses that border on what we might call the interplanetary sublime. Mars has the largest volcano in the solar system, Olympus Mons, which covers a territory the size of Arizona and rises thirteen miles above the datum (the Martian equivalent of sea level), and the Valles Marineris, a fault system that stretches for three thousand miles, dwarfing terrestrial analogues and making the Grand Canyon seem "a mere crack in the sidewalk" (Hartmann 2003, 315). Yet the spectacular landforms that bear witness to the planet's geophysical history are only part of a dynamic mosaic that scientists are trying to piece together. Orbital photographs show evidence of geologically recent lava flows and patterns of water erosion, indicative of processes that still may be occurring. The data returned since 1997 by the Mars Global Surveyor, Mars Odyssey, ESA orbiter, and the Mars Exploration Rovers have left scientists trying to reconcile seemingly incommensurate views of the planet and its history: a frozen snowball that has experienced only brief periods of warming versus a once warmer and wetter Mars that is still changing—resurfaced by volcanism, meteorite impacts, water and ice erosion, wind, and the effects of periodic climactic instabilities. Current scientific debates about Mars hark back to controversies that have existed, in a variety of forms, since the eighteenth century, and the historical, epistemological, and cultural implications of these debates are, in an important sense, the subject of my study.

With the advent of increasingly sophisticated technologies to collect and analyze data, analogies between Mars and Earth have become far more complex than they were even a quarter century ago. William K. Hartmann's excellent overview of the current (as of 2003) scientific understanding of Mars includes a dozen photographs he has taken of Mars-like analogues on Earth, from desert terrain in northern Mexico to gullies on volcanic hillsides in Iceland. As a planetary scientist, Hartmann describes the surface of Mars in terms of geological time; "recent" gullies, for example, may have been formed within the last ten million years. As a writer of science fiction, however, Hartmann deals with an imagined near-future, which brings Mars within the semiotics of historical and experiential time. This productive tension between geological and human time scales informs his photographs of Martian analogues on Earth. When he asks his readers to imagine what "a future party of astronauts" would see from the "summit of Olympus Mons," he is drawing on a complex fictional tradition in which the Martian landscape triggers a rethinking of the relationships between humankind and its environment(s) (Hartmann 2003, 313). Hartmann's invitation to his readers to imagine looking across the calderas of the solar system's largest volcano underscores the ways in which Mars exists as a complex multidisciplinary object.

To describe Mars as a "multidisciplinary object" is both to call attention to the historical, epistemological, and cultural relationships that I invoked above and to insist on the importance of the scientific debates that the planet has generated. Situating these debates within their historical and theoretical contexts is essential to understanding why any answer to the question "why Mars?" invariably crosses disciplinary boundaries. The study of the planet typically involves teams of scientists and engineers, who are needed to instrument experiments, collect data, interpret that data, and then contextualize the results and suggest their significance. Although almost all large-scale research projects in the sciences now involve multiple investigators, the interdisciplinary nature of planetary research involves complex interactions among specialists in various disciplines, including geology, chemistry, hydrology, meteorology, and microbiology. While researchers can obtain some provisional answers to specific questions about, say, the mineral composition of specific areas of the Martian surface, these answers invariably provoke new questions that, in turn, impinge on other disciplines, epistemologies, and discourses. As I suggest throughout this study, scientific debates about Mars since the canal controversy usually involve clashes between different disciplinary ideas about research protocols, the interpretation of evidence, and the nature of scientific reasoning: what is one researcher's legitimate inference about Mars is another's example of imprudent guesswork.

Precisely because it has generated such controversies, Mars has been at the center of the emerging, multidisciplinary science of comparative planetology—and of the still virtual sciences of exobiology and terraforming. These virtual sciences pose the big-ticket questions that both develop from and intrude on scientific discussions about the planet: Is, or was, there life on Mars? Can humankind colonize the planet and transform its forbidding landscape into a habitable biosphere? If so, when? In turn, these speculative questions have led scientists to reconsider theories about planetary formation and the origins and development of life. The very structure of analogy itself encourages the question posed by Hartmann's imagined astronauts: Where do we fit in?

This question has cultural, philosophical, and scientific dimensions and, by its very nature, encourages frequent crossings of the boundaries between science and science fiction. Hartmann's fictional climbers on Olympus Mons represent one strategy of trying to bring the billion-year time scales of Martian geological history within the imaginative frame of embodied experience, but they also testify to the desire to comprehend the vastness of planetary history. Since Schiaparelli, astronomers have attempted to extrapolate from what they know about Earth's climate and geological history to what they can surmise about conditions on Mars. Working within a set of scientific assumptions that still retain some currency, scientists in the 1880s and 1890s reasoned that since Mars was a smaller planet than Earth its molten core and then its surface would have cooled more quickly, and oceans would have formed soon afterward. Mesmerized by Darwinian theory, Percival Lowell and many of his contemporaries argued that life would have developed on Mars sooner than it did on Earth and also would have evolved more rapidly. But the geothermal engine on this smaller world would have run down more quickly, and, as the red planet cooled, it would have lost its oceans and much of its atmosphere. Evolutionary logic then suggested that any beings who could survive on this dying planet might far surpass humans in intelligence and be able to marshal their dwindling resources to stave off extinction. Even Lowell's critics accepted the broad outlines of this evolutionary theory and appropriated the metaphor of a dying world to describe their views of a Mars without canals. This quest to understand the vast time scales of planetary development has been at the forefront of comparative planetology since Lowell's time, and his canal theory owed much of its popularity to his skill in relating an impersonal geological history to the compelling picture he evoked of the struggle for existence on a desert world. In this respect, the canals allowed Lowell's readers to meditate on relationships between the human experience of scarcity and the larger cosmic forces that, in the late nineteenth century, seemed to indicate that Earth was "going the way of Mars" (Lowell 1909, 122). To imagine gazing across the Martian landscape in 1905 or 2005 is to confront the possibility that one is looking, in some still-to-be-defined way, at the future of the human species.

As the canal controversy demonstrates, the fascination with extraterrestrial life and, more broadly, extraterrestrial ecologies lies at the heart of humankind's obsession with Mars. The exploration of Mars as the site of past or present life, says Chris McKay of NASA-Ames, is a "fundamental science driver" for future exploration (Markley et al. 2001, "interviews" 5). Historically, questions about life on Mars have been bound up with changing scientific ideas about the limit conditions for terrestrial biology. In different ways, the science teams studying the possible nanofossils in Mars meteorites are asking more precise and more complex versions of questions debated by Lowell and his critics: Is the environment on Mars too harsh to support indigenous forms of life? If it exists or has existed in the past, could Martian biology be fundamentally different from life on Earth? To pose, try to answer, or reframe these questions is to realize that exobiology—the study of life beyond the Earth's biosphere—remains both a virtual and syncretic science. Like comparative planetology, the search for life on Mars is a collaborative effort based on intersecting specialties and subspecialties that together can offer only heuristic answers. The search for Martians since the 1890s has proved immensely controversial because the conclusions drawn, like the questions posed, depend on often competing assumptions about exobiology and the technologies and protocols that can be used to establish or rule out its existence. Nonetheless, the assertion that the discovery of life on Mars or elsewhere in the universe will alter profoundly humankind's perception of its place in the cosmos is an extraordinarily common one. Without disputing its significance, I suggest that this fascination with exoecologies rests on a host of extrascientific values and assumptions—at once theological, philosophical, social, and cultural. To explore these values and assumptions is to offer an ongoing narrative in response to the question, "Why Mars?" and to rethink the meaning of some key terms that invariably recur in discussions of the red planet: analogy, methodology, and history.

ANALOGIES AND METHODOLOGIES

Areology, the study of Mars, is defined by intersecting research programs: the coordination (and conflicts) among various disciplines (climatology, hydrology, petrology, extremophile biology, and chemistry); different experimental styles and practices; and different rhetorical and conceptual traditions. Most planetologists, whatever their philosophical leanings, emphasize the limitations of current knowledge about the planet. As Kevin Zahnle observes: "In some ways the debate [about Mars and the possibility of life] has really moved little since the days of Flammarion and Lowell. The most interesting information remains right at the limits of resolution, be it metres in satellite images of gullies, or nanometres in microscopic images of magnetite crystals. Always life on Mars seems just beyond the fields that we know" (2001, 213). The data on Mars, as Zahnle suggests, remains a partial mosaic that leads scientists to rely on what Victor Baker calls "analogies of form and context" between Earth and Mars to explain the latter's surface features. Rather than leading to hard and fast conclusions, however, such analogies, Baker argues, "suggest fruitful working hypotheses, thereby leading to completely new theories that bind together any newly discovered facts" (2001, 228). Even if such hypotheses may strike many in the scientific community as "outrageous," he continues, "it is the productive pursuit of such hypotheses that leads ultimately to new understanding, not only of Mars, but also of Earth" (228). Baker's "productive pursuit" is an ongoing process; new theories and newly established facts are dynamically as well as dialectically related.

His emphasis on the value of Mars for understanding Earth underscores the close connections that have evolved between ecology and areology. The study of both planets has moved from being governed by deterministic metaphors to a recognition that "multiple causal pathways" (Lewontin 1991, 147) have shaped their surfaces, atmospheres, and potential as life-bearing worlds (Bell 2003, 34–41; Zuber 2003, 42–48).

Scientific analogies between conditions on Earth and Mars, as Hartmann and Baker suggest, cut in two directions, and one of the reasons for this divide has to do with the nature of analogical thinking itself. Analogies have significant consequences in science, social science, and literary studies because they disrupt analytical methods based on the application of accepted principles to specific examples; analogical thinking, Ronald Schleifer argues (2000, 13-15), involves a reciprocal process of negotiation between two terms or objects. Because it works by induction, a point-by-point comparison of observed characteristics or phenomena, analogical thinking has the potential to call into question the very principles that allow such comparisons to be made.

Schleifer's discussion of analogy is helpful in understanding the dynamic interchanges that structure historically the relationship between Earth and Mars. Not surprisingly, even as Mars has been perceived and described scientifically as an earthlike planet, Earth has become, in effect, a Mars-like planet. One of the consequences of this analogical relationship has been an increasing wariness about assuming that the Earth represents general principles of planetary evolution and the ideal "solution" to the problems of sustaining a viable biosphere. The "danger of having so much planetary science grow out of the field of earth science," David Grinspoon argues, "is that we may be overly predisposed toward 'geocentric' interpretations: we sometimes assume Earth to be the standard against which other planets are measured, rather than simply one among many possible outcomes of planetary evolution." For this reason, over the course of the last century "the growth of planetary science has also produced an enlarged, less provincial perspective on Earth" (1997, 57). Bedrock conceptions of what a planet is have been called into question, and scientists have been forced to justify their assumptions and reasoning: comparative planetology frames questions that cannot be asked from a parochial perspective. This "less provincial perspective" about the geochemical history of the planets also suggests a questioning of fundamental assumptions about ecology; the range of conditions—available nutrients, energy sources, chemical reactions, water, and temperatures—under which life can exist.

In the multidisciplinary context of comparative planetology, Mars has become a bellwether in both scientific and broader cultural debates about the nature, legitimation strategies, and aims of science. Because planetary science is inductive and heuristic, it resists the all or nothing assertions, the "epistemological absolutism" (Shapin 1999, 13) that too often dominates, and impoverishes, debates in the philosophy of science (Smith 1997, 243–66; Markley 1999a, 47–70; Plotnitsky 2002). In describing the "cultural clashes within the hybrid discipline of planetary science," Grinspoon argues that "equations, and the conclusions they lead us to, are only as good as the assumptions that go into them" (1997, 173). Rather than deductive models based on mathematical predictions, planetary scientists characteristically offer inductive and heuristic accounts of probable interpretations of the available evidence, even as "cultural clashes" among astronomers, chemists, geologists, and microbiologists reflect often very different ideas of the reliability and significance of particular data and of the values and assumptions that underlie different disciplinary models of knowledge.

In an important challenge to the regime of "epistemological absolutism," biologist Richard Lewontin offers a historicist account of the methodologies of scientific inquiry: "If one examines science as it is actually carried out," he asserts, "it becomes immediately clear that the assertion that it consists of universal claims as opposed to merely historical statements is rubbish" (1991, 142). Lewontin maintains that even agreed-on, widely practiced methodologies are culturally and historically situated. In his view, scientific truth-claims have complex internal structures and complicated networks of external affiliations that cannot be explained solely by reference to internal standards of legitimation. Significantly, Lewontin's description of the contingency of representations of the physical universe suggests why he shares a heuristic vocabulary with practitioners of the cultural study of science, including Donna Haraway, Michel Serres, Bruno Latour, Steven Shapin, Katherine Hayles, and many others.[4] In different ways, these cross-disciplinary researchers are concerned with investigating what Andrew Pickering (1995) calls "the mangle of practice" and the contingent languages of scientific debate rather than with seconding or attacking specific pronouncements about the nature of "reality."

In discussing the ongoing controversy over the possible nanofossils in Martian meteorite ALH84001, Hartmann contrasts the operations of science to the discourses of persuasion, marketing, and competition that dominate popular media. Science, he argues, "works . . . by appeal to evidence. In ideal science, the glory goes . . . to the person who brings the best data to the table. All the data are spread out, and the best estimate of truth emerges from it, not from the rhetoric of the person who makes the best case" (2003, 267). The crucial terms in Hartmann's description—"ideal science" and "best estimate of truth"—do not suggest a hard-and-fast opposition between "evidence" and "rhetoric" but rather processes of negotiation, debate, and deliberation by which an "ideal" consensus can be reached. In practice, the "best estimate of truth" becomes generally accepted only by sorting through and debating exactly what constitutes "the best data," and new data invariably provoke more discussions and eventually a new, always provisional, consensus. As the current controversy about Martian nanofossils suggests, stalemates not only occur with regularity, they often result in bitter disagreements, including confrontations that Hartmann describes "where nominally fair-minded researchers . . . traded accusations of inept science [and] shook with red-faced rage" (266). The cultural study of science is interested in such debates not to glorify the "winners" and ridicule or explain away the "losers" but to investigate the ways in which consensus views have been debated, established, modified, and overturned. To a great extent, scientists and laypersons come to trust certain analyses precisely because both methodologies and conclusions have been subject to such scrutiny and revision.

The two parties of red-faced scientists that Hartmann describes cannot agree about the likelihood of nanofossils in meteorites because (as I suggest in chapter 8) they are relying on different methodological assumptions and different conceptions of probabilistic "proof." Such debates indicate why Lewontin insists that "facts in science do not present themselves in a preexistent shape. Rather it is experimental or observational protocol that constructs facts out of an undifferentiated nature" (1991, 147). "Constructs" in this sentence does not mean "make up out of whole cloth"; instead, the verb calls attention both to the practice of science (the designing, building, testing, and calibrating of instruments; experimental and observational protocols; running and rerunning trials; writing and rewriting papers) and to the complicated negotiations that go into producing a consensus about the "best data" and its significance. This process of constructing facts, Lewontin maintains, must be historicized; recognizing the contingency of these ongoing processes means that in physical sciences such as "biology there may be *general* statements, but there are no universals, and . . . actual events are the nexus of multiple causal pathways and chance perturbations" (147). By complicating notions of causality, Lewontin renders the practice of science both inductive and historicist because "experimental [and] observational protocols" must themselves be flexible and dynamic in order to describe accurately a dynamic physical reality.

Lewontin's comments indicate something of the resistance among scientists to universalizing assumptions and deterministic metaphors. As George Lakoff and Mark Johnson suggest in their study of metaphor, "truth is relative to our conceptual system, which is grounded in and constantly tested by, our experience with our cultural and physical environments" (1980, 193); consequently, scientific language is always troubled and enriched by the interference patterns of differing discourses (Bono 1990, 59–90; see also Woolgar 1988; Rotman 1993). This attention to the relationship between physical reality and systems of representation allows both scientists and cultural critics to offer complex and self-critical descriptions of the systems of signification and knowledge making that they, and others, employ. Rather than opposing objective truth to subjective belief, Lewontin insists that scientific systems of representation are historically and culturally mediated, and he advocates extending a methodological rigor and skepticism to analyzing their specialized languages and modes of representation rather than simply assuming that they are value-neutral reflections of a transhistorical reality. Neither scientists such as Lewontin nor cultural critics of science such as Shapin make the obviously absurd claim that the world is a hodgepodge of subjective impositions. Instead they maintain that because all systems of representation are contingent there can be no unmediated access to "reality."

Rather than providing easy paths to consensus, scientific debates grow heated and end in disputes about data, methodology, and even competence because different groups of scientists have different institutional and disciplinary investments, a point that I will develop at length in this study. Admittedly, it would be easier to believe, in the words of physicist Steven Weinberg, that truth and representation are reflexive, that "the correct [scientific] answer when we find it is what it is because that is the way the world is" (1996, 14). But this view offers no means to assess the plausibility of claims when scientists, red-faced with rage or otherwise, arrive at different interpretations of phenomena. The differences between Weinberg's language—based on four iterations of an intransitive verb—and Hartmann's language are striking, even though both men are describing ostensibly the same process: how scientists arrive at the "truth." Hartmann describes a historical situation in which the quality of the "best data" depends on all sorts of outside factors, ranging from the reliability of subcontractors, to political debates over scientific funding, to the assumptions behind the instruments being flown into space. Weinberg, in contrast, deals in epistemological absolutes. The limitations of such a reflection theory of scientific representation become evident when its proponents try to account for conflicts within science—such as those endemic to debates about Mars in the nineteenth, twentieth, and twenty-first centuries.

The study of the red planet since 1878 has produced a series of paradigm-shattering discoveries that time and again have rendered previously accepted wisdom about the planet spectacularly wrong. What makes this history so fascinating is that almost all efforts to form legitimate inferences about Mars from limited, disputed, and ambiguous data have been hotly contested. As generations of astronomers have studied the planet, new technologies have been developed, new data collected, and new interpretations offered that have allowed planetary scientists to arrive, at different points in time, at different versions of the "best estimate" of the truth about Mars.

Debates about life on Mars, from canal-building Martians to suspected nanofossils, reveal the ways in which scientific discussions are embedded in complex professional networks and in ideology—the assumptions, values, and beliefs that are taken for granted, half articulated, or defended vigorously as the self-evident parameters of good judgment and common sense. Ideology is not primarily a consciously held set of beliefs or a "false consciousness" that is diametrically opposed to the objective realities of science; instead it describes the culturally and technologically specific ways in which knowledge is put together and disseminated. Ideology is the cultural form of complex dynamics; and its negotiations of "knowledge" and "meaning" both redefine and constrain our horizons of expectation and intelligibility. In this regard, the cultural study of science does not maintain that "society," "ideology," or "politics" determine scientific practices or knowledge but demonstrates why the discourses and practices of science cannot be hermetically sealed from their sociocultural environments. Put simply, there is no way to separate "society" or "culture" from "science" and "technology," no standpoint from outside the cultural matrix from which one could make such distinctions. Therefore, while the practice of a particular science never can be reduced to a "nonscientific" belief system, it also can never exist independently of the systems of representation, belief, legitimation, and socioeconomic rewards that make possible its complex practices of knowledge making. This is the distinction that Shapin makes between "the [social and professional] bases of scientific authority and epistemological absolutism" (1999, 13). No one would argue against the proposition that the scientific study of Mars is cumulative and progressive: technological developments in telescopes and then in spacecraft design, photography, spectroscopy, and computer hardware and software have allowed scientists increasingly to collect more, and more accurate, data and to pose far more precise questions than their predecessors did in 1903, 1933, or 1963. At the same time, however, this history is embedded in larger scientific and nonscientific discourses and practices—in the complex processes of contextualization that I describe below.

MARS, EVOLUTION, AND THE ECOLOGICAL IMAGINATION

The analogies between Mars and Earth proposed in the nineteenth century depended in large measure on visual inference: the waxing and waning of the polar caps and the wave of darkening that swept across the planet suggested to almost all astronomers an earthlike ecology, even those who believed that Mars harbored only low forms of vegetation and animal life. For some scientists, as well as for many who followed the controversies in the popular press, the effects of environmental change on Mars's presumed inhabitants became a topic of intense speculation. The canal theory owed much of its appeal to the analogies it suggested to ecological concerns on Earth. Severe droughts in the late nineteenth century ravaged large areas in India, Africa, China, and Brazil, and the environmental writing of the period often dealt as much with problems of resource management, labor, agriculture, and pollution as it did with the beauties of a pristine nature (Davis 2001; O'Connor 1998; Bramwell 1989; Worster 1993, 1994). Mars became a prominent object of speculation because its presumed inhabitants, digging canals thousands of miles long to connect various "oases," seemingly provided an object lesson in harnessing political will and technological expertise in the service of a larger social good—staving off environmental collapse by husbanding dwindling supplies of water.

To conceive of ecological change on Earth occurring on a planetary scale, nineteenth- and twentieth-century observers resorted to imagining intelligent creatures on Mars looking at our planet. Visualizing how the Earth might appear from space involved a number of uncertain assumptions and best-guess analogies, many of them back formations based on what astronomers could observe on Mars. With its seasonal changes, the fourth planet provided a refracted image of what the Earth might look like from space, and both popular and scientific discussions of Mars between 1880 and 1964 therefore reveal some of the problems that attended efforts to understand our planet as an integrated set of biological, hydrological, and climatological systems. The Earth, seen from the eyes of putative Martians, became a site to speculate about the consequences of humankind's exploitation of its resources and the fragility of its biosphere. The understanding of the complex feedback loops between organism and environment that developed during the twentieth century stems, in part, from the ways in which Mars encouraged scientists to think about the planetwide conditions necessary to sustain life.

It is significant that Mars became an object lesson in planetary ecology at the same time that the discourses of physics and neoclassical economics seized on the concept of the conservation of energy as a way to understand the seemingly ironclad "laws" of the natural and socioeconomic world. Both domains employed what Philip Mirowski terms "a reciprocal metaphorical legitimation" in order to justify research programs that ultimately depended on "an ideal of unification" (1989, 108) of disparate phenomena. This ideal, in turn, was predicated on "a purely abstract, conventional standard"(115)—a belief in the epistemological absolutism of its representational schemes that defined the physical (and economic) universe by time-independent laws. As Mirowski demonstrates, the concept of the conservation of energy, based on a mathematical formalism, did not describe the natural world so much as it asserted "the very ideal of natural law: the mathematical expression of invariance through time, the verification of a stable external world independent of our activity or inquiry" (75). Mirowski's analysis is useful in explaining why nineteenth-century discussions of ecology paradoxically commit themselves to two different, even incommensurate, views of the relationship between the natural world and humans. The first is an economics of indefinite growth that identifies progress with an ad hoc principle of invariance; this view assumes that technological innovation and industrial expansion continually can develop new resources to replace those being depleted. The second view envisions the world as a set of closed systems, an ecology defined by limits and carrying capacities that can be temporarily extended only at the expense of sustainability (Ashworth 1995, 124–34). Paradoxically, the dominant metaphor for describing both worldviews was derived from the second law of thermodynamics; entropy could be appropriated to predict a downward spiral, over eons, of the physical universe and therefore the Earth itself: humankind could respond either by circumventing or transcending environmental limits or succumbing to resource exhaustion, scarcity, and social disintegration (Clarke 2001; Gold 2002, 449–64). Not surprisingly, Mars was invoked to buttress both of these views.

The idea that the universe was in the thrall of evolutionary laws that determined the life cycles of the planets was generally accepted throughout the nineteenth century and into the early twentieth (Brush 1987, 245–78). For some scientists, notably Lowell, the prospect of Mars as a dying world challenged theological notions of "man's" place in the cosmos by mocking humankind's pretensions to transcend ostensibly universal laws of planetary evolution and their inevitable consequences—the entropic heat death of all planets that would lead to drought, scarcity, privation, and ultimately extinction. In this respect, a roughrider optimism about economic growth as a means to overcome ecological limits constituted only one half of a dialectic; the alternative held that not only individual species but entire worlds were destined to suffer the eventual loss of water, heat, and atmosphere. Therefore the notion that intelligent beings on Mars were grimly crisscrossing their world with canals seemed, to some, both a harbinger of humankind's fate and a universal Darwinian response of intelligence to the problems posed by planetary desiccation.

In some striking ways, bringing Mars within the orbit of "environmentalist" principles anticipated one of the fundamental principles of current ecological thinking: the idea that landscapes manifest "ongoing dialectical relations between human acts and acts of nature" (Crumley 1994, 9). Rather than reinforcing romantic notions of living in "harmony" with nature, Mars figured into scientific discourses that conceived of these "dialectical relations" in terms of the costs and consequences of *intensification.* Intensification is defined by the cultural anthropologist Marvin Harris as "the investment of more soil, water, minerals, or energy per unit of time or area." While such escalating investments are humankind's "recurrent response to threats against living standards," intensification invariably proves counterproductive over time because "the increased effort sooner or later must be applied to more remote, less reliable, and less bountiful animals, plants, soils, minerals, and sources of energy" (1976, 5). As resources become increasingly scarce, living standards decline, usually with disastrous consequences for the most vulnerable populations, until cultures "invent new and more efficient means of production which sooner or later again lead to the depletion of the natural environment" (5). Such a view of intensification underscores the complex interactions among climactic variations, population pressures, technological developments, rates of resource extraction, environmental degradation, the need to maintain living standards, food production, working conditions, political power, transportation requirements, economic structures, and social customs (Markley 1999b, 817–37).

During the years of the canal controversy, as I argue in chapter 2, the consequences of intensification were cast in the ontogenetic analogies that dominated nineteenth-century biology.[5] In turn, these analogies encouraged cosmological narratives, such as theories of planetary life cycles, based on the universalizing concepts that Mirowski describes. As ecology and physical sciences underwent profound changes throughout the twentieth century, so too did the inflections of the analogies between Mars and Earth. The Mars that has emerged since 1965 has been informed by conceptual and methodological changes within the disciplines of ecology and microbiology. Historical ecologists such as Elizabeth Graham thus describe the study of Earth in language that has affinities with both the discourses of planetology and the language and methodology of the cultural studies of science. These affinities might be described as a shared epistemological heuristic, one that challenges the supposedly fundamental distinctions that are said to exist between the sciences and the humanities (Haraway 1991; Bryld and Lykke 1999; Plotnitsky 2002; Brande 2006). Rather than emphasizing stability in ecosystems and holistic and deterministic notions of "system," many ecologists base their analyses on models of heterarchy, which according to Graham is "a system in which elements are unranked . . . or ranked in a variety of ways depending on conditions," or "scalar hierarchies" in which any level of organization can affect or temporarily control others (1998, 124). On Earth, at least, such heterarchies imply scientific methodologies that work "by embodying contradiction" and therefore "allow [scientists] to move more freely among the three realms [of nature, society, and meaning] because [historical ecology] does not require a macrotheory to integrate them" (125). Mars, as yet, lacks an indigenous social science, but the complex, nondeterministic methodology that Graham describes foregrounds an attention to complex interactions among science, meaning, and culture rather than an unwavering belief in determinant causes and predictable effects.

The analogies I have suggested among the methodological languages of ecology, areology, and the cultural study of science cannot disguise the fact that scientists and cultural critics typically hold very different notions of causation: the "best" explanations for scientists are the simplest; the "best" interpretations for cultural and literary critics are those that register the complexity that evades macrotheories. Yet simplicity itself, as I argue throughout this study, is always a contested term within science. In different contexts simplicity can be, and has been, invoked to defend mutually exclusive hypotheses about life on Mars. Scientists on both sides of the canal debate, the interpretation of the Viking life detection experiments, and the current controversy over nanofossils in ALH84001 invoke classical ideals of parsimony—Occam's razor: the idea that the simpler the explanation the better—as their trump card to justify their interpretations of available data. Yet such appeals to simplicity invoke aesthetic standards of epistemological elegance. In turn, these standards involve larger conceptions of order—aesthetic, philosophical, and theological as well as "purely" scientific. While communities of scientists and artists may share broadly acknowledged aesthetic standards, these standards are themselves historically contingent, embedded in complex cultural assumptions, practices, and values. Consequently, they remain open to competing interpretations (Gauch 1993, 468–78). Employing Occam's razor, in short, requires unpacking a whole shaving kit. Consequently, although one of the interpretations of the carbonate structures in the Martian meteorite, at some point in the future, will be proved "right," all of the interpretations of these structures—right and wrong—are shaped by cultural, historical, and disciplinary assumptions, by different ways of deploying the languages of epistemological simplicity.

SCIENCE AND SCIENCE FICTION: THE LURE OF THE RED PLANET

With the release of the presidential commission's June 2004 report recommending eventual human voyages to the Moon and Mars, debates about the future of planetary exploration are raging on the Internet, in the popular press, and among scientists. The commission's recommendations, like all such scenarios, depend on projecting into the future a sequence of events and likely outcomes that can be simulated but never predicted with any certainty. The further into the future that scientists project scenarios for harvesting solar energy and minerals from asteroids, the Moon, and Mars, the more imaginative such speculation becomes (Lewis 1996; Schmidt and Zubrin 1996; O'Neill 2000). At the boundaries between extrapolations from existing data and speculations that project contemporary cultural, political, and economic assumptions into the future lies the realm of science fiction. What Carl Sagan terms "the continuing dance between science and science fiction" (1994, 340) involves the kinds of complex interactions between physical reality and modes of imaginative expression that, in different ways, Lewontin and Graham describe. Science fiction, however, does not extrapolate from given facts so much as it simulates possible or alternative realities. For cultural critics, investigating the affiliations between science and science fiction does not mean simply passing aesthetic judgments on particular texts but investigating the reasons why some works, like Burroughs's novels, prove extraordinarily popular and influential, even when they are panned by literary critics.

To study the significance of Mars as a dying planet in popular culture is to explore the political implications of science fiction. Most practitioners and critics of the genre take pains to distinguish its aims and methods from those of canonical literature. As "an unconscious and figurative projection of some more 'realistic' account of our situation" (Jameson 1989, 283), science fiction does not reflect a familiar reality but suggests ways in which alternative realities or future histories might take shape given other initial or governing conditions. Rather than representing what is or has been, science fiction offers a simulation of what might be; it changes the postulates of realistic fiction and then runs simulations based on these premises. In contrast to postmodern conceptions of representation predicated on lack (the absence of the signified) or deferral (the gap between the signifier and the signified), simulation, Steve Shaviro argues, "precedes its object: it doesn't imitate or stand in for a given thing, but provides a program for generating it. The simulacrum is the birth of the thing, rather than its death" (1996, 17). In Shaviro's sense, science fiction works by convincing its readers of the internal consistency of its simulation so that its obvious differences from reality engender both "cognition," the recognition that an imagined world can be compared to the reader's experience of her or his reality, and "estrangement," the recognition that its differences from "reality" provide a means to analyze the ideological conditions of existence that the reader otherwise takes for granted (Suvin 1979, 7–9). That said, anyone who has read a lot science fiction realizes that much of it is pretty bad. Most science fiction novels do not explore the complex interactions between cognition and estrangement but reduce simulation to caricature, imagining that while technologies will change, human (or inhuman) nature never evolves beyond hoary stereotypes. Some simulations, in this regard, are more convincing and thought provoking than others precisely because they ask to be judged both by the standards of aesthetic consistency and by their ability to challenge, question, or defamiliarize the historical experience of their readers.

If estrangement marks the difference of these imagined worlds from historical reality, as Fredric Jameson and Darko Suvin argue, these differences allow readers to imagine that their socioeconomic and political reality could be otherwise: the "homemade qualities and amateurishness" of much science fiction, Jameson contends, are not marks of their aesthetic limitations but indicate "their political function . . . *to bring the reader up short against the atrophy of the utopian imagination and of the political vision in our own society*" (1987, 54). This political function of science fiction offers a means to "dramatize this contradiction" between history as an open-ended process and the formal constraints of narrative closure; therefore, Jameson argues, because "the vision of future history" can neither know nor predict the future, history itself becomes a "work in progress," defined by its potential rather than by its representation of current social, economic, and political conditions (Jameson 1982, 148; see also Philmus 1970; Penley 1997). Science fiction registers possibilities; it does not make predictions. If, as Suvin suggests, the genre exists "between the utopian and anti-utopian horizons" (1979, 62), it is because science fiction resists formal—and ideological—closure.

Suvin's and Jameson's definitions suggest why science fiction seems to oscillate between ideological critique and escapist fantasy. The relationship between "reality" and the logic of simulation is paradoxical because the two realms are both opposed and mutually constitutive. "Realistic" projections for, say, the future explorations of Mars—the idea, for example, that the planet can provide abundant resources to sustain and expand human civilization (Lewis 1996; Schmidt and Zubrin 1996)—depends explicitly and implicitly on the traditions of science fiction to provide a narrative mapping for an uncharted conceptual space; conversely, science-fiction authors frequently defend the genre's literary and social value by emphasizing the significance of its "realistic" extrapolations from current science and technology. The novelist and anthologist Judith Merril, several decades ago, claimed that " 'realistic fiction,' rather than speculative or science fiction, was the transient oddity—[a] grotesque . . . product of nineteenth-century super-rationalism and mechanistic philosophy" (1971, 61). In defending science fiction, she argues that literary realism narcissistically inflates individualistic concerns to the status of "universal" human problems, thereby ignoring or marginalizing the profound effects on human and social identity of scientific and technological innovation. For this reason, Merril maintains that "the literature of the mid twentieth century can be meaningful only in so far as it perceives, and relates to, the central reality of our culture: the revolution in scientific thought which has replaced mechanics with dynamics, classification with integration, positivism with relativity, certainties with statistical probabilities, dualism with parity" (54). Although she describes these transitions in oppositional terms, she also insists that science fiction foregrounds irrevocably complex relationships between science and society. During the 1940s and 1950s, she argues, science fiction helped to raise "the acceptance threshold of society in general and of the scientific/technical segment in particular" for the paradigm shifts she describes. As Constance Penley (1997) suggests, it is the fantasy at the heart of "realism" and the "realism" at the heart of fantasy that explains our fascination with the genre.

Merril's comments anticipate, in some respects, both the theoretical concerns of cultural critics of science, such as Latour and Haraway, and later analyses by science-fiction writers who distinguish their genre from the aesthetic, social, and political values of realistic fiction. Samuel Delany shares Suvin, Jameson, and Merril's conviction that the genre's willful distortions of "reality" offer possibilities for a social criticism that rejects "the monologic aesthetic" of modernism "in which art itself denies all dialogue, contest, agonism, and history to become an individual subject's representation *of* an individual subject (or a series of individual subjects) *for* an individual subject, with the exciting, material, impinging social object relegated to a wholly secondary position" (1994, 192, 194). By inverting, complicating, or disrupting the conventional relationship between subject and object, the "genre-effect" of science fiction blurs distinctions between "high" and "popular" art (194). Rather than placing humankind at the center of a coherent representational narrative, science fiction allows other kinds of "subjects"—cyborgs, software programs, and aliens—to speak for themselves. The imagined technologies and strange environments of science fiction become ways to simulate in provocative ways the dialogic relationship that exists between interpenetrating "subjects" and "objects" in historical reality.

Delany's "exciting, material, impinging social object" has affinities with Latour's "actant" (1987), his designator for material objects—flesh, blood, silicon, plastic, steel, copper, or wood—that cut across and blur distinctions between mind and matter, subject and object, and nature and culture. According to Latour and Delany, the laptop on which I am writing this sentence is not a passive object or a reification of design specifications, human expertise, production techniques, semiskilled labor, and operating and word processing systems that "I" use; rather it redefines what I, a human subject, can do and think, radically altering my techniques of composition and my understanding of my relationship to print and electronic media (Markley 1996, 55–77). Technologies and identity are mutually constitutive: things "use" humans, Latour argues, as much as humans use them. Science studies and science fiction share this fascination with reconceiving the relationship between humankind and technology and, more broadly, between humankind and a complex material reality—including the twentieth century's favorite site for interplanetary science fiction, Mars.

For a planet that no human has visited, Mars plays a surprisingly significant role in defining the intersections between individual and sociocultural identities and the physical universe. Since H. G. Wells published *War of the Worlds* in 1898, Mars has imposed the ecological constraints of a dying planet on the imaginations of writers and readers. Science fiction about Mars is obsessed with ecological issues, insistently foregrounding the problems of survival for humans and Martians in a fragile or exhausted environment. The notion of a dying planet paradoxically spurs the cultural imagination precisely because it imposes severe environmental constraints on human explorers or Martian civilizations. Consequently, as Mars becomes a favored site to depict the consequences of resource exhaustion—battles against rivals and against a hostile or indifferent nature—the planet offers refracted visions of political economy on Earth. Lowellian Mars served as a crucial impetus in the early twentieth century for scientists, novelists, and social commentators to view Earth itself not as a patchwork of countries, tribes, and localities, or a boundless reserve to be exploited for adventure and profit, but as a global ecosystem in crisis.

Paul Alkon credits science fiction with "offer[ing] our most powerful literary defense against unthinking collaboration with the impulses behind our worst nightmares" (1987, 4), and quite often Mars is the setting for ecological versions of these nightmares of planetwide exploitation. In projecting evolutionary narratives onto the natural history of the solar system, science-fiction writers followed Lowell in seeing Mars as an older and more decrepit version of Earth. In his study of twentieth-century pulp fiction, Paul Carter suggests that "Mars as a nineteenth-century frontier America and Venus as a nineteenth-century frontier Africa had a mythic appeal that pushed aside all demands of scientific exactitude" (1977, 65), but the two planets served different functions and generated different kinds of narratives. Cloud-shrouded Venus may have given writers free rein to fantasize about jungle beasts and warring tribes, but its featureless disk imposed no ecological constraints on the imagination and suggested no tragic narrative foreshadowing the fate of the Earth. In contrast, Mars determined a course of sociocultural as well as planetary history: the response of an advanced civilization to a dying world. Venus was the abode of primitive life-forms or the kinds of the "backward" humanoids who Europeans and Americans already had subjected and colonized. In contrast, even "primitive" Martians (those imagined by Leigh Brackett and Philip K. Dick, for example) had degenerated from a master race of canal builders. In describing ecological devastation on a planetary scale, science fiction about Mars offers a refracted image of humankind's efforts to live on an Earth disfigured by industrialization, pollution, and resource depletion.

The persistence of this image of a dying planet in science fiction reveals something of the dark underside of modern myths of technological and social progress. Dying, after all, is a liminal state; it is both a process and a border between human experience and the absolute, whether that absolute is figured as spiritual transcendence or utter annihilation. Dying also implies a transition through time and history, and the use of this metaphor to describe planetary life cycles projects onto Mars explicit values and implicit assumptions about the passing of civilization. The planet offered readers the vicarious experience of inhabiting an always liminal world. In popular culture, Mars is both a ghost planet haunted by past life-forms—from the vanished civilizations of twentieth-century pulp fiction to the putative nanofossils of exobiological controversy—and a world waiting to be inhabited, or already haunted by, future generations of human explorers and colonists. At the end of Ray Bradbury's *Martian Chronicles,* the last family of colonists on Mars see their reflections in a canal. The Martians have died off, although they move spectrally throughout the vignettes that comprise the novel; humankind has destroyed itself in a nuclear holocaust. When one of the children asks his father where the Martians are, the man points to their reflections in the water: "The Martians stared back up at them for a long, long silent time from the rippling water" (1950, 181). Metaphorically, the imagined colonists of the future are identified with the ghosts of a long vanished past, and, for Bradbury and many others, Mars remains a planet haunted by both its past and its future.

The spectres of a once-living world that haunt a now-dead or dying planet recur frequently in science fiction because they represent the paradoxes of temporal dislocation: a past and future that alike disturb the present. The languages of popular culture and planetology are full of the ghosts of Mars, ghosts who seem uncanny harbingers of the fate of the

Earth. Planetary scientists such as John Lewis often deploy such images to describe the loss, over eons, of the planet's water and atmosphere: "Mars today," he writes, "is but a ghost of the planet it once was" (1996, 150). The line between science and fiction in such comments blurs because our understanding of the surface of Mars is haunted by evidence of a dynamic, if not a biological, past. As Lowell's canal builders receded from science fiction, they were replaced by the new ghosts of long-vanished aliens who have left behind ancient cliff dwellings (Ben Bova's *Return to Mars* [1999]), mysterious artifacts (William K. Hartmann's *Mars Underground* [1997]), and the so-called face on Mars, the mile-long plateau supposedly carved to resemble a humanoid face by a mysterious civilization (Allen Steele's *Labyrinth of Night* [1992]; Ian Douglas's *Semper Mars* [1998]). In the best of these novels, such as Hartmann's or Terry Bisson's *Voyage to the Red Planet* (1990), no ultimate or originary truth underlies or explains the uncanny artifacts left by billion-year-old civilizations. They are, in one sense, ghostly markers for a virtual future of human exploration and colonization. Like Bradbury's Martians, the alien artifacts of 1990s science fiction bring disparate, even contradictory, views of the planet into the same fictional space. In turn, this fictional future of human settlements and terraforming technologies serves as the ultimate goal for enthusiasts in the Mars Society, for the authors of the 2004 presidential commission report on the United States space program, and for a number of well-respected scientists.

· • ● • ·

In this study, the chapters on science and those on science fiction alternate and interweave. My purpose (as this Introduction suggests) is not to divide "science" from "fiction" but to pursue the internal logic of developments in each genre as well as to explore the ways in which their concerns overlap and interpenetrate. The chapters that follow, then, do not indicate hard and fast conceptual boundaries but explore the territories that science and science fiction share.

Chapter 1 traces debates about Mars and its imagined inhabitants from the seventeenth through the late nineteenth century. After Kepler and Galileo, belief in the plurality of worlds played an important role in seventeenth-century intellectual history; in addition to the well-known works of Bernard le Bovier de Fontenelle and Christiaan Huygens, debates about life on other planets intrigued some of the major literary figures of the period, notably Aphra Behn (Fontenelle's translator) and Daniel Defoe. As telescopes improved in the eighteenth and nineteenth centuries, Mars became increasingly important in debates about extraterrestrial intelligence, particularly in the work of William Herschel, who placed Mars at the center of exobiological speculation. By the time Camille Flammarion produced his massive history of the planet in 1892, a consensus had emerged that was based on analogies between Mars's seasonal changes and terrestrial climatology. In turn, however, such analogies also reshaped conceptions of Earth. The major challenge to visions of an earthlike Mars came from Giovanni Schiaparelli, who saw and drew a complex network of straight lines, *canali,* on its surface. These canali touched off a half century of debate: some observers denied their existence, others confirmed Schiaparelli's observations and argued that they were natural geological features; and still others, including Flammarion, entertained the idea that they were the artificial products of intelligent Martians. Schiaparelli remained reticent to explain the significance of his canali, and his own commentaries focus on the scientific and philosophical problems of analogical reasoning.

The second chapter examines Lowell's canal theory in the contexts of nineteenth- and early-twentieth-century scientific, environmental, social, and political thought. A master stylist and popularizer, Lowell offered a plausible narrative that drew effectively on two major scientific theories of the period: the nebular hypothesis of planetary formation and Darwinian evolution. For many readers, he provided eloquent defenses of scientific progress and secular objectivity by decentering humankind in the universe. In contrast, his major critics—among them E. W. Maunder and Alfred Russel Wallace—argued from theological as well as scientific postulates that "man" was alone in the universe. These critics, however, were often divided among themselves, some accepting his maps of the canals, others claiming that the straight lines he mapped were illusory. Yet even many of these skeptics accepted key aspects of Lowell's thesis—that cold and dry Mars offered a harbinger of Earth's future. In this respect, Lowell's view of Mars as a dying world both drew on contemporary ecological concerns—particularly the loss of arable land in equatorial deserts—and encouraged his readers to imagine a technologically advanced civilization confronting dire environmental conditions. For all his political and social conservatism, Lowell's grim narrative of the evolution of worlds from life-bearing planets to desiccated husks challenged metanarratives of technological progress and manifest destiny. While E. E. Barnard, Maunder, and Wallace offered trenchant critiques of the canal thesis, Lowell and his allies launched vigorous counterattacks. Beginning in 1905, the photographs of Mars by Lowell's assistant, V. M.

Slipher, in the eyes of most observers, confirmed the general character of the planet's surface that Lowell described. The canals of Mars attracted serious attention from literary figures, scientists, philosophers, and political theorists as a site of speculation about the interactions among a hostile environment, evolutionary pressures, and the sociopolitical organization of intelligent beings forced to cope with an ecological catastrophe.

Chapter 3 deals with the outpouring of science fiction about Mars written between 1880 and 1910. Before Lowell, Mars was a popular setting for utopias that offered different visions of the technoscientific and sociopolitical ideals of advanced civilizations. After 1895, however, science fiction became concerned with depicting the social, political, and economic consequences on Lowell's dying planet. Although H. G. Wells's *War of the Worlds* is the best known of the turn-of-the-century science-fiction novels, works by Kurd Lasswitz, Alexander Bogdanov, and Alexei Tolstoy offered popular, often compelling visions of the significance of advanced Martian civilizations. Wells's alien invaders allow the novelist to critique nineteenth-century imperialism and human pretensions to scientific mastery and metaphysical significance. In depicting humanoid Martians, Lasswitz and Bogdanov extend popular analogies between the two planets to challenge the values and assumptions of European civilization. What unites these three novelists is the belief that ecology is destiny—that the fate of intelligent Martians is tied irrevocably to the evolutionary decline of their planet.

Chapter 4 explores the afterlife of Lowell's canal thesis in midcentury planetology. Between Lowell's death in 1916 and the Mariner 4 mission in 1965, most scientists accepted some aspects of Lowellian Mars—a dry, cold world still characterized by some form of vegetation—while dismissing his canal-building Martians. Yet temperature measurements by William Coblentz in the 1920s led to a brief revival of the canal theory; simmering controversies about the canals hinged on the varying interpretations of results from thermocouples, new spectrographic techniques, photographs, and, after World War II, radio astronomy. In charting the ups and downs of the canal thesis through the 1930s, I examine a 1928 *New York Times* article that surveyed twelve prominent American astronomers and found them divided in their beliefs about the canals. Even Lowell's critics conceded the existence of some kind of linear markings on the surface and acknowledged the possibility that Lowell may have been right. Such interpretations, as Henry Norris Russell concluded, did not constitute proof, and he and several other astronomers withheld judgment. By 1933, however, the canals had begun to fade when researchers failed to detect water vapor in the planet's atmosphere. Nonetheless, studies of Mars for the next thirty years struggled to reconcile the seeming existence of vegetation with evidence that the planet was so cold and dry that it could not support terrestrial life.

Chapter 5 discusses the half century of Martian science fiction between Lowell's death and the first Mariner missions to Mars. Beginning before World War I, Edgar Rice Burroughs crafted a series of ten Martian novels that proved enormously influential. Burroughs invented the first mass-media American superhero, John Carter, while exploiting and popularizing the vision of a dying world. In depicting this Earthman's adventures among various human and nonhuman races struggling to survive on Mars, Burroughs defined the action-adventure genre in terms that both draw on and challenge myths of the American frontier. Burroughs's followers in the 1930s through the 1950s adapted these conventions to dramatize the ecological trade-offs that characterize existence on a dying planet. In studying novels and stories by P. Schuyler Miller, C. S. Lewis, Leigh Brackett, C. M. Kornbluth and Judith Merril, Lester del Ray, and Ray Bradbury; Howard Koch's 1938 radio adaptation of *The War of the Worlds* for the Mercury Theatre; and two 1950s science-fiction films, I explore the dialectically related logics of cynicism and paranoia that shape the ideational landscape of midcentury science fiction. Mars becomes both a site for serious political protest against no-holds-barred capitalism and a camp setting for matinee redactions of Wells's invaders. It allows writers and readers to deal with the paradox of imagined colonization: liberal-utopian ideals of cooperation fostered by the need to manage scare resources set against the industrial and corporatist models of planetary exploitation. For Bradbury, Brackett, and others, Mars became a vehicle to critique the self-destructive militarism and mindless conformity of postwar America.

Chapter 6 examines the Mariner and Viking missions between 1964 and 1976 that substantially altered scientific views of Mars. The photographs taken by Mariners 4, 6, and 7 revealed a cratered, lunarlike surface that ended speculation about canals and stretched the limits of longstanding analogies to Earth. But in 1972 Mariner 9 returned over seven thousand photographs of the planet that revealed massive shield volcanoes, a canyon system three thousand miles long, and evidence that water had once flowed across the surface. This mission literally remapped the surface of the planet and raised a host of questions about its chemical composition and geomorphic history. In 1976, the Viking missions placed the first two landers to function successfully on Mars. While the

Viking orbiters photographed the planet's surface and studied its atmosphere and geology, the landers conducted three experiments to search for microbial life. The results, at first ambiguous, eventually were interpreted by most scientists as evidence of a lifeless, arid world bathed in ultraviolet radiation. The life-detection experiments and the orbital photographs marked important advances in interplanetary exploration, but their results also provoked new debates about the usefulness of continuing the search for life on Mars.

Chapter 7 examines the response of science-fiction authors, movie directors, and readers to the red and apparently dead Mars of the post-Mariner era. In the pre-1964 novels and short fiction of Robert Heinlein and Philip K. Dick, Mars retained some of its associations with ancient races and the commercial exploitation of a new frontier; after Mariner 4, it became the site for pointed critiques of the myths of new frontiers and new beginnings. In the late 1960s and early 1970s, British and European writers, such as D. G. Compton and Ludwig Pesek, painted bleak pictures of humankind's future on the planet. After the Viking missions, American authors, including scientists such as Hartmann, Gregory Benford, and Robert Zubrin—depicted fictional voyages to Mars that extrapolated from existing knowledge to speculate about what astronauts eventually may find. Even before the missions of the 1960s, however, other science-fiction writers, including Arthur C. Clarke, Isaac Asimov, and Walter M. Miller, speculated about terraforming the planet to make it habitable for human colonists. To conclude this chapter, I discuss Paul Verhoeven's *Total Recall,* a significant film in both popularizing the idea of terraforming and shaping that myth as a conservative fantasy of a new frontier safe for capitalist exploitation.

Chapter 8 traces changing perceptions of Mars since the Viking missions. Even as planetologists proposed various theories about the planet's geological past, controversies over the possibility of fossilized nanobacteria have provoked debates about what constitutes probable evidence of past or current life on Mars. In both geology and biology, seemingly anomalous data has led to reassessments of the methods and assumptions governing areology. In tracing ongoing debates about Mars and Martian meteorites, I explore the ways in which evidence of geologically recent water erosion, the detection of large amounts of subsurface water, and the success of the 2004 rover missions have affected planning for future missions. Mars remains a catalyst for the development of both innovative technologies in planetary exploration and new theories within comparative planetology. It is also a site for scientists, historians, philosophers, and cultural critics of science to explore the ways in which anomalous data redefine dynamically contemporary technoscience.

Chapter 9 examines Kim Stanley Robinson's trilogy, *Red Mars* (1993), *Green Mars* (1994), and *Blue Mars* (1996), the most influential Mars novels since Bradbury's *Martian Chronicles.* Drawing on scientific speculation about the possibility of terraforming, Robinson rethinks conceptions of planetary ecology, the interlocking systems that create and sustain the conditions that allow life to flourish, and political economy, the distribution of scarce resources among competing populations and interests. Beginning with two short stories, "Exploring Fossil Canyon" (1982) and "Green Mars" (1985), and continuing in the trilogy and his collection of stories, sketches, and poems, *The Martians* (1999), Robinson explores the utopian possibilities of depicting alternatives to the ideological conflicts and economic and ecological crises of the late twentieth century. In dealing with the fictional possibilities of terraforming Mars, the novelist suggests that the antagonism of environment and exploitation can be subsumed in what his characters call "eco-economics," a means of calculating value in order to minimize the degradation of the planet and to ensure social justice. In this regard, Robinson's Mars novels call into question the exploitative logic of late-twentieth-century capitalism to suggest alternatives to ever-increasing cycles of intensification and environmental degradation. *Red Mars, Green Mars,* and *Blue Mars* draw on and reconfigure the utopian tradition of science fiction represented by Lasswitz and Bogdanov.

CODA: AGNOSTICISM

At a prelaunch press conference in 1975, seventeen Viking scientists "were asked for a show of hands on whether they believed there was life on Mars. At first no hands went up; then 2 or 3 were raised; and after about a minute there were 11" (Cooper 1980, 12). The scientists' reticence to commit to a belief in even microscopic Martians harks back to the agnosticism of Schiaparelli and Behn and suggests as well the lure of a science-fiction tradition that still imagines Mars as the abode of strange life. In the mid-seventies, the differences among Viking researchers extended to questions about the usefulness of trying to detect alien life and to the philosophical and methodological differences among the three teams that conducted the Viking biology experiments. More generally, the scientists' delayed reaction is a representative moment in the history of our understanding: the eight or nine hands that hesitantly went up testify both to the problems of searching for exotic life-forms and to the agnosticism about exobiology that recurs throughout the scientific literature. The difficulty that Mars presents in 2005 is not so much imagining a single life-form that might eke out an existence on or under its surface, but in imagining the complex environmental systems—water, energy sources, transportation, reproduction, protection against ultraviolet radiation, and reciprocal effects on the environment—that would have to exist as well. If and when sixteen future scientists are ready to touch down on Mars and begin examining the most promising sites where life may have existed, a similar "show of hands" is likely still to produce a similar reaction: hesitant, hopeful, and agnostic.

ONE

"A Situation in Many Respects Similar to Our Own": Mars and the Limits of Analogy

CRUSOE ON MARS

In the third volume of *Robinson Crusoe,* the little-read *Serious Reflections,* Daniel Defoe's hero describes a conversation with a friend "upon the common received notions of the planets being habitable, and of a diversity of worlds." This discussion, Crusoe tells the reader, leaves him "for some days like a man transported into these regions myself" (Defoe 1903, 268). But unlike many of his contemporaries, Defoe rejects the popular view that other planets are "qualified for the subsistence and existence of man and beast" (273). Although Crusoe's imagined flight through the solar system is devoted to describing his "Vision of the Angelic World," his comments about Mars are both scientifically prescient and ideologically revealing. On Mars, says Crusoe, "the light [from the sun] is not above one-half, and its heat one-third of ours [on Earth]." Nevertheless, "this planet is hot and dry, and would admit of no habitation of man, through the manifest intemperance of the air, as well as want of light to make it comfortable, and moisture to make it fruitful; for, by the nature of the planet, as well as by clear-sighted observation, there is never any rain, vapour, fog, or dew in that planet" (274). Crusoe, of course, is an expert on what it takes to survive in a wilderness, and his Mars is more barren and inhospitable than any desert island. Hot, dry, and dark, the planet lacks an earthlike ecology and consequently the resources—water, flora, and fauna—necessary to sustain human existence.

Defoe's description of Mars registers some of the crucial concerns of eighteenth- and nineteenth-century thinkers who speculated about the possibility of life on the planet, and his emphasis on evidence obtained by "clear-sighted observation" anticipates the terms of subsequent scientific debates. Well into the twentieth century, controversies about conditions on Mars, and therefore the usefulness of analogies between that planet and Earth, centered on the problems of identifying the limit conditions for extraterrestrial biota. For Defoe and his contemporaries, habitability translated into questions about the resources that could be exploited by indigenous inhabitants. In a preindustrial world, dominated conceptually by agricultural metaphors, the analogy between Earth and Mars (or any other world) depended less on abstract speculations about the possibility of life than on the lived experience of environmental pressures and resource extraction—the complex interactions necessary to sustain what Defoe calls "the vegetative and sensitive life" (273). In this respect, "clear-sighted observation" implies not only atmospheric clarity and an unbiased observer but an economy of inference that translates visual data into hypotheses about planetary ecology.

The idea of a diversity of worlds consequently has georgic connotations well into the nineteenth century because "intelligence" is defined implicitly and explicitly in terms of resource extraction, agricultural production, and energy consumption. Although we hardly think of Crusoe as an astronomer, his description of a lifeless Mars suggests paradoxically why the red planet fascinated generations of fiction writers and their readers. As the urtext of Western "man's" conquest of the wilderness, *Robinson Crusoe* offers its readers a thought experiment to ponder the resourcefulness and technological ingenuity that an individual needs in order to thrive in an alien environment. Though Defoe's novel has been read as an exemplar of the Protestant ethic, an adventure tale, and a colonialist parable, it is also a tale of European "man" transforming an island ecology into a protoeconomy, exploiting the indigenous resources necessary to live in comfort. A "plantation," Crusoe calls his island, with cultivated fields, domesticated animals, foodstuffs, utensils, storage vessels for surplus goods, protective fencing, a summer house, and cottage industries in butchering goats, tanning their hides, and producing goatskin garments. For Crusoe, the island is a reservoir of abundant resources; it is, in a sense, the antithesis of his vision of Mars. Yet as the "common received notions" of his time suggest, many of his contemporaries were willing to entertain the idea that similar resources could be found and put to use on the fourth planet. By the end of the eighteenth century, the earthlike ecology of Mars was accepted generally as a scientific fact.

The tale of Crusoe's survival on a desert island hovers over a century of interplanetary science fiction and scientific speculation. In 1964, film director Byron Haskin updated Defoe and sent an American astronaut to the red planet in *Robinson Crusoe on Mars.* Shot on location in Death Valley, the film depicts Mars as a terrestrial desert. As a visual analogue for three centuries of areological speculation, Death Valley seems the logical extension of Defoe's vision of a planet without the water that would allow indigenous life-forms, or space-age colonists, to prosper. By restaging the Crusoe myth on a hostile, barely habitable planet, Haskin draws on a long tradition of speculation about the Martian environment and the life it might harbor. The limits of the film's analogy define the ways in which our fascination with the red planet depends, as Defoe intuited, on finding the food, water, and air necessary to make it a "fruitful" abode of intelligent life. Scientists studying Mars in the 250 years between Defoe's novel and Haskin's film refined questions about life on the planet, but could provide no definitive answers.

In this chapter and in chapters 2 and 4, I explore the scientific implications of the questions about the ecology of Mars in the era before spaceflight, and in chapters 3 and 5 I discuss a long tradition of science fiction that rejected Defoe's dead world and let loose provocative fantasies of a dying planet and its tenacious inhabitants. The need for "clear-sighted observation" encouraged astronomers to improve telescopes and develop new instruments to try to settle debates about conditions on the fourth planet. For three hundred years, Mars has remained a key site for debates about the nature of visual evidence in planetary astronomy, the role of inference and probability in scientific speculation, and the ways in which traditional value systems—Christian theology and anthropocentric philosophy—respond to the Copernican decentering of humankind in the universe. While Defoe remained a skeptic, most of his readers in the nineteenth and twentieth centuries rejected his reasoning and perceived Mars as an older and smaller version of Earth.

THE COPERNICAN REVOLUTION AND THE PLURALITY OF WORLDS

For a few weeks every twenty-six months, Mars and the Earth are aligned on the same side of the sun in their elliptical orbits.[1] During these periods of opposition, Mars is visible through comparatively small telescopes, and, since the mid-seventeenth century, scientific observations of the planet's surface and atmosphere have clustered during these periods. Because the orbit of Mars (compared to Earth's) is an elongated ellipse, its distance from the sun—and from Earth—varies significantly. The eccentricity of its orbit means that Mars may approach as close as thirty-four million miles to Earth when the two planets are aligned on the same side of the sun (as it did in 2003) or remain as far away as sixty-three million miles (as it did at the opposition of 1948). During its nearest perihelic oppositions Mars appears almost twice the size (25.0 seconds of arc) as it does at its most distant (13.8 seconds of arc).[2] As the ancient civilizations of Europe, Asia, and the Americas knew well, Mars appears as one of the brightest objects in the sky, and slowly traces a great loop through the constellations as the Earth overtakes and passes the slower orbiting planet; a Martian year at 687 days is almost twice as long as Earth's. Against the relatively fixed backdrop of the stars, Mars appears to move backward, and this retrograde motion fascinated the astronomers and stargazers of the ancient world.[3] The Chinese called Mars Ying-huo, the fire planet; the Greeks identified it with Ares, the god of war; the Babylonians termed the planet Nirgal, the god of the underworld; and the Aztecs saw in the fiery and erratic Mars the deity Huitzilopochtli, the destroyer of people and civilizations. For these cultures, the trajectory and appearance of the red planet seemed to symbolize the violence and uncertainties of existence (Burgess 1990, 5–7).

The retrograde motion of Mars historically posed difficulties for models of the solar system because it seemed to violate human notions of how a fixed empyrean, an ordered universe, should behave. The Ptolemaic system placed Earth at the center of the cosmos and tried to account for Mars's retrograde motion as well as the less-pronounced retrograde motion of Jupiter and Saturn by adding epicycles to the perfect circles of the planetary orbits. As observational data became more precise, Ptolemaic systems became increasingly complex in their efforts to explain the planets' apparent motion and to reaffirm geostationary models of the universe. By the late middle ages, the geocentric solar system was overwritten with epicycles, and yet it still failed to conform precisely to the observed motion of the planets (Randles 1999, 32–57; Koyré 1957; Grant 1994). But the ideological significance of an ordered universe dictated that the church uphold its age-old vision of a geocentric universe with humankind at its cosmological and moral center. In 1514, Nicholas Copernicus (1473–1543) circulated an anonymous pamphlet that challenged the Ptolemaic universe by placing the sun at the center of the solar system and relegating the Earth to merely another orbiting planet. Although he served as a canon at the cathedral of Fromborg, and therefore was ostensibly bound to uphold the church's position, Copernicus devoted much of his life to developing a heliocentric model of the solar system in an effort to simplify the byzantine complexities of Ptolemaic epicycles into a coherent model of planetary motion. Central to his argument was a rejection of the Ptolemaic explanation of retrograde motion. Copernicus described the backward track of the outer planets, particularly Mars, as an effect of the relative speeds at which they and the Earth orbit the sun. In 1542 he published *The Revolutions of the Heavenly Spheres*, which presented his heliocentric model, emphasized the "vast magnitude" of the visible universe (with other stars at the centers of their own solar systems), and suggested that gravity was not centered solely on a stationary Earth. His nervous publisher prefaced the treatise with a disclaimer that labeled his work a mathematical speculation, but Copernicus himself seems to have believed in the reality of his theory.

The Copernican theory was not an overnight success, but it provided the impetus for Johannes Kepler (1571–1630) to describe precisely the orbits of the planets. Committed to a quest for cosmic harmonies that would demonstrate the workings of divine providence, Kepler devoted years of his life to the elaborate mathematical calculations that led ultimately to his three laws of planetary motion.[4] In *The Mysteries of Cosmography* (1596), the first wholeheartedly Copernican treatise on the solar system since Copernicus's death, Kepler argued that the orbits of Mercury, Venus, Mars, Jupiter, and Saturn could be inscribed in the five regular polyhedrons. The universe, he believed, must exhibit the geometrical proportions that he and his contemporaries saw as evidence of divine order. Although he was a Protestant, Kepler reiterated a logic that dominated planetary astronomy during the seventeenth century: any calculations of celestial motions had to fit preordained assumptions about the ultimate harmony of the cosmos; in turn, that harmony confirmed a heliocentric—and divinely ordered—solar system (Paxson 1999, 105–23). Having succeeded Tycho Brahe as imperial mathematician in 1601, Kepler was directed by the Holy Roman emperor to prepare a massive chart of planetary positions. He spent several frustrating years trying to find a mathematical way to account for the orbit of Mars, the most eccentric among the then known outer planets. In *The New Astronomy* (1609), Kepler admits he "was almost driven to madness" by the years of calculation before he recognized that the paths of all the planets were ellipses with the sun at one foci. His work was subtitled *Commentaries on the Motions of Mars.*

Kepler's understanding of planetary motion marked a significant advance over the Copernican model and challenged the Ptolemaic universe conceptually as well as mathematically. While locating the sun at the center of the solar system, Copernicus viewed the heavens as a distinct realm from Earth; he made no connection between the forces that kept the planets in their orbits and the everyday effects of gravity on Earth. In contrast, Kepler sought a universal physical law to explain planetary motions—a law based on familiar principles of terrestrial mechanics. His recognition that planetary orbits were determined by gravitational forces liberated the planets from the crystalline spheres of medieval astronomy and gave them a corporeal reality. Even though Kepler's lifelong quest to unify mathematics, astronomy, music, and theology ultimately fell short of a grand synthesis, his commitment to the ideas that the solar system embodies a perfect, theocentric, order and that God does nothing in vain guided subsequent thinking about the physical nature of the planets as well as their motions. By vastly extending the dimensions of the universe, Kepler, like Copernicus, raised the possibility that the stars were other suns and that other planets, too small to detect from Earth, might be orbiting other heliocentric systems. The cosmos, then, might well harbor a plurality of inhabited worlds.

By the early seventeenth century, the Moon, Mercury, Venus, Mars, Jupiter, and Saturn had become refracted images of Earth, possible worlds rather than mere lights in the sky. Giodorno Bruno was burned at the stake in 1600 for his heretical beliefs, including his faith in a scientific knowledge "which freeth us from an imagined poverty and straitness to the possession of the myriad riches of so vast a space, of so worthy a field, of so many cultivated worlds" (quoted in Guthke 1990, 69). A century before Defoe, Bruno redefined the geocentric conception of the planets into a dynamic model of resource-using, intelligent beings. It is significant that Bruno defines "worlds" in terms of "myriad riches" and "cultivation"; the transcendence from "poverty" to "riches" is material as well as metaphysical, ecological (in the broadest sense) as well as theological. "Cultivated worlds" clearly implies rational beings to do the cultivating, and these beings posed a threat to a religious orthodoxy that placed "man" at the moral center of creation. A just and merciful Supreme Being who created intelligent life elsewhere in the universe, Bruno reasoned, presumably would supply those beings with the capacities, resources, and opportunities to transform raw materials into material comforts, as humankind had done on Earth. The alternative—lifeless planets that had no discernible function in a divinely created universe—would threaten the bedrock principle of seventeenth-century natural philosophy: each aspect of creation fulfills a divine, if mysterious, purpose.

With the development of the telescope in the seventeenth century,

Mars slowly began to be perceived as the most likely candidate in the solar system for the "cultivation" and "riches" that would define an extraterrestrial civilization. For telescopic astronomers, Mars had distinct advantages over the Moon, Venus, Jupiter, and Saturn as an object of study and contemplation. The surface of the planet was both visible and dynamic; those of the Moon and Venus were not. Mars's surface features, though, were consistent enough to offer pointed contrasts to the atmospheric turbulence of the gas giants. As early as the mid-seventeenth century, the growth and shrinking of the Martian polar caps fascinated planet watchers and led them to reason analogically that they were observing earthlike seasons. Between 1654 and 1892, most oppositions witnessed efforts by astronomers to scrutinize the planet and to understand, as best they could, its climate, geography—or, more accurately, areography—and indigenous forms of life. To a great extent, however, these astronomers were hampered by the limitations of their telescopes; and, as William Sheehan argues, until the late nineteenth century "the story of Martian exploration [is] intertwined with the optical improvement of the telescope" (1996, 57). Yet even as telescopes improved, astronomers often complained that they could see little of value, and celebrated those rare moments when the planet came sharply into focus and they could distinguish features on its surface. The British natural philosopher Robert Hooke found during the opposition of 1666 that the surface of Mars was too indistinct to make useful observations. "I could find nothing of satisfaction [and] I could not conclude upon anything" (cited in Sheehan 1996, 23), he reported to the Royal Society of London.[5] Although Hooke persisted and had better luck a few nights later, this report by the author of *Micrographia* and a meticulous observer of the natural world is characteristic of early efforts to describe the planet: no resolution or "satisfaction," and then, during brief periods of comparative clarity, suggestive glimpses of another world. These moments allowed early observers the opportunity to sketch prominent features on the planet's surface and offered the possibility that Mars could be described accurately enough to meet the developing criteria for the objective observation of natural phenomena.[6] At best, though, the most distinctive surface features of Mars, such as Syrtis Major, could be seen clearly only at irregular intervals, and Hooke and his contemporaries, constrained by the limitations of their telescopes, had to content themselves with rough drawings.

Mars is a difficult object to observe under the best conditions, and many of the controversies that have swirled around the planet since Hooke's time have centered on the interpretation of ambiguous visual evidence, filtered through the dense, moist, and roiling atmosphere of Earth. Conditions of seeing can change from moment to moment: temperature differentials, the interactions among density layers in the Earth's upper atmosphere, humidity, and air currents all contribute to make Mars—even when it is observed under nearly ideal conditions—appear like a quarter shimmering at the bottom of a swimming pool. Hooke noted in 1664 that his difficulties in observing the planet could be attributed to the fact that "the Inflective veins of the Air . . . have a greater or less Refractive power, than the air next adjoyning, with which they are mixt, [and] . . . make [the image] confus'd and glaring" (cited in Sheehan 1996, 23). For Mars to come clearly into focus, the air currents immediately above the telescope and higher in the atmosphere must be comparatively still to minimize visual distortion; conditions of good or excellent seeing are often described in minutes, seconds, and even fractions of seconds. As Percival Lowell wrote in a commissioned article in *Nature* in 1908, "Living as we do under a gaseous ocean in constant turmoil, no image from beyond it stays perfect for long" (1908,402). Not surprisingly, then, the history of the observation of Mars is characterized by often acrimonious debates about which telescopes can best penetrate or compensate for the distorting envelope of the Earth's atmosphere. In addition, by the nineteenth century astronomers were well aware that the observation of the planets was bound up with the problems of environmental degradation—smog and light pollution as well as humidity and turbulence. The quest for geographical locations that offered the best conditions for planetary observation thus embroiled many astronomers in controversies about the air quality on Earth.

Given the combination of factors required for good seeing, the siting of telescopes became a crucial issue in the nineteenth century as air pollution and city lights rendered many observatories in metropolitan areas in Europe and North America almost useless for planetary study. Hooke observed Mars from Greenwich at a time when the "smokes" of London already were driving some members of the upper classes out of the city for months on end. As late as the nineteenth century, important observations of the planets were made from sites in low-lying urban areas; Schiaparelli mapped his canals from the Brera Observatory in Milan in 1877 with a 8¾ inch refractor (Moore 1999, 50). But telescopes situated at high altitudes have less of the Earth's atmosphere (and often less pollution) to contend with, and, if the air above them is comparatively still, offer more frequent opportunities for seeing through the "gaseous ocean" that inhibits observations of planets and stars.[7] In an important sense, the history of the observation of Mars can be described as a movement away from centers of population, light, and pollution to mountain-top or high plateau retreats. The large astronomical observatories that were built in the United States in the late nineteenth century—such as Mt. Lick and Mt. Wilson—were situated in the dry and still air of the mountains of Southern California, and in the early 1890s Lowell traveled from the Sahara to South America in search of the best location for observing Mars. He finally settled on Flagstaff, Arizona, in 1894, and he also funded expeditions to Arequipa in Peru to observe Mars from the southern hemisphere. The California and Lowell Observatories were farther south than the important European observatories such as Meudon in France, and Mars appears higher in the sky and remains visible for longer periods from these sites—facts that the defenders of the canal thesis emphasized well into the twentieth century.

Until the Mariner missions of the 1960s, historically important observations of Mars and ensuing debates about the conditions on the planet tended to cluster during perihelic oppositions such as those of 1781, 1894, 1909, and 1956 when the planet comes closest to the Earth; planetologists routinely kept nightly vigils at telescopes to seize every occasion of good-seeing. Paradoxically, however, as telescopes improved in the nineteenth century and more detail on the surface was observed, a two-hundred-year-old consensus about Mars began to break down, and individuals' conclusions about what they had or had not seen on the planet's surface became the subject of increasingly intense debate. The emergence of this consensus between roughly 1656 and 1877, the year of Schiaparelli's first observations of the Martian "canali," reveals some of the ways in which astronomical observations are bound both to histories of technological developments and to complex scientific, philosophical, theological, and sociopolitical narratives.

HUYGENS OBSERVES MARS

By the mid-seventeenth century, telescopes had progressed to the point that they could reveal features on the Martian surface, and Mars emerged as a kind of Latourian quasi-object that revealed as much about technology, cosmological values and assumptions, and the politics of contemporary natural philosophy as it did about its "true" nature (Latour 1987). In 1664, the Italian astronomer Giovanni Cassini (1625–1712), who then held the chair in astronomy at the University of Bologna, observed, drew, and published the first detailed images of the Martian surface (Markley et al. 2001, "early views" 5). Cassini was noted for the precision of his astronomical observations and for his work in improving and building telescopes, including one with a focal length of 136 feet. Despite the limitations of his unwieldy telescope, his observations of Jupiter and Mars during the 1660s were accurate enough to allow him to determine their periods of rotation. By studying the reappearance of surface features in the same relative position over the course of thirty-six or thirty-seven days, Cassini concluded that a Martian day was twenty-four hours, forty minutes, only three minutes longer than its actual rotational period (Sheehan 1996, 21–25; Flammarion 1981, 11–57 [3–35]). His drawings of Mars indicate twin dark patches that bear some resemblance to the albedo features that were mapped as "seas" in the mid-nineteenth century and as "oases" and "canals" a half century later. But the accuracy of Cassini's determination of the length of the Martian day indicates that he had a good enough view of the planet to recognize distinctive features and to measure their positions accurately. Mars was no longer a wandering light in the sky but a world that could be brought tentatively within the regime of scientific measurement.

The Dutch scientist Christiaan Huygens (1629–1695) began observing Mars and the other planets in the 1650s (he discovered Saturn's moon Titan in 1656), and his drawings from the opposition of 1659 (which were not published until the end of the century) record the distinctive feature of Syrtis Major. Independently of Cassini, Huygens determined (within two minutes) the length of the Martian day, and he began to speculate on the similarities between Mars (and other planets) and Earth. Establishing that Mars had a rotational period similar to Earth's and intuiting that the polar caps and dark areas were permanent features on the planet's surface, Huygens speculated about a larger analogical relationship between the two worlds. Writing near the end of his life in the liberal intellectual atmosphere of his native Holland, Huygens seconded the views of the French deist philosopher, Bernard le Bovier de Fontenelle, who had published *Entretiens sur la pluralité des mondes* (*Conversations on the Plurality of Worlds*) in 1686.[8] Although Huygens distanced himself from Fontenelle's Cartesian mechanism, he drew on his "experience as an observational astronomer" to offer a materialist as well as philosophical rationale for his defense of the many worlds hypothesis (Dick 1982, 128). While Fontenelle was well known for his nonchalance about established religion (as I shall discuss below), Huygens followed a long-established tradition in tying the hypothesis of a plurality of worlds to the argument from design—the idea that God's handiwork could be read in the structure of the physical universe (Markley 1993, 56–61, 122–23).

In his posthumously published *Cosmothereos,* translated into English as *The Celestial Worlds Discovered* (1698), Huygens extended some of the arguments advanced by Fontenelle, but brought them firmly within a theocentric rhetoric of natural philosophy.[9] Fontenelle speculates about a plurality of inhabitable worlds without committing himself to a theological interpretation of the cosmos; Huygens, in contrast, reasons from a theological absolute, extending the argument from design to the question of what order of intelligent beings might exist on other planets. Early in *Cosmothereos,* Huygens reiterates the idea, familiar to contemporary readers of Robert Boyle and Isaac Newton, that all aspects of the physical universe are exquisitely structured according to a divine plan: "For every thing [in the natural world] is so exactly adapted to some design, every part of them so fitted to its proper uses, that they manifest an Infinite Wisdom, and exquisite knowledge in the Laws of Nature and Geometry" (1698, 20). Humankind, however, is not the only species capable of appreciating this cosmic order. Unless other planets, including Mars, are inhabited by intelligent and God-fearing beings, Huygens suggests, the universe can have no theological or philosophical rationale. In *Cosmothereos,* the implications of the Copernican revolution are extended into the galaxy.

Yet even as Huygens considers the Earth as "one of the Planets of equal dignity and honor with the rest" (1698, 62), he envisions these other worlds in anthropocentric terms. Extraterrestrial intelligence, he maintains, must manifest itself in the wise use of divinely created resources: other planets offer idealized images of pious and prosperous (Christian) civilizations on Earth.

> For all this Furniture [plants, animals, and water] and Beauty the Planets are stick'd with seem to have been made in vain, without any design or end, unless there were some in them that might at the same time enjoy the Fruits, and adore the wise Creator of them . . . otherwise our Earth would have too much the advantage of [other planets], in being the only part of the Universe that could boast of such a Creature so far above, not only Plants and Trees, but all Animals whatsoever: a Creature that has a Divine somewhat within him, that knows, and understands, and remembers such an innumerable number of things; that deliberates, weighs, and judges of

> the Truth: a Creature upon whose account, and for whose use, whatsoever the Earth brings forth seems to be provided. For every thing here he converts to his own ends. With the Trees, Stones, and Metals, he builds himself Houses: the Birds and Fishes he sustains himself with: and the Water and Winds he makes subservient to his Navigation . . . It is therefore the Principle we before laid down be true, that the other Planets are not inferior in dignity to ours, what follows but that they have Creatures not to stare and wonder at the Works of Nature only, but who employ their Reason in the examination and knowledge of them, and have made as great advances therein as we. (Huygens 1698, 37–38, 61)

Huygens has a very specific idea of what a "planet" is, an idea that excludes the uninhabited worlds of Defoe's vision but that makes Mars a prime candidate for speculations about extraterrestrial intelligence. By definition, a planet embodies seventeenth-century ideas of a theologically sanctioned balancing act among physical needs, rational desires, and available resources. A just Creator, Huygens maintains, would not put intelligent beings on a world where they could not feed, clothe, and improve themselves. Paradoxically, then, he relies on terrestrial standards of intelligence, piety, and reason to decenter geocentrism. The other planets are marked by the assumptions and values of a Baconian natural philosophy: intelligence is a function of use-value and the exploitation of extraterrestrial "Trees, Stones, and Metals" an indication that the "Planetarians" possess that "Divine somewhat" that elevates humankind above the beasts. Such beings are coequal to humankind, Huygens emphasizes, because they possess the principles of "Reason" both to comprehend the agency of God in the physical universe and to convert "every thing . . . to [their] own ends" (1698, 78). Like a golden-age Earth, the other planets are storehouses of God-given resources waiting to be exploited. In one respect, then, Huygens invests the plurality of worlds with the values of late-seventeenth-century republicanism. The rational order of the universe becomes an idealized projection of the good life as understood by a Dutch scientist near the end of his country's golden age: civilizations across the universe are distinguished by their architecture, trade, commercial fishing, and "Navigation."

The myriad inhabited worlds that were conjured into existence in the eighteenth and nineteenth centuries project into the solar system and beyond Huygens's fundamental assumptions: no resources, no intelligence; no intelligence on other planets, no divine order to the universe.

To assume or argue that the planets were inhabited was to invoke images of industry and cultivation. Yet if Huygens's planets are populated by beings who exhibit the same mental faculties and desires as humankind, these other worlds are also idealized extensions of the blank spaces on contemporary European maps of the Americas, Africa, and the Pacific. In Huygens's time the dominant technoscientific industries of western Europe—shipbuilding, navigation, cartography, finance, ballistics, and even optics—were offshoots of Dutch, Portuguese, English, Spanish, and French commercial adventurism; like the Spice Islands or the West Indies, Huygens's planets are stocked with seemingly infinite resources waiting to be exploited.[10] Huygens's "Planetarians" exist in an ideological realm that owes much to values and assumptions underlying the seafaring commercial empires of the British and Dutch. In his *Second Treatise of Government,* Huygens's correspondent John Locke argues that the basis for a constitutional alternative to absolute monarchy is a natural world open to endless cultivation: "In the beginning," he declares, "all the World was *America*" (1960, 2, ¶36, 292)—that is, all the world was open to exploration and exploitation. Nature is not nature, Locke implies, unless it is in the process of being mined, refashioned, and turned to "use." Yet this idealized world suffers no ecological degradation and no decrease in either value or productivity. Locke's pristine "*America,*" in short, is the model for, and a terrestrial analogue of, Huygens's planets "stick'd with" natural resources and "Beauty." Looking at the heavens, Huygens projected onto the surface of Mars conditions that could sustain intelligent life and provide evidence that the principles of seventeenth-century political economy were truly universal.

FONTENELLE AND BEHN

Cosmothereos was conditioned by a half century of speculation about extraterrestrial life, much of it prompted by the invention of the telescope (Nicholson 1935, 428–62; 1940, 83–107). The advent of the telescope made the rocky surface of the Moon a favorite site for fantasies about extraterrestrial life-forms and a familiar setting for utopias relocated from Earth or, as Karl Guthke characterizes them, ventures in speculative fiction, such as Kepler's *Somnium* (1634), Francis Godwin's *The Man in the Moon* (1638), John Wilkins's *Discovery of a New World,* and Cyrano de Bergerac's *The Other World* (1657–72).[11] In contrast to the dark blotches on Mars or the bright, featureless disk of Venus, the unchanging face of the Moon offered a convenient image of timeless perfection, an ideal site on which to place virtuous beings who live in peace and prosperity (Dick 1982, 176–83; Guthke 1990, 144–58). The idea that sociopolitical and ecological conditions on the Moon had to be better than on Earth filters into satires such as de Bergerac's and imaginary voyages such as Godwin's. Like other utopias, ideal societies on the Moon offer a standard against which the shortcomings of contemporary European societies can be judged, mocked, or condemned. But even as they retain generic traits familiar to readers of Thomas More or Francis Bacon, they redefine the relationship between intelligent beings and the worlds they inhabit. Rather than being confined to isolated islands, these lunar utopias offer planet-wide visions of rational economy, moral perfection, and unfallen nature.

The denizens of utopian societies on the Moon never foul their nests or degrade their always abundant resources. Godwin, for example, sends his hero, Domingo Gonsales, to a Moon inhabited by gigantic, God-fearing Christians who are without mortal sin. Their virtue, not surprisingly, is a function of living amid conditions of abundance: "There is no want of any thing necessary for the use of man," the reader is told, because "Food groweth every where without Labour" (1638, 102). There is no luxury or surplus value in their society, no unsatisfied desire; use-value and resources are forever in balance. Without fear, envy, greed, or the hardships of agricultural labor, the Moon "seemeth to be another paradise" (1638, 104). If this balance between human needs and a beneficent nature is a familiar topos in utopian literature, Godwin's Moon differs significantly from the islands of More's *Utopia* and Bacon's *New Atlantis.* While these utopias are socially, economically, and ecologically contained, Godwin's "paradise" imagines an Edenic *planetary* ecology—a world infinitely productive without labor. At a time when Europeans had begun to recognize the ecological limits of the tropical islands that they had colonized, the fantasy that new worlds could offer limitless possibilities for improvement and enrichment became a crucial aspect of the scientific speculation prompted by the plurality of worlds (Grove 1995). The infinite exploitability of resources on such worlds remained, for Huygens and the utopianists (if not for Defoe), an article of religious as well as scientific faith.

The literary and scientific speculation about extraterrestrial intelligence has suggested to some historians a dialectical response to the implications of Copernicanism: either credulous belief or a conservative rejection in the name of religious orthodoxy. In practice, however, the reactions to the prospect of "new worlds" were more ambiguous and complex. If Defoe could use dissenting Protestantism to depopulate the planets and thereby preserve humankind's privileged position in the cosmos, other writers could employ different strategies.

One of the most important early responses to the plurality of worlds is the preface by Aphra Behn to her English translation of Fontenelle's *Entretiens sur la pluralité des mondes* as *A Discovery of New Worlds* (1688).[12] Remembered primarily for being the first professional women writer in England and an influential poet, playwright, and novelist, Behn was one of the shrewdest cultural and political commentators of her era; her preface to Fontenelle reveals some of the complexities within seventeenth-century speculation about the plurality of worlds. In contrast to Huygens, Behn tries to limit the implications of Copernican cosmology.[13] Rather than a license for wanton speculation, the plurality of worlds, in her view, is useful for expanding the limits of the human imagination and confirming rather than challenging Christian belief.

Behn's preface to Fontenelle's *Entretiens Conversations*, as Lisa Schnell (1992, 109) suggests, is historically significant for maintaining that women have the same intellectual capacity to understand science as men. Its claims for gender equality, however, are only one aspect of Behn's commentary on cosmology. Like Newton, Behn offers an extended defense of the Copernican system as being the most compatible with the account of the heavens in Genesis; yet her voluntarist belief in a version of the argument from design renders her skeptical of Fontenelle's theories and wary of his motives in promoting a materialist defense of extraterrestrial life. In distancing herself from his Cartesianism, Behn attacks Fontenelle because he "ascribes all to Nature, and says not a Word of God Almighty, from the Beginning to the End [of the *Entretiens*]; so that one would almost take him to be a Pagan" (1992, 77). Moreover, his cosmology is unconvincing. In his efforts to persuade his readers "that there are thousands of Worlds inhabited by Animals, besides Earth, [Fontenelle] hath urged this Fancy too far" and engaged in wild speculations that have no basis in either theology or observable fact. Cleverly inverting Fontenelle's heterodox intentions, she declares, "I shall not presume to defend his Opinion, but one may make a very good use of many things he hath expressed very finely, in endeavouring to assist his wild Fancy; for he gives a magnificent Idea of the vastness of the Universe, and of the almighty and Infinite Power of the Creator" (1992, 77). Behn effectively turns Fontenelle the stylist against Fontenelle the iconoclast. His prose creates—three centuries before Carl Sagan—a cosmological sublime: the "vastness of the Universe" becomes comprehensible not through Newtonian mathematics or "wild Fancy" but through a mode of literary voluntarism: she invokes a grandeur and mystery that humans can intuit but never truly understand.

Behn's decision to translate Fontenelle's three histories, including his anticlerical treatise, *The History of the Oracles and the Cheats of the Pagan Priests,* highlights the historical significance of her agnostic response to the hypothesis of a plurality of worlds. If speculations about extraterrestrial life in the seventeenth century occasion a variety of responses, these responses do not necessarily resolve themselves into hard and fast distinctions between belief and skepticism. Between obvious fantasies such as Godwin's and impassioned defenses of the plurality of worlds, such as Huygens's lie the complex strategies of ironic distance, hypothetical thought experiments, speculative reasoning, and invocations of the limitations of human knowledge. In translating Fontenelle, Behn appropriates his arguments to support her own view—an agnosticism that, on the one hand, refuses to license pure speculation and, on the other, rejects rigid dogmatism. Nowhere in her preface does she attack Fontenelle's thesis by denying the possibility of extraterrestrial life; instead she defends a minimalist or "primitive" Christianity that reasons from the argument from design to celebrate the divine order in the universe. Yet Fontenelle is obviously an attractive author for her to translate because his work allows her to satirize—through the very act of translation—the corruption of organized religion, particularly the power and privileges of the clergy.[14] Accordingly, Behn places arguments about the Copernican system and the plurality of worlds within the realm of individual conscience: debates about heliocentrism have no bearing, she contends, on the validity of scripture. In contrast to Huygens, Behn paradoxically lowers the stakes involved in extraterrestrial life debates. In this respect, her preface anticipates a prevalent attitude toward the possibility of life—on Mars and elsewhere in the "vastness of the Universe"—that persists into the twenty-first century: higher intelligence is neither disproved nor rendered probable; it exists instead in the liminal spaces of speculations that might yet be proved. In the three centuries since Fontenelle and Huygens, astronomers and philosophers have hedged their bets, deferred judgments, offered hypothetical scenarios, and engaged in speculations that resist precise generic classification as either "science" or "fiction." Behn's response to the plurality of worlds thesis, in this respect, is significant not because it foreshadows modern conceptions of scientific objectivity but because it anticipates a rhetoric of deliberate oscillation between skepticism and belief, a rhetoric that still characterizes scientific speculation about extraterrestrial intelligence.

The development of refracting telescopes in the eighteenth century allowed a new generation of astronomers to get better views of Mars during oppositions and to improve on the drawings of their predecessors. Cassini's nephew, Giacomo Filippo Maraldi, observed Mars for almost half a century; his most detailed observations and sketches date from the perihelic oppositions of 1704 and 1719 (Flammarion 1981, 58–65 [35–38]; 65–79 [39–46]; Sheehan 1996, 29–30). Planetary observation, however, was a part-time enterprise for the handful of astronomers at European observatories, and there was little significant work done on the planet after Maraldi until that of William Herschel (1738–1822) over a half century later. No one exemplifies the nascent professionalization of astronomy more than Herschel, who was born in Hanover and came to England in 1757. A noted musician, he began buying telescopes and learning to grind lenses in 1773; within a few years he was the most celebrated maker of telescopes in England. Herschel established his reputation as an astronomer by discovering Uranus in 1781, then spent much of his time cataloging thousands of nebulae and star clusters. He became the official astronomer to George III and president of the London Astronomical Society. The author of seventy papers that appeared in the *Transactions of the Royal Society* between 1781 and 1818, Herschel determined the length of the Martian day, conducted the first extensive survey of the planet's polar regions, made relatively accurate observations of its polar and equatorial diameters, and noted its obliquity (its rotational deviation from the vertical axis). Herschel's drawings of Mars, like those of his predecessors, trace irregular shaded or dark patches as the planet rotates and mark the seasonal shrinking and growth of the polar caps (Markley et al. 2001, "early views" 6–7).

Carefully tracking changes in the northern polar cap during the opposition of 1781, Herschel reinforced and extended seventeenth-century analogies between Mars and Earth. His paper, "On the Remarkable Appearances of the Polar Regions of the Planet Mars, the Inclination of Its Axis, the Position of Its Poles, and Its Spheroidical Figure; with a Few Hints Relating to Its Real Diameter and Atmosphere," read before the Royal Society in London on March 11, 1784 (and published later that year) established the parameters that were to dominate the study of the planet for the next century and a half. In summing up the state of knowledge about Mars, Herschel offers scientific support for the speculations that Huygens had advanced in *Cosmothereos.*

> The analogy between Mars and the earth is, perhaps, by far the greatest in the whole solar system. Their diurnal motion is nearly the same; the obliquity of their respective ecliptics, on which the seasons depend, not very different; of all the superior planets the distance of Mars from the sun is by far the nearest alike to that of the earth: nor will the length of the martial year appear very different from that which we enjoy, when compared to the surprising duration of the years of Jupiter, Saturn, and the Georgium Sidus [Uranus]. If, then, we find that the globe we inhabit has its polar regions frozen and covered with mountains of ice and snow, that only partially melt when alternately exposed to the sun, I may well be permitted to surmise the same causes may probably have the same effect on the globe of Mars; that the bright polar spots are owing to the vivid reflection of light from the frozen regions; and that the reduction of those spots is to be ascribed to their being exposed to the sun. (Herschel 1912, 148)

Herschel's paper presents itself as a model of scientific deduction: his data result from painstaking observations, he reproduces a sequence of sketches of the Martian surface as it changes with the seasons, and his "surmise[s]" are measured and logically coherent. Mars exhibits earthlike seasons, and therefore similar effects on the two planets safely can be presumed to arise from similar causes. As Flammarion wrote a century later, "Herschel's explanation of the polar patches of Mars has been adopted since his time as the most natural, the simplest and the most logical answer, since it is identical with the explanation of our own polar patches. . . . Since it *completely* satisfies the observed phenomena, there is no need to look around for alternatives" (1981, 86 [50]). Occam's razor can shave no closer.

The Martian polar caps, for Herschel, become the key to a comprehensive view of the planet's climate and hydrological cycles. Because its seasons mirror Earth's, and the surface temperature must be warm enough for the polar cap to melt, Herschel concludes that the fourth "planet has a considerable but moderate atmosphere, so that its inhabitants probably enjoy a situation in many respects similar to our own" (1912, 156). He ascribes albedo changes in the planet's equatorial regions to cloud cover. In the same paper, however, he also demonstrates that the Martian atmosphere is comparatively thin. In 1783, he observed two stars that were occulted by Mars. As they came close to the rim of the planet, Herschel noted that their light did not fade and correctly deduced that a dense atmosphere on Mars would have obscured their light substantially. Herschel's recognition, however, did not affect his views of the planet or its presumed inhabitants. However thin Mars's atmosphere, it had to be substantial enough to retain sufficient heat to melt the polar caps. The analogy to Earth, then, suggested that if the climates of the two planets were similar, their life-forms must be similar as well. Although he concedes that astronomers often may be frustrated by the lack of hard data, Herschel makes a persuasive case that they can extrapolate from what they know about Earth to fill in the gaps of their knowledge about Mars. The plurality of worlds hypothesis, it seemed, had been put on a sound scientific footing.

Herschel was followed in the late eighteenth century by the German astronomer Johann Hieronymous Schroeter, also patronized by George III. Schroeter built a telescope with lenses ground by Herschel and recorded a series of observations of Mars from 1785 to 1803. Praised by historians of astronomy such as Flammarion (1981, 107–34 [62–83]) and Sheehan (1996, 36–41), Schroeter's work lay neglected until the end of the nineteenth century. Although convinced that his difficulties in perceiving markings on the planet's surface were the result of cloud cover, he seconded the notion that Mars is an earthlike planet and made more accurate estimates than Herschel did of the planet's polar and equatorial diameters. He also sketched over 230 images of the planet (see the reproductions of his drawings in Sheehan 1996, 40), and while these drawings clearly show recognizable features such as Syrtis Major, Schroeter missed their significance: while he notes that "we must be cautious in our conclusions because we lack definite proof" (cited in Flammarion 1981, 109 [63]), he suggests that the clarity of the atmosphere is illusory. The planet's albedo changes, he maintains, are caused by atmospheric clouds rather than changing surface features. Accepting Herschel's reasoning about the polar caps, Schroeter begins to fill in the gaps in a supposed hydrological cycle: melting snows and ice at the poles offer de facto evidence of the cloud cover that obscures his view.

Schroeter's caution about what he had seen marks the end of an era when the telescopic observation of Mars was restricted to easily identifiable features, notably the polar caps. The differing interpretations of the dark and light areas on the planet's surface by Herschel and Schroeter focused attention in the nineteenth century on determining which markings were permanent features and which, if any, were transitory. The improvement and rapid dissemination of refractor telescopes allowed for unprecedented clarity in viewing the Moon and the planets, and astronomers soon began trying to determine what the albedo changes and differing shades of color observed on the surface of Mars might signify.[15] A century after Schroeter, the astronomer and historian Camille Flammarion called attention to the ongoing problems of recording and interpreting such changes: "All observers who draw Mars know that it is extremely difficult to make a faithful drawing of what can be seen, because the forms are nearly always indefinite, diffuse, vague, without sharp outlines, and sometimes quite uncertain. The aspects are vague, feeble, dubious, and hard to draw; the instruments used are different, the observers' eyes and methods of observation are perhaps even more different" (1981, 133 [83]). Despite these difficulties, by the middle of the nineteenth century astronomers had begun to map the planet's surface, converting "vague, feeble, and dubious" markings into a "knowledge of Martian geography, or *Areography*" (183 [118]). Mars began to assume the characteristic features of terrestrial geography—oceans, continents, lakes, and rivers. Water on the surface was accepted almost as a given, and consequently astronomers presupposed the existence of some form of life on an aqueous planet with a "moderate" climate.

MARS IN THE NINETEENTH CENTURY

The most influential of these nineteenth-century areographers were Wilhelm Beer and Johann Heinrich Mädler, who translated their years of observation into the first detailed map of the Martian surface in 1840. Beginning in 1828, they observed Mars through numerous oppositions and, like Herschel, concentrated on the polar caps. Beer and Mädler noted differences in the behavior of the polar caps: the south polar cap grows larger in winter and shrinks in the summer to a much smaller size than its northern counterpart. They also sketched surface features and suggested that the dark areas might be marshes that grow with the melting of the ice caps (Sheehan 1996, 45–50; Moore 1999, 41–42). Comparing Beer and Mädler to Christopher Columbus, Flammarion hails them as "true pioneers" for their "first methodical attempt at studying Martian geography" (1981, 158 [101]; 163 [104]). Even though Mädler later came to doubt the stability of some of the dark surface features, the map he and Beer produced brought the fourth planet within the framework of nineteenth-century geographical understanding. It is, however, a geography stripped of many of the sociopolitical overtones that characterized European mapmaking: without nations, borders, or colonies, Mars appears as a kind of environmental thought experiment on which presumed landforms and bodies of water can be observed or partially deduced from the behavior of the polar caps. The waning of the polar caps drives climatological and hydrological cycles: melting ice produces what might be marshes, fills low-lying areas, and strongly suggests the presence of vegetation. The planet's surface takes on, by analogy, aspects of terrestrial maps.

By the opposition of 1856, telescopes had improved and the number of observers who were able to sketch the surface of the planet increased dramatically. These included Warren de la Rue, Angelo Secchi, the director of the Observatory of the Collegio Romano in Rome, Lord Rosse, Sir Norman Lockyer, John Phillips, William Lassell, Frederik Kaiser, and in the 1860s William Dawes (Sheehan 1996, 49–52). The opposition of 1862 brought a flurry of new drawings and reports. As Flammarion notes, "almost all astronomers who had good instruments at their disposal made observations of the planet" (1981, 239 [170]). The accounts of these men, many of them published in scientific journals, were marked by debates about the nomenclature best suited to describe what were considered permanent features on the planet's surface. The related processes of mapping and naming drew on and reinforced a descriptive vocabulary of terrestrial terms—seas, channels, and continents—to describe Mars, a vocabulary that suggested far more certainty about observational data than the cautious language that Flammarion, among others, employed. The sketches made by an English clergyman, William Rutter Dawes, according to Flammarion, "brought a new precision to the studies of Mars" (288 [204]). Dawes's sketches also became the basis for the map produced by Proctor, and were republished with some changes in several of his books, including *Other Worlds than Ours* (1870) and *Essays on Astronomy* (1872).[16] Proctor named the various landforms and seas that he drew after his predecessors: Beer, Mädler, Herschel, Lockyer, and so on. His nomenclature, in an important sense, both shapes and is shaped by a value-laden view of planetary science: astronomical understanding progresses, vague suppositions are transformed into hard knowledge, and maps become more detailed and precise. Through hard work and judicious induction, Proctor maintains, Mars can be brought within the paradigms defined by terrestrial cartography.

By 1877, Flammarion counted "391 different drawings of the planet, made by all the observers" (1981,350 [242]) over the course of two centuries. Writing at the end of the nineteenth century, he was in a position difficult for twenty-first century readers to imagine: he recognized the heuristic nature of these drawings and yet remained convinced that areography had become a hard science rather than a speculative genre.

"Each drawing," Flammarion writes, "represents a personal equation, an individual interpretation, and as the details of a globe seen from the distance of Mars through two atmospheres are always vague and excessively delicate, there can be no single drawing which gives a rigorous and exact representation of Mars as it would appear to an observer close to the surface." Yet despite this recognition of the limitations of individual observers and their telescopes, Flammarion believes that "our knowledge of the planet gradually advances, year by year, with the progress of observations" (350 [242]). This progressivist narrative of the history of astronomy overrides Flammarion's methodological caution; he concludes that "Mars is the only planet whose geographical configurations are known accurately enough to be put on a map" (290 [205]) This cartographic knowledge, however, rests on transforming fleeting impressions and "enigmatical variations" (185 [120]) into identifiable surface features: continents, seas, gulfs, and bays.

As recent historians of cartography argue, mapping is never a purely objective rendering of a static landscape; it is always marked by competing practices, discourses, and interests. The historiography of map-making on Earth demonstrates the complex ways in which naming features and discerning political boundaries help to construct ideas of nationhood, property, and sociopolitical identity (Wood 1992). At the same time that Britain, France, and Germany were mapping their overseas empires, mapping Mars presented a different set of challenges for aerographers. And yet these nineteenth-century astronomers too were subject to broader ideological conceptions of what maps represent and the functions that they serve.

If two centuries of geocentric analogies had brought Mars within the discourses of topography, climatology, and hydrology, they also implied relationships between surface features and indigenous networks of resource extraction and distribution, commerce, communication, and travel. Yet astronomers gazing at the red planet and sketching what they saw had no political boundaries to mark, no frontiers to chart, and no distinctions to enforce between civilized areas and unexplored continents. Consequently, the mapping of Mars focused primarily on global cycles of temperature, hydrology, and presumed biological activity. Above all, Mars was conceived as a planetary whole, and efforts to differentiate putative land masses from bodies of water and to chart, with only a consensual, probabilistic basis in observable "fact," the features of the landscape kept turning to questions about the interactions of land, water, and climate. In this respect, Mars offered a means to think about general principles governing planetwide relationships between environment and biology. Yet if planetary astronomers in the second half of the nineteenth century routinely projected terrestrial characteristics onto Mars, invoking the age-old analogy between the two planets had its limits. Debates about the color of the dark areas on the surface—variously described as blue, green, and blue-green—led to questions about the composition and nature of presumed Martian seas. Flammarion suggested that "these aqueous stretches appear to be in a different physical state from our own seas: less dense (?), less liquid (?), or covered with viscous fogs (?)" (1981, 352 [243]). Even if Flammarion's efforts to rethink physical chemistry on the fourth planet stretch the boundaries of the two-planet analogy, his fascination with water was bound up with speculations about life on its surface. The existence of intelligent Martians, in turn, could be deduced only by inferring a planetary ecology that was stable enough to be mapped, categorized, and named. Victorian controversies about whether dark blotches on the planet's surface should be named after astronomers, classical figures, or geographic analogues on Earth may seem, in retrospect, a quaint preoccupation, but the terrifying droughts in India, South America, and China from 1877 to the end of the century made scientific efforts to describe the complex relations among climate, hydrology, food, population, and networks of distribution anything but an academic exercise (Davis 2001). The fascination with water on Mars developed, in part, from an awareness of the consequences of drought on Earth.

The tendency to impose on Mars an earthlike geography of waterways and continents was corroborated, to some extent, by the first scientific efforts to measure the chemical composition of the planet's atmosphere. Flammarion, like many of his contemporaries, was swayed by spectrographic evidence of water on Mars that was offered independently by Jules Janssen in France, Sir William Huggins in England, and Hermann Vogel in Germany in the 1860s (Sheehan 1996, 114; Moore 1999, 70–71; Horowitz 1986, 84; Markley et at. 2001, "canals of mars" 18–20). A spectrograph is a photographic recording device that measures the dispersal of radiation or light into bands that indicate either emission (bright lines) or absorption (dark lines). Because each element has a characteristic signature of such bands, spectrographic analysis can identify and record the chemical composition of each radiating or reflective source. Like other bodies in the solar system, Mars reflects light from the sun, but when earthbound observers look at Mars, they see rays of light that have passed twice through the Martian atmosphere, once on their way from the sun to Mars and again on their way from Mars to Earth. Janssen, Huggins, and Vogel realized that they could attach spectrographs to telescopes in order to study the chemical composition of the light that had been filtered through the Martian atmosphere. Such readings, they reasoned, would provide hard and fast evidence about the composition of the planet's atmosphere.

In 1867, Janssen conducted the first spectroscopic analysis of Mars by comparing its spectrum to the Moon's. Because the Moon has no atmosphere, its spectrum is a feeble version of the sun's unfiltered light. Any differences between the two spectra, Janssen reasoned, could be attributed to gases in the Martian atmosphere. He took his measurements from the top of Mount Etna to minimize interference from oxygen and water vapor in the Earth's atmosphere, though he does not seem to have been successful: working with a new and experimental technology, he detected water vapor in the Martian atmosphere. Huggins and Vogel confirmed his findings (Flammarion 1981, 277–81 [180–183]). Well into the twentieth century, some spectrographic readings seemed to indicate the presence of water vapor in the Martian atmosphere. The absorption bands were difficult to read and difficult to interpret; for a century, planetary astronomers struggled to distinguish the characteristic absorption signatures of gases in the Earth's atmosphere from those that could be attributed to the gases surrounding Mars (Schorn 1971, 223–36). By 1892, Flammarion could write that "spectral analysis . . . establishes that the waters [on Mars] are analogous to ours in chemical composition" (351 [243]). "Aqueous" Mars, as it turned out, was an artifact of the interference of the Earth's atmosphere, and as early as 1894, W. W. Campbell at the Lick Observatory failed to find any evidence of water vapor in the Martian atmosphere, although his findings were disputed for decades (DeVorkin 1977, 37–53). While astronomers recognized that it was difficult to interpret spectrographic readings taken through the Earth's atmosphere, many of them accepted the view (at least until 1933) that Mars had measurable amounts of water and oxygen. Even after improved instruments and techniques indicated no traces of water and oxygen on Mars, the tendency to describe the dark areas of the surface as some sort of vegetation persisted into the 1960s, a century after Janssen and Huggins had published their findings.

SCHIAPARELLI AND THE CANALI

At this juncture, most historians of planetary astronomy begin a new chapter on Schiaparelli and the canal controversy. But I want to consider Schiaparelli's *Astronomical and Physical Observations of the Axis of Rota-*

tion and the Topography of the Planet Mars, 1877–78 as a work very much in the tradition of his predecessors and contemporaries, such as Flammarion. In one sense, Schiaparelli's observations of a network of lines on the surface of Mars seem more an extension by visual inference of generally accepted characteristics than a radical departure from the views of previous astronomers. In another, however, they defamiliarize the planet by redefining the terrestrial analogies imagined by Herschel, Dawes, and Proctor. Paradoxically, Mars becomes both a world made strange by its differences from Earth and a near twin that can provide significant evidence for understanding terrestrial climatology. Schiaparelli, as Sheehan argues, was a gifted planetary observer who, until his eyesight began to fail, "recorded traces of real features lying at the threshold between visibility and invisibility, and [these] bear out E. M. Antoniadi's assessment that as 'a record of fleeting impressions, Schiaparelli's stands unrivalled' " (Sheehan, in Schiaparelli 1996, i). Although by 1905 almost all commentators, including the anticanalists, preferred Lowell's drawings to Schiaparelli's, the Italian astronomer remained an authoritative figure who put Mars on the front pages of newspapers across Europe and the Americas. The "fleeting impressions" that he recorded took a geometrical form that called into question the values and assumptions encoded on the maps of his contemporaries. To a greater extent than his predecessors, Schiaparelli transformed glimpses of the planet through a small telescope into a kind of Rorschach test that had profound effects on planetary astronomy and on debates about extraterrestrial intelligence.

Although enmeshed in the canal controversy almost from the start, Schiaparelli was a self-conscious agnostic about what his canali signified. As much as any of his critics, he recognized that in sketching or describing the planet's surface "interpretations necessarily find their way into the very descriptions themselves" (1996, 9). Like Flammarion, he carefully qualifies both the evidence he considers and the conclusions he reaches. Emphasizing the "clear analogy [of Mars] with Earth," Schiaparelli nonetheless calls attention to the conventional, analogical nature of the terminology he employs. "Do not brevity and clarity," he asks, "compel us to make use of words such as *island, isthmus, strait, channel, peninsula, cape,* etc.? Each of which provides a description and notation of what could otherwise be expressed only by means of a lengthy paraphrase" (47, 9). In this context, the controversy over the translation of Schiaparelli's canali as either "channels" (Sheehan's preference) or "canals" (the headline grabber of the 1890s) may be understood as an effort to pin down a term that the astronomer seems to have left deliberately ambiguous. The self-critical nature of Schiaparelli's descriptive vocabulary, to some extent, distinguishes his work from that of the die-hard canalists who followed him, particularly Lowell. Yet paradoxically the "brevity and clarity" of his terminology allows Schiaparelli to push the analogy between Mars and Earth farther than many of his contemporaries were willing to go. He asserts that Mars "is a small version of the Earth, with seas, an atmosphere, clouds and winds, and polar caps; and it promises, in this regard, a good deal more" (52). The "good deal more" skirts the issue of extraterrestrial life as neatly as Behn had done two centuries earlier. Mars, "our neighbor and almost brother," offers a mirror on which astronomers may "examine closely the analogous phenomena [climate and weather] . . . [and] take in at a glance the meteorology of a whole hemisphere" (52). This "meteorology" is driven by a hydrological cycle that requires large bodies of open water. Accordingly, Schiaparelli declares that the "dark areas of the planet" are "seas" and suggests that "if the Earth were seen from a similar distance, it would present much the same appearance as Mars does to us" (48). This hypothetical view of our planet from space inverts the perspective of the two-planet analogy. Earth, in effect, becomes a Mars-like planet: its features and ecologies are understood as a back formation from the "complex embroidery of colors" that are visible on the Martian surface.

But even as Mars reveals its earthlike character, it remains an elusive and mysterious object. Schiaparelli's heuristic vocabulary reflects the tenuousness of his evidence. "Such is the number of the details [on the surface], and so fugitive the duration of [brief moments of good seeing]," he acknowledges, "that it was impossible to form a clear and coherent idea of the things seen, so that what remained was only the confused impression of a dense network of thin lines and minute spots" (51). Mapping Mars, for Schiaparelli, is not an exact science but an effort to register the contours of possibility for further investigation. Schiaparelli's diagrams of the canali both represent a "dense network of thin lines and minute spots" and call into question the bases of such representations. This self-critical aspect of Schiaparelli's description of Mars is a key reason why his reputation (in general) has avoided the ups and downs of Lowell's over the past century. The expressions of uncertainty and the qualifications that characterize his areography make few claims for the significance of his data. His observations, particularly in contrast to Lowell's grand narrative of a dying planet, seem a model of scientific restraint.

Nevertheless, Schiaparelli's drawings of his "confused impression[s]" resolve themselves into networks of straight lines that often doubled, or geminated, so that Mars became sectioned, in Flammarion's words, by "a geometrical reseau of straight lines, crossing each other at angles" (8 [viii]). These lines, in an important sense, lead nowhere; they offer no self-evident explanation, no narrative to explain their existence. For Flammarion, the puzzle of Schiaparelli's canals lies in their resistance to the very analogies that structured contemporary descriptions of Mars. "On Earth we have nothing comparable to guide us in explaining these features," the French astronomer declares. "We are dealing with a new world, incomparably more different from ours than the America of Christopher Columbus differed from Europe" (8 [viii]). This difference cannot be contained by the analogical relationship between Earth and Mars. Even as Schiaparelli invokes a familiar analogy to Earth, his canali paradoxically undo many of the assumptions that had defined areography since Herschel. His "dense network of thin lines" requires precisely the kind of "lengthy paraphrase" that he foregoes—that is, a narrative explanation of their nature, significance, strange doublings, and disappearances. All efforts to characterize these lines as physical features rather than optical illusions are necessarily interpretive and raise more questions about the geology and climatology of Mars than they answer.

Given this state of uncertainty, Schiaparelli falls back on the logico-deductive narrative born of the nebular hypothesis (see chapter 2). With one-twelfth the volume of Earth, Mars would have cooled more quickly than its planetary neighbor; it "is further along toward the period of absolute senescence of its internal [volcanic] forces" (52). This heat loss indicates that the planet is older than Earth, and that it is on the downside of its entropic journey toward planetary death. Yet even as Schiaparelli lays out some of the basic principles of a theory of the canals, he studiously refrains from relating his canali to the larger issues raised by the nebular hypothesis. In an article originally published in Italy, Schiaparelli rejects "inorganic" explanations for the gemination of the canals, including the view that the doublings were mere "optical illusions," and declares that it was "far easier to explain" the phenomenon by invoking an "organic" cause. But he quickly adds that "the field of plausible supposition is immense" (1894, 90). Waves of vegetation spreading across the surface of the planet is the explanation to which he devotes most of the rest of the article. This, however, is as far as he goes. Without hard data to limit possible explanations and to allow him to frame a plausible narrative, further conjecture would be unscientific. "The great liberty of possible supposition," Schiaparelli concludes, "renders arbitrary all explanations" (90). While encouraging further investigation and hoping for a "ray of light" to illuminate the problem of the doubling canals, he leaves his readers without a conceptual framework that would guide inquiries about this phenomenon. Even granting the established analogy between Earth and Mars, the canali have an unearthly, even uncanny quality that resists precise description.

THE STORY THUS FAR: FLAMMARION ON MARS (1892)

The difficulties and opportunities that Schiaparelli posed for nineteenth-century readers can be gauged, in part, by the response of Flammarion, an important figure in his own right in the history of astronomy.[17] In his translator's foreword to Flammarion's 1892 history of areography, dated 23 August 1980, Patrick Moore suggests that this "great book is the most complete study of the history of the observations of Mars ever written" (Flammarion 1981, n.p.), and it remains unrivalled in its copious analysis of two hundred years of sketches of the planet's surface. Having surveyed this vast body of material, Flammarion is in a position to argue that Schiaparelli's is "the greatest work" (421 [288]) of the lot. Yet while he acknowledges that the canali "represent some mobile liquid element," he hesitates to embrace the canal theory wholeheartedly (880 [577]). "Well aware of the difficulties" posed by the canals, Flammarion steers a middle course between Schiaparelli and his critics, who contended that the "thin lines" were optical illusions (882 [579]). Flammarion considers but ultimately rejects the possibility "that the idea of lines had been put into observers' minds, by the maps in front of their eyes and a preconceived idea which they came to by a kind of auto-suggestion" (883 [579]); too many qualified observers have confirmed the general character of Schiaparelli's "thin lines." But Flammarion remains guarded in his response: "It seems very difficult to accept the full accuracy of Schiaparelli's observations," he argues, because they defy explanation and not all the double canals can be confirmed by independent observation. Weighing the evidence, he concludes that astronomers "should regard this strange reseau of lines as real, at least in its general canvas" (884 [580]). His skepticism leads Flammarion into an extended discussion of the limits of the analogies that lie behind the canal theory. Rather than the unrestrained enthusiasm that he later exhibited for Lowell's account, Flammarion offers a near-textbook example of the rhetoric of scientific reasoning, repeatedly emphasizing the limitations of speculation when dealing with limited data. "There is nothing analogous [to the canals] on the Earth," he asserts, "and we have, unfortunately, only terrestrial ideas from which to reason" (1981, 884 [580]). His restraint both echoes and questions Schiaparelli's own metacritical commentary on the two-planet analogy.

As a historian, however, Flammarion has generic options that the Italian astronomer cannot pursue in trying to fit the latter's data into a still-developing narrative of nineteenth-century areography. He tries out different explanations for Schiaparelli's lines, based on the geological understanding of the time, testing the limits of analogic reasoning. As opposed to myriad canals, Flammarion sees rivers running north and south that require no canals or canal builders. Yet even the issue of water, long accepted as a distinguishing feature of the Martian surface, raises problems. What troubles him are "the variations of extent and tone [of Schiaparelli's canali], the doublings, the disappearances and reappearances according to the seasons, and all the innumerable changes in aspect which are admittedly hard to account for by water in the same state as ours" (1981, 892 [585]). Consequently, he suggests that liquid water may be "chemically and physically" different on Mars and even theorizes a "sixth state" of water—a "vapour which is so dense it is almost liquid" to explain the changes observed along the dark areas of the planet's surface. This "sheet of thick mist, viscous, sombre, dark" might even take the form of "a viscous fluid, heavier than air, and subject to forces other than gravity[,] . . . for instance electricity and planetary magnetism, as well as unknown forces" (890–91 [584]). Flammarion labels these ideas as pure speculation and, throughout the discussion that follows, emphasizes the limitations of human experience and scientific knowledge. His "viscous fluid" is a thought experiment; so too is the possibility that Mars "could be inhabited by human species whose intelligence and methods of action could be far superior to ours" (894 [586]). No astronomer in 1892, however skeptical of Schiaparelli's observations, could rule out either of these speculations.

In a rhetorical move that anticipates the literary strategies of Carl Sagan seventy years later, Flammarion argues that, given our "absolute ignorance," to reject the possibility of canal-building Martians "would be unscientific" (895 [586]): "Our knowledge is inadequate. It would be a great error to claim that Science gives us the last word, and that we are in a position of knowing everything; it is no less puerile to claim that we are familiar with all the forces of Nature. On the contrary, the Known is a tiny island in the midst of the ocean of the Unknown. Moreover, our senses are very limited; our power of perception is still lacking; our science remains, and will always remain, fatally incomplete" (901 [589]).

Flammarion's rhetoric verges, self-consciously, on the sublime—a response characterized by awe tinged with both terror and excitement—in the face of that which remains beyond representation. Significantly, Flammarion does not depict a scientific future in which the secrets of Mars will be revealed; his emphasis on a knowledge that will remain "fatally incomplete" leaves the interpretation of Schiaparelli's lines to the realms of the imagination—that is, to mystery, faith, and fiction. Ultimately, Flammarion can offer only a probabilistic assessment of Schiaparelli's work. A sixth state of water, a viscous mist, straightened rivers, mysterious electromagnetic forces, an absolute lack of "information to speculate about the possible forms of the 'human,' animal, vegetable and other types of life on Mars" all temper his conclusions. Yet this lack of information provokes speculation, and for a popular writer such as Flammarion it is difficult to resist erring on the side of extraterrestrial life and higher book sales. For the most knowledgeable historian of planetary astronomy of his era, "the habitation of Mars by a race superior to ours seems . . . very probable" (905 [591]). By the time that Lowell was putting the wheels in motion to build an observatory for the opposition of 1894, the burden of disproof had fallen on skeptical astronomers such as Edward Holden, E. E. Barnard, and W. S. Campbell to explain away the headline-friendly reseau of straight lines. Since Herschel, Mars had stimulated theorizing about planetary ecology, and with the publication of Schiaparelli's *Observations*, that speculation at the very limits of vision intensified.

TWO

Lowell and the Canal Controversy: Mars at the Limits of Vision

We Earth-dwellers, accustomed to judge according to the evidence of our eyes, and unable to imagine the unknown, have extreme difficulty in explaining phenomena which are strange to our own planet, and consideration of them can even plunge us into hopeless embarrassment.—CAMILLE FLAMMARION, *The Planet Mars*

LOWELL AND THE HISTORIANS

The canal controversy is one of the most notorious episodes in the history of science, and the consensus view of Percival Lowell—an erratic, if talented, amateur who let his imagination run wild—is unlikely to be eradicated. At best, Lowell is treated as a popularizer who came to the study of Mars with rigid ideas, stubbornly held onto them in an unscientific fashion, and was soon discredited.[1] Most historians boil down Lowell's three books and dozens of articles on Mars to a simple thesis—canals meant intelligent Martians—and emphasize his isolation from the small community of professional astronomers. William Sheehan and Richard McKim champion his opponents—E. E. Barnard, A. M. Antoniadi, and Alfred Russel Wallace among them—who fit, or can be made to fit, into a familiar narrative of scientific progress: scientists objectively considered, tested, and then rejected the canal thesis (Sheehan 1995, 1996; McKim 1993, 164–70, 219–27). Even Lowell's biographers, William Graves Hoyt and David Strauss, who chart his contributions to the study of Mars as well as the lengths to which he went to defend his thesis, maintain that the canal controversy "was largely of Lowell's own making" (Hoyt 1976, 212; Strauss 2001). These historians, however, tend to gloss over the persuasiveness of Lowell's rhetoric in its own day and the continuing fascination of his thesis for more than a few scientists well into the 1950s. At least until 1910 and for a time in the late 1920s, controversies about the surface temperature on Mars, the presence of oxygen and water vapor in its atmosphere, and the habitability of its surface did as much to confirm as to dislodge Lowell's reputation in the popular imagination and in some scientific circles. One of the ironies that many historians overlook is that Lowell presented himself as, and seemed to many of his contemporaries, the champion of a secular, progressive science doing battle against God-fearing conservatives—including many noted astronomers—who clung to the belief that "man" was the be-all and end-all of creation. The canal thesis reignited debates about humankind's place in the cosmos and thrust Mars into the forefront of controversies about ecology, evolution, and social organization.

In this chapter I place the canal controversy within the scientific, ecological, and sociopolitical contexts of the late nineteenth and early twentieth centuries in order to explain its intuitive appeal. In building on the work of Hoyt, Stephen Dick, and Strauss, I argue that until the 1930s the canal thesis had enough currency within the scientific community to reinforce a widespread agnosticism about the possibility of intelligent life on Mars. I want to emphasize, though, that I am not out to "defend" Lowell, to claim for him the status of a "major" scientist, or, still less, to excuse his conservative politics. But any attempt to chart the cultural and scientific significance of Mars as a "dying planet" both needs to explore the ways in which the canal theory insinuated itself into the discourses of cosmology, evolutionary biology, photographic technology, ecology, and sociology and to analyze the difficulties his critics faced in trying to dislodge the canals from popular consciousness. Contrary to the assertions of some critics and historians (see Crossley 2000, 297–318), Lowell's theory operated within the bounds of accepted scientific practice and offered a compelling narrative to many of his contemporaries committed to Darwinian evolution. If the reticence of Schiaparelli and Flammarion to explain the significance of the Martian canali offered Lowell a convenient opening to promulgate his theory, he used effectively the rhetoric of scientific objectivity to challenge the values, assumptions, and methods of his opponents.

I begin by examining Lowell's fascination with the nebular hypothesis and his efforts to describe "planetology" as "the connecting link in the long chain of evolution from nebular hypotheses to the Darwinian theory" (Lowell 1909, 2). By linking his observations of Mars to the two great secular theories of nineteenth-century science, Lowell was able to portray himself as the heir to scientific pioneers—Copernicus, Galileo, and Newton—who struggled against the prejudices and conservatism of their contemporaries.[2] I then describe the probabilistic argument for the canals that he advanced in *Mars* (1895). From the start, Lowell's trump card in the canal controversy was his ability to offer a coherent narrative in which his observations and conclusions "fit into one another [like] a mortised whole" (Lowell 1906b, 165). His critics often disputed his conclusions and rejected his logic, but they also frequently reinforced his description of Mars as a dying world: none of them challenged his accomplishments as a literary stylist, his scientific training, or his dedication. To suggest something of the range of responses to his theory, I discuss two prominent reviews of Lowell's first book, by Agnes Clerke and William Lockyer, which defined the contours of the canal debate for some years.

In 1903, the British scientists E. W. Maunder and J. E. Evans dealt a potentially devastating blow to the canal thesis by publishing a study of schoolboys who, when asked to reproduce the markings on a disk the approximate size of Mars at opposition, connected discrete dashes into straight, canal-like markings. Their paper marks a definitive turning point for most historians of the canal controversy, but, I suggest, the Lowell Observatory's publication of photographs of Mars taken by C. O. Lampland in 1905 and 1907 marked a significant—though often ignored—counteroffensive that revealed inconsistencies among the assumptions and strategies of the anticanalists. Lowell's *Mars and Its Canals* (1906) and *Mars as the Abode of Life* (1909) bracket Alfred Russel Wallace's attack on the canal thesis, *Is Mars Habitable?* (1907); the debate between Lowell and Wallace highlights their radically opposed conceptions of ecology, politics, and theology and indicates the stakes that were in play in this controversy. Although many historians suggest that Lowell was ostracized by the scientific community, this view simplifies a complex relationship among Lowell, the observatory he founded, and the small community of professional astronomers who attacked his theory. To give some sense of the popular as well as scientific response to the canal debate, I examine over two hundred entries about Mars and Lowell between 1894 and 1916 in the British weekly *Nature* (including articles, short notices, unsigned summaries of breaking news, and readers' letters). I concentrate on those years between 1905 and 1910 when the canal theory supposedly was falling out of favor. Then, as now, *Nature* presented both specialized articles and pieces aimed at scientifically literate but nonspecialist readers. Although many of these entries take no stand on the canal controversy, the vast majority of them treat Lowell with more respect and accord him more scientific legitimacy than they do his critics.

Finally, I examine the reasons why Lowell tended to find support among biologists and sociologists. Repeatedly invoking Darwinian theory, Lowell cast himself as the first exoecologist, insisting that the environment of Mars would shape its biota in nonterrocentric and nonanthropocentric ways. While the canal controversy began to die down after the opposition of 1909, none of Lowell's adversaries had advanced a widely accepted rejoinder to explain the wave of darkening spreading from the Martian poles to the equator. Against the backdrop of the droughts of the late nineteenth century, the canal theory drew on and informed different versions of ecological theory; it dovetailed both with a managerial ethic that sought to harness nature for humankind's economic benefit and with a darker entropic vision born of a biosphere in seemingly irrevocable decline. Mars, in brief, provided a complex metaphor for understanding planetary ecology; it could be and was used to support other metaphors for the laws that governed the operations of nature (Kingsland 1985, 207–8). In the decades after Lowell's death in 1916, as I argue in chapter 4, the canal theory was neither believed nor disproved, yet it continued to inform conceptions of planetary ecology.

THE NEBULAR HYPOTHESIS AND THE EVOLUTION OF WORLDS

Lowell's enthusiasm for Mars and the possibility of canal-building Martians antedated his first series of observations. He scrawled "Hurry!" in his copy of Flammarion's *Mars* in 1892 and scrambled to have his observatory built in time for the opposition of 1894. But this injunction can be read in more than one way. Flammarion and Schiaparelli were cautious about the possibilities of intelligent life on Mars, and throughout his career Lowell resisted being drawn into speculating about the nature of his canal builders. Instead, he emphasized the significance of Mars within an overarching theory of planetary evolution. In one of his last lectures, Lowell reiterated the premise that had guided his work for twenty years: "The study of [Mars] is as important to cosmic understanding in general as it is to knowledge of the planet's self" (1916b, 18). The red planet was not a singular curiosity but a crucial piece of evidence in comprehending the evolution and fate of the solar system. The "physical laws" of "cosmic understanding" that Lowell invokes are those described by the nebular hypothesis.

The nebular hypothesis was the brainchild of Pierre Simon Laplace and went through several refinements during the nineteenth century. In *Exposition du système du monde* (1796) Laplace theorized a natural rather than divine origin of the solar system. He replaced Newton's voluntaristic universe, one shaped and reshaped by the mysterious will of God, with a mechanistic system that obeyed regular laws.[3] Imagining a primordial, rotating nebular mass, Laplace suggested that as the mass cooled, its speed of rotation increased. The centrifugal force acting on particles in the plane of its equator eventually came to equal the gravitational attraction of the central mass. A large diffuse ring of nebulous matter then gradually coalesced into a ring (like those of Saturn), and the process repeated itself each time the centrifugal force became great enough to counterbalance gravity. In the rare instance of a uniformly distributed mass, the rings would continue to orbit (as they do around Saturn), but most would break up into rotating masses that would agglomerate and eventually form planets (Numbers 1977, 10). In 1802, Laplace met William Herschel, who had discovered what he called "nebulous" fluid around distant stars; the French mathematician seized on Herschel's findings and claimed them as "empirical evidence" to support his theory of the origin of the universe (Herschel 1912, 1: 415–16; Numbers 1977, 11). Subsequent refinements in Laplace's theory provided a coherent, mechanistic narrative of the origin of the solar system—no divine creation, only the immutable laws of physics.

By the mid-nineteenth century, the nebular hypothesis was widely accepted by scientists of opposing philosophical and theological views. Between 1840 and 1860, William Thompson universalized the second law of thermodynamics into a vision of the "heat death" of the known universe; in thirty million years, he calculated, entropy would bring the solar system to a halt (Clarke 2001, 5). Paradoxically, other theorists used similar logic to justify a theological explanation for the origin of the solar system. At Harvard, Lowell studied mathematics under the eminent American mathematician Benjamin Peirce (Hoyt 1976, 16–17). A stalwart defender of the nebular hypothesis, Peirce praised the "untiring genius of Laplace" but envisioned the universe going through successive ages of chaos, nebular formation, planetary maturity, and then a reordering of creation by a benevolent deity, with one universe endlessly succeeding another (Peirce 1881, 171). Peirce thus adapted the nebular hypothesis to support a new version of the argument from design and attributed the coherence and beauty of the natural world to divine intervention.[4] By invoking the deity, he could overcome the implications of the second law of thermodynamics: without such divine intervention, entropy led to a fascination—in science fiction, philosophy, and mainstream literature—with the ultimate destruction of Earth and its inhabitants (Clarke 2001, 2–3; Gold 2002, 449–64). As Ronald Numbers argues, the dynamic model of the universe offered by the nebular hypothesis "encouraged Victorians to envision all creation as an evolutionary progression" and consequently helped to foster "in both the scientific and religious worlds, that climate of opinion that made possible the rapid assimilation of [Darwinian] evolution" (1977, 109, 108). The nebular hypothesis therefore papered over the philosophical divisions between two radically different versions of the "evolutionary worldview": "guided" development engineered by a supreme being versus immanentist theories, including entropic heat death, committed to "unidirectional change, to finite domains of space and time; to mechanistic causation" (Brush 1987, 248, 268). In either case, "evolutionary progression" could be pressed into the service of conservative views of natural and social history that protected "man"—particularly white, middle- and upper-class "man"—from the unsettling implications of Darwinian theory.

Lowell wrote his undergraduate thesis on the nebular hypothesis and, in contrast to Peirce, committed himself to a relentlessly mechanistic interpretation of its evolutionary logic. If Peirce's solar system offers evidence of divine intention working through natural law, Lowell turns the nebular hypothesis into an entropic master narrative: decay is inevitable and the evolution of worlds heads toward an irrevocable—and lifeless—senescence. Rather than Peirce's anthropocentric cosmology, he describes a fundamental antagonism between human intelligence and an indifferent nature. Given Lowell's efforts after 1894 to distance himself from Harvard, his rejection of his former teacher's theology can be seen, in part, as an intellectual rebellion against arguments from design as well as an effort to yoke Darwinianism and entropy.[5] At the heart of his work lies an evolutionary and ecological antiromanticism. Lowell sees himself as the heir to a tradition of scientific secularism, and his Martians, whatever their engineering feats, are themselves victims of inexorable evolutionary laws. The canals are his proof that the nebular hypothesis explains all.

Even in his most strident defenses of the canal thesis, Lowell places Mars within this larger theory of planetary evolution. He defines "planetology" concisely as "the study of the laws governing the evolution of bodies revolving around a controlling sun in the long history of their careers from molten masses of matter, through world stages, to old age, decrepitude and decay" (1905a, 2). In *Mars as the Abode of Life,* Lowell identifies six stages of planetary evolution: "I. The sun stage. Hot enough to emit light. II. The molten stage. Hot, but lightless. III. The solidifying stage. Solid surface formed. Ocean basins determined. Age of metamorphic rocks. IV. The terraqueous stage. Age of sedimentary rocks. V. The terrestrial stage. Oceans have disappeared. VI. The dead stage. Air has departed" (1909, 12). The known planets of the solar system are then placed within one of these categories: the Moon and Mercury are at the "dead stage"; Earth obviously is at stage four; Venus, beneath its cloud cover, is at three or four; the gas giants, Jupiter and Saturn, are at stage two; and Mars is at stage five. Given this evolutionary taxonomy, Lowell insists that if sufficient carbon, hydrogen, oxygen, nitrogen, phosphorous, and sulfur exist "under suitable temperature conditions," then "it seems inevitable that life will ensue" on these planets because stage four dictates the appearance of water. In turn, water indicates that "the necessary temperature" is present for the evolution of a complex ecology because, he concludes, "the gamut of life [is] coextensive with the existence of water" (1909, 38–39). Although his critics debated what kind of life might arise on other planets, most accepted the broad outlines of Lowell's evolutionary taxonomy. By the late nineteenth century, Mars was considered, almost without exception, a smaller world that had cooled more quickly than Earth and therefore could be seen as a harbinger of its geological, climatological, and biological fate. Although the nebular hypothesis after 1900 was challenged by the Chamberlin-Moulton theory of the accretion of "planetesimals," the idea that Mars was an older world farther along in its planetary evolution was not seriously contested (Strauss 2001, 154–59). As Mary Proctor wrote in 1929, "Mars . . . illustrat[es] the fate awaiting our planet at some remote period, when it will have reached the stage of decrepitude now so plainly visible on our neighbour world" (1929, 85). Because their alternatives to the canal thesis assumed some version of this evolutionary cosmology, Lowell's critics were forced to counter his theories using the same assumptions on which his case rested.

While the nebular hypothesis provided the theoretical basis for Lowell's study of Mars, much of its popular appeal stemmed from his ability to turn the fourth planet into a landscape that could be experienced, at least imaginatively, in a way that Schiaparelli's straight lines and Flammarion's "viscous fogs" could not. Although trained as a mathematician and dedicated, during the last years of his life, to locating the mysterious Planet X that orbited beyond Neptune, Lowell insisted that planetology was an inductive science; astronomers could extrapolate from what they knew of the geology, climatology, and evolutionary biology of Earth to investigate the "nature, constitution, and history" of other planets (1905a, 2). Mars was thus fascinating in its own right and it foretold the fate of the Earth. "Pitiless as our deserts are," he assured his readers, "they are but faint forecasts of the state of things existent on Mars" (1906b, 157). Once Lowell had helped disprove the nineteenth-century commonplace that Mars had open bodies of water on its surface, the study of the planet's "nature, constitution, and history" became, in his mind, a function of its desertification—its slow but inexorable transformation from a "terraqueous" world to a stage-five planet, its oceans gone and its lifeforms engaged in a life and death struggle to obtain dwindling supplies of water.

In this respect, Lowell's conviction that Earth was beginning its inevitable decline to a Mars-like desiccation underwrites his analogies between the two planets. In his mind, the "lambent saffron" of the deserts of northern Arizona are reminiscent of "the telescopic tints of the Martian globe" (1906b, 149). Such analogies are emotive rather than intellectual: Lowell does not so much investigate the Martian ecologies he imagines as invoke a version of the scientific sublime, an admixture of terror and awe that cannot be represented by the familiar languages of terrestrial experience: "Deserts already exist on the earth, and the nameless horror that attaches to the word in the thoughts of all who have had experience of them . . . is in truth greater than we commonly suppose. For the cosmic circumstance about them which is most terrible is not that deserts are, but that deserts have begun to be. Not as local, evitable evils only are they to be pictured, but as the general unspeakable death-grip on our world. They mark the beginning of the end" (1909, 124). Lowell's insistence on the "nameless horror" of desertification in some measure belies his scientific interest in studying the semiarid ecology in the mountains near the observatory he founded.[6] Nonetheless, his fascination with the "unspeakable death-grip" of the nebular hypothesis is crucial to understanding his popularity. His sublime, even tragic, depiction of a dying world suggested that astronomy involved more than mere data gathering, and the glaring lapses that he made in underestimating the significance of greenhouse effects on planetary temperatures could be overlooked by even scientifically literate audiences who were swayed by the aesthetic and philosophical implications of his insistence that "the earth . . . is going the way of Mars" (1909, 122).

Lowell's success in popularizing his vision of Mars turned the seemingly abstruse nebular hypothesis into a plausible, even compelling portrait of a political as well as ecological world. In a story headlined "Tragic Struggle of the Dying Martians," the *London Daily News* in 1907 imagined Martians," physically and intellectually ages ahead of [humans], . . . fighting a Titanic battle against the inexorable laws of the Cosmos in an effort to stave off for a few centuries the annihilation of the race" (*London Daily News* 1907, Lowell Observatory Archives [LOA]). As melodramatic as these dying Martians may seem, they served important ideological functions in deflecting questions about conflicts over scarce resources on Earth. For some ecologists early in the twentieth century, such as Paul Sears, the problems of encroaching deserts, particularly in the western United States, were a consequence of human mismanagement and greed (Sears 1935; Mitman 1992; Worster 1993, 157–61). In contrast, rather than blaming human exploitation for drought conditions, Lowell and his followers argued that deserts were the result of inexorable evolutionary laws.

Lowell's vision of dying planets recasts the history of late Victorian imperialism in universalist and evolutionary terms, depoliticizing horrific conditions of drought and mass starvation. Mike Davis argues persuasively that the "great famines" in India, China, and Brazil at the end of the nineteenth century were the result of the "extraordinary succession of tropical droughts and monsoon failures" brought about by El Niño events between 1877 and 1902 and were exacerbated by the inequities of distribution, price gouging, and hoarding during "the golden age of Liberal Capitalism" (2001, 10, 9). Blame for these widely reported famines and the millions who died divided along political fault lines: apologists for European imperialism shrugged their shoulders or castigated the inability of traditional agricultural societies to adapt to the demands of a market economy; socialists and their allies decried the consequences of colonialism. A decade before he attacked Lowell's canal thesis, Alfred Russel Wallace, the noted naturalist and a committed socialist, characterized the famines in India and China, along with the slum conditions in European cities, as the worst failures of nineteenth-century civilization (Davis 2001, 8). In marked contrast to Wallace, Lowell describes desertification (and by implication famine) as an irrevocable consequence of planetary evolution. The tragic struggle of the dying Martians suggests, by analogy, that no blame for the "nameless horrors" of drought and starvation on Earth need be attached to capitalism or colonialism. Confronted by the prospect of Earth decaying to the "terrible reality" of the Martian deserts, middle- and upper-class readers of Lowell could contemplate ecological collapse and starvation from the safe distance of scientific speculation. In the wake of the famines of the 1870s and 1880s, invoking the "survival of the fittest" could insulate well-to-do readers from recognizing their complicity in the suffering caused by colonial exploitation. The implied kinship between heroic Martians struggling to survive on a dying world and privileged Europeans and Americans intent on justifying their way of life reinforced the argument that the millions who starved were victims of nature's laws, not human greed, indifference, and cruelty.

THE EVIDENCE FOR THE CANALS AND LOWELL'S "CHAIN OF REASONING"

After sending assistants across the globe to find a site that had the best possible conditions for planetary observation, Lowell established his observatory in Flagstaff, Arizona Territory in time for the opposition of 1894 (Strauss 1994, 37–58). Flagstaff was located farther south than the major observatories in Europe, and consequently during this and subsequent oppositions, Mars remained higher in the sky for longer periods of time.[7] Over the course of several months, Lowell and his assistants, William H. Pickering and A. E. Douglass, made 917 sketches of Mars, a number that far surpassed the efforts of other astronomers, including Schiaparelli. These observations of the canals confirmed what Lowell believed: Schiaparelli's canali were the herculean engineering project of a race struggling to survive on a world that had lost most of its water. In *Mars* (1895), Lowell articulated the "chain of reasoning" that led him to conclude that it was "probable that upon the surface of Mars we see the effects of local intelligence" (201). The conditions on Mars, though harsh, did not preclude "some form of life." The melting of the polar caps during the Martian spring indicated that temperatures were warm enough to melt water ice; the wave of darkening that swept from the poles toward the equator and then faded during the Martian autumn and winter could be explained simply and logically by the life cycle of vegetation. This counterintuitive growth pattern that began during the Martian spring at the coldest areas of the planet and then moved toward the equator suggested that water was being transported from the poles. The geometrical regularity of the dark lines noted by Schiaparelli made sense if they were connected to the problem of conveying water to irrigate the parched areas of equatorial Mars. The convergence of these lines at dark areas indicated the presence of "lands . . . artificially fertilized," and the "network of markings covering the [surface] precisely counterpart[ed] what a system of irrigation would"—or should—"look like" (201). The engineers who had masterminded this system of canals, Lowell later added, had to be "necessarily intelligent and non-bellicose," willing to put aside "their local patriotisms" in order to distribute scarce reserves of water across the planet's surface (1906b, 377; 1909, 207). Lowell sketched, named, and mapped these canals, creating on the face of his representations of Mars the outlines of an implicit narrative: a race's heroic efforts against the forces of nature (figure 1).

For many of his contemporaries, the plausibility of Lowell's argument compensated for its boldness. Lowell and his collaborators had conducted the most intensive and systematic study of Mars to date, and they argued that their findings were the result of superior technology, the excellent seeing at their observatory, their rigorous professionalism, and their experience as dedicated observers. Like Schiaparelli, Lowell was a firm believer in using small telescopes to observe planetary detail, and he insisted throughout his career that the key to observing Mars lay in "the combined performance of three factors: the observer, his instrument and the air through which he is compelled to look" (1905a, 7). Since the eighteenth century, astronomers had recognized that the larger the telescope, the greater the width of the column of air that has to settle to allow the planet's light to pass relatively undisturbed to the eyepiece, and Lowell hammered home the point that his critics lacked the advantages afforded by his new observatory: an excellent instrument, still air, and trained and patient observers (Lankford 1981a, 11–28; Sheehan 1988, 1996, 7; Strauss 1993, 159). To observe Mars, Lowell used a twenty-four-inch telescope, usually with the aperture damped down to reduce distortion: he argued repeatedly that "detail which would remain hopelessly hid with the full aperture of the 24-inch, starts forth to sight" when the aperture is reduced to eighteen or twelve inches (1905a, 25).[8] He derided those astronomers trying "to look for the canals with a large instrument in poor air," and asserted that such an endeavor "is like trying to read a page of fine print kept dancing before one's eyes" (1895, 140). The still air of Flagstaff and his small telescope were advantages that informed all of his claims about the canals.

Lowell insisted as well that his methods were superior to those employed by occasional observers of the planet. He claimed that he had

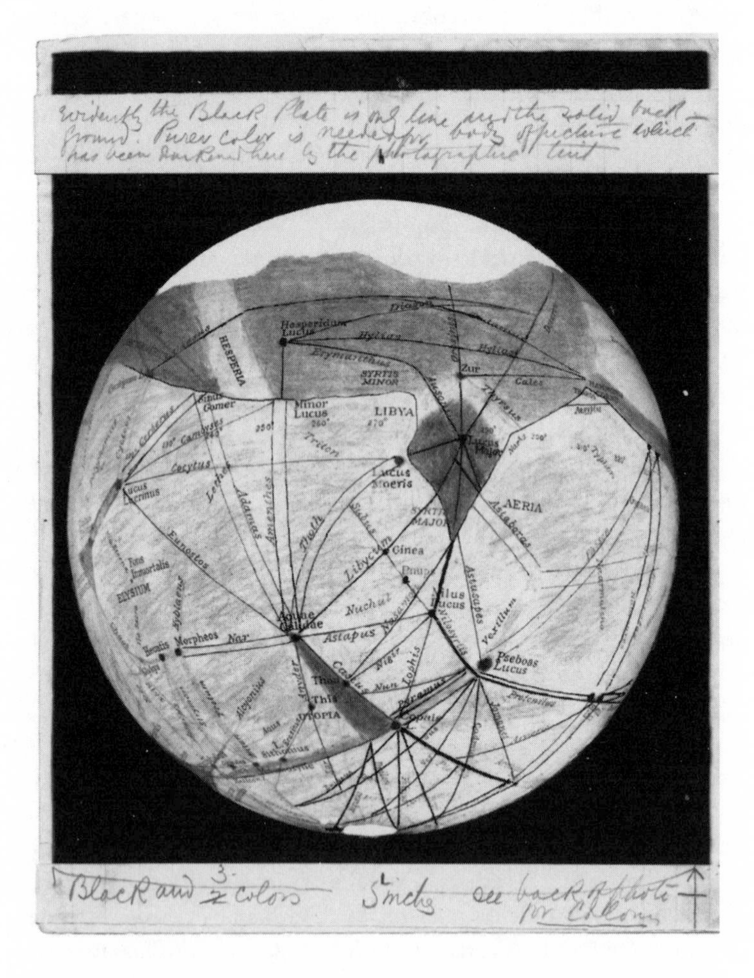

1. LOWELL'S SKETCH OF MARS, C. 1895, WITH INSTRUCTIONS TO THE PRINTER. COURTESY OF LOA.

trained his eye to unusual acuity and this training enabled him to see and record the canals. Continued observations, in effect, honed one's ability to recognize the fine, linear detail that could be discerned on the Martian surface: "success" in seeing the canals, he wrote, "depends upon the acuteness of the observer's eye and upon the persistence with which he watches for the best moments in the steadiest air" (1906b, 175). The archives at the Lowell Observatory include numerous sketches that Lowell revised minutely before publication (figure 2). Because he and his assistants seldom missed an opportunity to observe Mars during the oppositions between 1894 and 1909, they had seen Mars at many of these "best moments," and their experience outweighed the hit-and-miss observations made by stellar astronomers who happened to train their large telescopes on Mars in between other projects. Lowell never failed to paint his critics as hidebound conservatives who did not understand planetary science or who violated the basic premises of inductive science: painstaking observation over an extended period of time and the independent confirmation of the observed phenomena. An astronomer's "first opposition" Lowell told his audiences, is "always spent in learning how to see," and he questioned the commitment of his critics to the painstaking observation that he believed essential to planetary astronomy: to understand the significance of the canals "systematic, continuous, study is absolutely vital" (1905a, 33, 36). While many of his critics waited for comparatively rare periods of good or excellent seeing, Lowell and his collaborators observed, sketched, and produced riveting maps of an alien world.

Having established that "there is an apparent dearth of water upon the planet's surface," Lowell hinged his canal theory on a conditional clause: "If beings of sufficient intelligence inhabited it, they would have to resort to irrigation to support life" (1895, 201). The "if" lay at the heart of his disputes with many of his religiously inclined opponents. When some of his critics, like Wallace, maintained that intelligent beings had not evolved on Mars, Lowell countered that they were neglecting the evidence of both the nebular hypothesis and Darwinian theory. He challenged them to explain why life on the planet had stopped short of producing an intelligent species. When Peirce and later Wallace reasoned scientifically that human beings were unique in a theocentric cosmos, Lowell mocked human pretensions to significance: "Man is but a detail in the evolution of the universe. . . . That we are the sum and substance of the capabilities of the cosmos is something so preposterous as to be exquisitely comic" (1895, 210, 212). By decentering man in the universe, Lowell lay claim to the mantle of Copernicus, demystifying theological arguments by presenting himself as the standard-bearer for scientific secularism. His evocative prose underscores his effectiveness as a scientific popularizer.[9] Imagining what Earth would look like from space, Lowell reinvented an analogy that would dominate both science fiction and scientific speculation during the first half of the twentieth century: "From the standpoint of forty millions of miles distance, probably the only sign of [humankind's] presence here would be such semi-artificialities as the great grain-fields of the West when their geometric patches turned with the changing seasons from ochre to green, and then from green to gold. By his crops we should know him" (1895, 129). Before spaceflight, astronomers pondered endlessly what Earth would look like from outer space in an effort to understand what they were seeing when they observed other planets, particularly Mars. With no satellite images to draw on, Lowell's "semi-artificialities" of regimented crops turn scientific speculation into an evocative analogy predicated on the wave of darkening on Mars: seasonal changes on Earth are understood as back formations from Mars. And Lowell's parodic biblical echo brings human pretensions down to a material level—the agricultural transformations that civilization has wrought on the Earth. Human endeavor comes down to amber waves of grain.

As his "chain of reasoning" suggests, Lowell the philosopher of science operates outside of the Popperian falsification hypothesis: it is not science if it cannot, in theory, be proved wrong. He offered a probabilistic argument for the canals and challenged his opponents to overturn the arguments that he advanced. "Proof," Lowell declared, is "a supposition advanced" that "explains all the facts and is not opposed to any of them." Because he consistently casts his theory of the canals in terms of "the balance of probability" rather than ironclad certainty, he presents himself as eminently reasonable, open to competing views, and even-handed in his presentation of the evidence (1906, 160). "The point at issue in any theory," Lowell wrote, "is not whether there be a possibility of its being false, but whether there be a probability of its being true. . . . and the possibility that a thing might be otherwise, [offers] no proof whatever that it is not so" (1895, 6–7). Anticanalists are thus put in the position of arguing from negatives, and well into the mid-twentieth century astronomers had to contend with repeated sightings of the canals that kept alive the possibility that Schiaparelli and Lowell had seen actual markings on the surface of Mars. In large measure, the canal controversy hinged on the initial assumptions that astronomers made: if one accepted the premise of the canals, the evidence refuting their existence looked subjective, even

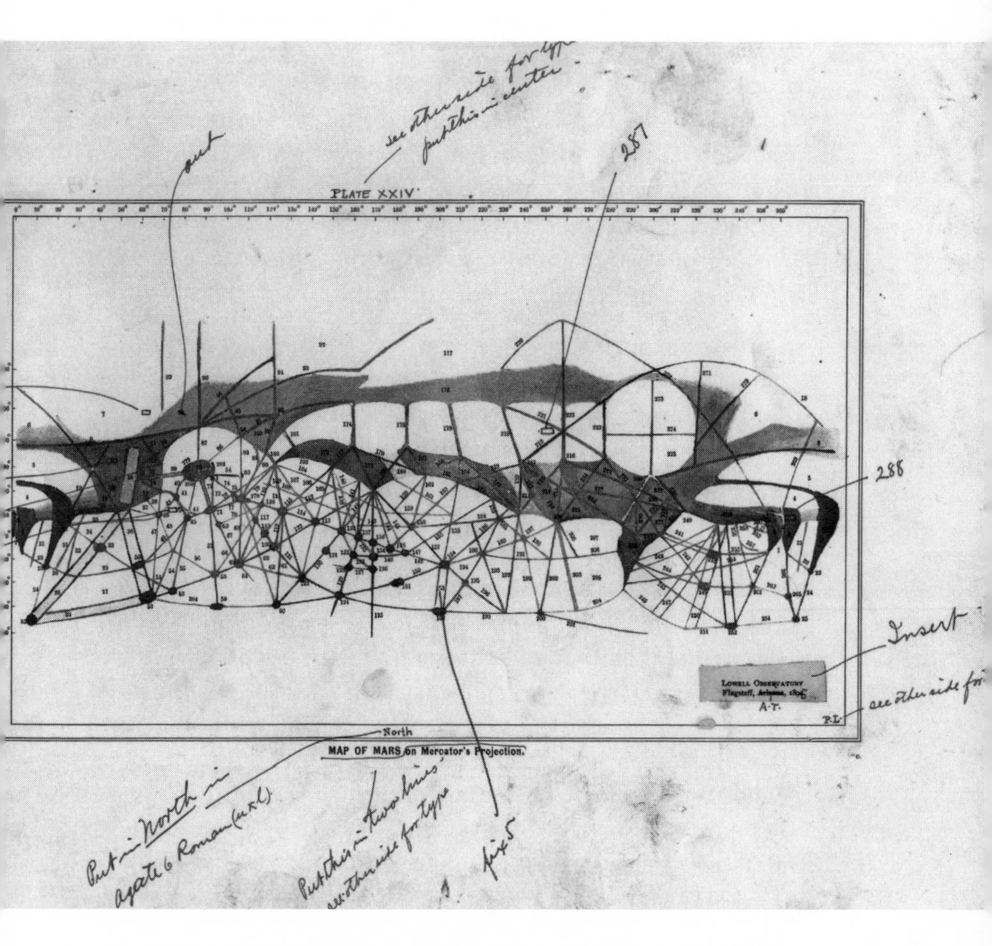

2. PROOF PAGE OF LOWELL'S SKETCH OF THE CANALS, DATED 1896, WITH HIS CORRECTIONS. COURTESY OF LOA.

skimpy; if one rejected the findings on which the canals were based—visual evidence or the spectroscopic data provided by Lowell and his associates—then the terms of the controversy could be reversed: the canals probably were not there and Lowell's thesis seemed as thin as the Martian atmosphere.

Lowell's yoking of the nebular hypothesis and Darwinian theory, though, allowed him to cloak his suppositions and hypothetico-deductive leaps of faith in the rhetoric of inductive methodology and the excitement of a paradigm-shattering discovery. In his most explicit statement of his philosophy of science, Lowell describes astronomy in the language of Victorian detective fiction:

> Discovery of a truth in the heavens varies in nothing, except the subject, from the discovery of a crime on earth. The forcing of the secrets of the sky is, like the forcing of a man's, simply a piece of detective work. . . . In astronomy, as in criminal investigation, two kinds of testimony require to be secured. Circumstantial evidence must first be marshaled, and then a motive must be found. To omit the purpose as irrelevant, and rest content with gathering the facts, is really as inconclusive a procedure in science as in law. . . . Unless we can succeed in assigning a sufficient reason for a given set of observed phenomena, we have not greatly furthered the ends of knowledge and have done no more than the clerkage of science. A theory is just as necessary to give a working value to any body of facts as a backbone is to higher animal locomotion. . . . Coordination is the end of science, the aim of all attempt at learning what this universe may mean. And coordination is only another name for theory. . . . (1909, 184–85)

This passage suggests why Lowell succeeded as a popular writer. "Coordination," "purpose," "motive," and "sufficient reason" all point to logical and narrative structures that persuade by integrating "facts" into a structure with a beginning, middle, and end. Lowellian Mars is, in part, the product of a well-turned story. "Theory," as its Greek etymology suggests, is a way of seeing, but theory, for Lowell, produces a literary effect beyond that of "clerkage," the mere "gathering of facts" that can lead to no discovery. However uncertain Lowell's observations, however problematic his insistence on the acuity of eyesight as a determining factor in seeing the canals, his narrative gave compelling imaginative shape to his speculations. Two decades before Edgar Rice Burroughs, Lowell had etched his dying planet in the public imagination.

However controversial the canal theory, *Mars* made widely acknowledged contributions to planetary astronomy. By 1900, Lowell's argument that there were no open bodies of water on Mars, except for the seasonal, shallow polar seas that resulted from the melting of the polar caps, was generally accepted, and his description of a planet dominated by vast deserts, "world-wide in [their] extent, girdling the planet almost completely in circumference" was unchallenged (1909, 134). None of his opponents could come up with a better explanation for the seasonal wave of darkening that swept from the poles toward the equator during the Martian spring; most believed that melted snows percolated toward the equator without massive irrigation. Many of his critics, in short, challenged the canal theory but did not fundamentally question Lowell's description of a dying world. The narrower the grounds of the canal debate, the more Lowell could lay claim to a central position in the small field of areology.

Rather than a dogmatic theory, then, Lowell's canal thesis emerged dynamically in conversations and debates with his predecessors, adversaries, readers, and audiences. In their various manifestations, the canals of Mars are dialogical models of imagined ecological and cultural responses to planetary catastrophe. They are the products of discourse—of writing about and debating the response of evolutionary "intelligence" to the prospect of a dying planet. As discursive constructions that are always in the process of being redefined, the canals spill across disciplinary boundaries between astronomy and ecology, observation and speculation, science and science fiction. They served a variety of purposes for Lowell and his followers, and they retained their hold on the popular imagination precisely because they were an evocative symbol that represented the power of science to transform perceptions of humankind's place in the cosmos.

LOWELL'S CRITICS

From the start, Lowell had critics among professional astronomers; as David Strauss suggests, the canal debate "was shaped by a number of changes in American astronomy in the 1890s," including "the rise of astrophysics as the most prestigious form of astronomical activity, the related emergence of the factory observatory with its emphasis on hierarchy and specialization, and the increasing dominance of professional as opposed to amateur astronomers" (1998, 97). As a rule, stellar astronomers, particularly in the United States, kept Lowell and the canals at arm's length. Yet the adjective "professional" applied to astronomers in the late nineteenth and early twentieth centuries is itself a complex term. Professionalization was, as John Lankford notes, "a contingent, historical, and political process" (1981a, 11), and this process—particularly in the small field of planetary astronomy—often had less to do with peer review than with the conditions of employment. Professional astronomers were paid salaries at institutional observatories with large telescopes. These instruments were designed to maximize the light-gathering capacities necessary for stellar astronomy, and the research undertaken at the Yerkes, Lick, and Mt. Wilson observatories concentrated on mapping interstellar space. The noted historian Agnes Clerke observed a long-standing division as far back as the mid-nineteenth century between stellar astronomers and "observers who spent their nights in scrutinising the faces of moons and planets rather than in timing their transits" (1902, 2). In 1900, both groups of astronomers numbered in the dozens rather than the hundreds. Professional astrophysicists were a small group within the larger scientific community, and Lowell's isolation from this "new generation of practitioners" cannot be equated unproblematically with a widespread rejection of the canal hypothesis either by scientists in other fields or by educated nonspecialists (Strauss 1998, 101). Many of Lowell's antagonists—Simon Newcomb, Edward S. Holden, William Wallace Campbell, George Ellery Hale, Antoniadi, and Barnard—either held only undergraduate degrees or, in the case of the latter two, never completed university educations. In brief, Lowell's qualifications to make pronouncements about Mars were on par with those of his critics. After graduating from Harvard with a degree in mathematics, he was offered a professorship at MIT, which he declined (years later, he became a nonresidence professor at that institution). None of Lowell's contemporaries, including those who questioned the canal thesis challenged his diligence, methods, or the world-class status of the Lowell Observatory. Perhaps more important, his critics were often divided among themselves—some contending that the canals did not exist, others that they were natural features, perhaps volcanic fissures, on the surface. The professional community of astronomers remained skeptical or agnostic about the canals but never launched a concerted counteroffensive—at least before 1916—that demolished Lowell's paradigm.

Between 1892 and 1909, four basic, though often interanimating (and in one case conflicting), lines of argument were advanced to refute Lowell. Several prominent astronomers who observed Mars never saw any canals; others saw streaks or linear formations on the surface but interpreted them as natural formations. Both of these critiques found support in the work of W. W. Campbell. In 1892 and again early in the twentieth century, Campbell published spectroscopic results that showed no water or oxygen in the atmosphere of Mars (DeVorkin 1977, 37–53). His results, however, were contested by Lowell's colleague, V. M. Slipher, who repeatedly got positive results for water vapor, confirming the chemical signatures obtained in the nineteenth century by Sir Norman Lockyer, Jules Janssen, and others. (The latter stages of this debate were played out in *Nature* and in other scientific publications, as I shall discuss below.) The most serious challenge to Lowell's thesis, though, was advanced in 1903 by the British astronomer E. W. Maunder, who argued that the canals were tricks of the eye and mind that led otherwise competent scientists to play connect the dots among discrete surface features or dark patches on the disk of Mars. Lowell and his allies responded to each of these criticisms, and the debates reveal a good deal about Mars as a culturally as well as scientifically contested landscape.

As early as the early 1890s, some astronomers questioned Schiaparelli's observations and suggested that the Italian astronomer was reducing complex natural formations to an angular geometry. Rather than networks of straight lines, the astronomer and artist Nathaniel Green saw on the surface of Mars "the most delicate mottling of shapes and streaks of shade, and it requires a long training of the eye to see and follow the extent of these shades, and a considerable training of the hand to imitate them when seen" (cited in Sheehan 1996, 109; see also 105–8). Green's emphasis on the difficulty of interpreting and then drawing the patterns one sees makes planetary astronomy an aesthetic as well as scientific endeavor. Objectivity, in this regard, becomes a function of artistic as well as scientific "training," an acquired ability to represent, with a minimum of distortion, a surface characterized by "shades and streaks." Schiaparelli's Mars, for Green, simplifies a mottled, seemingly more earthlike landscape.

During the same opposition of 1892, E. E. Barnard voiced similar reservations about the canals. Using both the twelve- and thirty-inch refractors at the Lick Observatory, Barnard saw complex details on the Martian surface but no linear markings (1892, 680–84). Two years later, Edward S. Holden, the first director of the Lick Observatory, reacted to a newspaper description of Lowell's search for life on Mars as "very misleading and unfortunate" (1894, 160). Like Holden, Barnard held firm to his skepticism even as Flammarion and other astronomers became convinced of the reality of the canals (Sheehan 1995, 195–96). In a letter to Simon Newcomb in September 1894, Barnard described "the surface of

Mars" in starkly different terms from those employed by Lowell and his assistants: "To save my soul I can't believe in the canals as Schiaparelli draws them. I see details where he has drawn none. I see details where some of his canals are, but they are not straight lines at all. When best seen these details are very irregular and broken up—that is, some of the regions of his canals; I verily believe—for all the verifications—that the canals depicted by Schiaparelli are a fallacy and they will be so proved before many favorable oppositions are past . . . It is impossible to adequately draw all that can be seen" (quoted in Sheehan 1995, 246). To twenty-first century readers, Barnard's description seems extraordinarily prescient, a dead-on anticipation of what we now know the surface of Mars to be. Schiaparelli's straight lines decompose into artifacts of the imagination or "irregular" detail. Barnard finds "mountains" and "plateaus" where Lowell, painstakingly studying the Martian terminator, concluded that its irregularities were the result of transitory cloud formations, and that therefore the surface of Mars had to be flat.[10] Yet as Barnard's language suggests, his critique of Schiaparelli is based on "belief" as much as it is on objective standards of visual acuity. With a thirty-six inch refractor, no matter how good the conditions of seeing, the most dramatic features of Martian topography—the three-thousand-mile-long Valles Marineris and the huge shield volcanoes—are not visible. In this respect, no matter how astute an observer Barnard may have been, he was not describing self-evident features on the surface of Mars that Schiaparelli and Lowell should have seen, but offering an interpretation of "irregular" detail based on analogies to terrestrial topography. At a time when many observers, including Schiaparelli, persisted in calling the dark areas on Mars "seas," Barnard describes albedo differences in terms of other familiar geographic features—mountains and plateaus. His "details," as he admits, are "impossible" to reproduce accurately and consequently do not lend themselves to the kind of dramatic narrative that Lowell was concocting in Flagstaff. In this respect, Barnard's skepticism about Schiaparelli's canali neither disturbs the age-old analogy between Mars and Earth nor demands a revaluation of the theoretical contexts in which Mars, prior to Schiaparelli, was described.

Before the controversy over Maunder's experiments, Lowell already had staked out a rejoinder to suggestions that Schiaparelli had converted a profusion of detail into a network of interconnected lines. Predisposition, he argued, is a two-edged sword: "A few years ago it was the fashion not to see the canals of Mars, and nobody except Schiaparelli did. Now the fashion has begun to set the other way, and we are beginning to have presented suspiciously accurate facsimiles of Schiaparelli's observations" (1895, 160). As he does elsewhere in *Mars*, Lowell assumes the role of the skeptic, trying to guard against the effects of predisposition and expectation. Characteristically, he argues that the differences among various observers' drawings of the canals actually confirm the validity of his and his associates' methods:

> In any observation, the observer is likely to be unconsciously affected in some way or other *pro* or *con*, which, from the fact that he is unconscious of it, he is unable to find out. The only sure test, therefore, is the seeing what no one else has seen, the discovery of new detail. Next to that is not too close an agreement with others. Inevitable errors of observation, to say nothing of times and seasons, distance and tilt, are certain to produce differences, of which one has ample proof in comparing his own drawings with one another. Even too close agreement with one's self is suspicious. In the matter of fine detail, absolute agreement is therefore neither to be expected nor to be desired. (1895, 160)

"Fashion," for Lowell, goes beyond the power of suggestion to encompass what cultural critics would now call "ideology"—the complex of values, assumptions, and beliefs that are deeply imbricated in, and in complex ways coextensive with scientific, philosophical, and cultural worldviews. Ideology remains "unconscious," then, because there is no analytical vocabulary that can locate itself outside of its cultural and historical contexts in order to issue objective pronouncements about "reality."[11] Because perceptions of Mars and its canals are shot through with assumptions and values about ecology, humankind's role in the universe, and the legitimacy of speculation in science, Lowell paradoxically must try to press his inductive methodology into the service of his hypothetico-deductive "chain of reasoning." The canals are not the product of one man's vision but a collaborative enterprise that depends on accumulated observations and a process of mutual correction through disagreements about what has been seen. Even Lowell's dubious assertion that the "sure test" of visual acuity is seeing what has not been seen previously involves an inductive and self-critical process of refining and recalibrating sketches of the Martian surface. In this respect, the 917 drawings made by Lowell, Pickering, and Douglass during the 1894 opposition are a data archive against which subsequent observations can be tested, refined, and used to redraw the latticework of canals. Because "absolute agreement" is neither "expected nor . . . desired," Lowell can subsume his critics' attacks within a collaborative model of scientific progress: the more observations are made, the more the accumulated evidence will reveal a consensual model of Mars and its canals.

The broad outlines of the canal debate were drawn by 1896, and the implications of Lowell's arguments clearly grasped by his contemporaries. Two influential reviews of *Mars* helped to shape the contours of that debate for the next decade: a negative response by Agnes Clerke in the *Edinburgh Review* and a positive reaction from the son of the editor of *Nature,* W. S. Lockyer. Taken together, these reviews suggest both the range of responses that Lowell elicited and the ways in which his critics and admirers responded to his "chain of reasoning."

The author of the standard history of nineteenth-century astronomy, Clerke provided the framework for most of Wallace's arguments a decade later, as the biologist acknowledged. She begins her review by acknowledging that *Mars* offers "a marked advance in [understanding] Martian topography" and praising it as "eminently readable," but then turns its very readability against its author: "As a contribution to science," she writes, "this pleasant and clever work can scarcely be taken quite seriously" (1896, 371). In her mind, the key scientific problem that Lowell fails to explain adequately is the seasonal melting of the polar caps. Given its distance from the sun and low albedo, Mars should be frozen into a permanent winter, but the melting and reforming of the polar caps clearly indicates that a greenhouse effect is raising the planet's temperature considerably: "Some ingredient" in the Martian atmosphere, Clerke reasons, must be "acting more powerfully than even water-vapour" in retaining solar energy (373). Yet the rapid disappearance and reappearance of the polar caps indicates that they must be "quite flimsy structures" containing at most the equivalent of one foot of water over a surface area of 2.4 million miles, hardly enough to irrigate the dark areas that Lowell believes are under cultivation. If this estimate is close to the mark, and one assumes that some water must evaporate from the melting polar caps, Lowell's hypothesis becomes "hopelessly unworkable" (379, 380). Even if this meager water supply were used for irrigation, there would be no means for it to return to the polar caps each winter: the canals would disrupt the planet's hydrological cycle and the polar caps would disappear permanently. Consequently, while Clerke admits that the canals—"the very strangest of planetary phenomena"—do exist, she denies that their "extraordinary complexity" could be used for planetwide irrigation (374). Where Lowell challenges his critics to disprove his "chain of reasoning," she inverts his logic to argue that he has not proved his case. If his theories cannot be shown "in all shapes, demonstrably false," they nonetheless "open the door to pure license in theorising" (384). Because the wave of darkening on the surface of Mars can be explained by primitive vegetation, Clerke concludes that the irrigation hypothesis becomes "superfluous" and therefore "inadmissible" as scientific evidence. More fundamentally, she suggests that Lowell's "inferences" from the nebular hypothesis that Mars "is far advanced in senile decay" are "questionable": the planet in 1896 remains "a sealed book" (382–83).

As compelling as Clerke's argument seemed to Wallace in 1907, it led her to reiterate the dubious theory that the dark areas on Mars were open bodies of water. If they were merely Lowell's oases, the planet's hydrological cycle, as she understood it, would cease to function because there would be no way for the polar caps to be replenished during the Martian winter. The "Martian seas cannot be abolished," she maintains, because "their presence [is] indispensable to the systemic and rapid circulation of water" (384). She is left, then, in the awkward position of accepting the reality of the linear markings—recognizing that they must be related to a hydrological cycle that eventuates in shallow, seasonal waterways and the restoration of the polar caps—and yet unable to describe the relationships among these phenomena. At one point Clerke compares the supposed canals to what she imagines the Gulf Stream would look like if humans could view it from Mars, but admits that any theorizing about the planet in the absence of "definitive evidence" must remain "nebulous conjecture" (384). Even as she calls into question Lowell's irrigation hypothesis, her own "conjectures" proved less durable than his as the "Martian seas" soon vanished into science fiction. His narrative "chain of reasoning" offered philosophical and aesthetic consolations that her description of Mars as a "sealed book" did not.

Writing with the implied editorial imprimatur of *Nature,* Lockyer took a more positive view of *Mars,* though he stopped short of wholeheartedly endorsing the canal theory. *Nature* had been founded by his father, Sir Norman Lockyer, in 1869, and this prominent solar astronomer edited it until his death in 1919. By 1896, it was widely recognized as the leading journal in Britain for both specialized articles as well as more accessible columns, letters, and short features. For the younger Lockyer, *Mars* is a model of scientific professionalism that describes "the most important" series of observations ever undertaken of the planet (Locker 1896, 625). In contrast to Clerke, Lockyer takes the canal thesis seriously. Even if it seems "rather premature for [Lowell] to draw such decided conclusions," the American astronomer remains well within scientific standards of observational rigor, and he is praised for his openness to alternative explanations. Whatever presuppositions Lowell may have brought to the study of Mars, Lockyer notes that "his own words show that . . . these views may be considerably changed by future observation, and he has not, therefore, tied himself too fast to them" (Lockyer 1896, 627). Rather than depicting the diplomat-turned-astronomer as an amateur pursuing a dubious thesis, Lockyer echoes Lowell's probabilistic language to describe a thesis that is still a work in progress. His review concludes by praising Lowell for having "added greatly to our knowledge of the surface-markings" and concludes that "astronomical science owes him a debt of gratitude" (627). Although Lowell had not proved his case, he had, for Lockyer, put the burden of proof on the anticanalists. As the Martian waterways of the nineteenth century disappeared, so did one obstacle to a tacit acceptance of Lowell's description of the planet. In the fourth edition of her *Popular History of Astronomy during the Nineteenth Century,* Clerke muted her previous critique of Lowell's canals: "Fantastic though the theory of their artificial origin appears, it is held by serious astronomers" (1902, 280). Some "serious astronomers," however, soon offered more pointed critiques.

IMAGINARY LINES

E. W. Maunder, the founder of the British Astronomical Association, had seen "canals" on Mars in the 1880s but believed that they were shadings within the complex of surface markings that Green and Campbell had noted. As early as 1894, he had done "rough little experiments" to demonstrate that a "narrow dark line" is visible "when its breadth is far less than the diameter of the smallest visible dot." More ominously for the canal enthusiasts, he argued that "a line of detached dots will produce the impression of a continuous line" when the dots are too small or too close together to be seen individually (1894, 251). In 1903, he and J. E. Evans sought to confirm experimentally the tendency of the mind's eye to see lines where there were only irregular details. Schoolboys from the Royal Hospital School at Greenwich were asked to copy "a canal expurgated picture of the planet [and they] supplied the lines which had been preceptorily left out" (Lowell 1906b, 202). Because most of the students sketched straight lines, Maunder and Evans argued that the mind was accustomed to seek linear order where there was none. The British astronomers then maintained that Lowell and Schiaparelli were similarly tricked by this perceptual tendency: "They have drawn, and drawn truthfully, that which they saw" yet "the canals . . . have no more objective existence than those" imagined by the "Greenwich schoolboys" (Maunder and Evans 1903, 488). The longer that Lowell and his followers stared at the disk of Mars, the more likely discrete, irregular features would assume the characteristics of canals.

For many historians of astronomy, Maunder's argument heralds the beginning of the end of the canal controversy, even if it took two generations for the last of the canalists to die out. The initial reaction by many scientists to the schoolboy experiment put Lowell on the defensive; W. H. Pickering, who never doubted the existence of the major canals, conceded that Maunder and Evans's experiment had "materially weakened" Lowell's hypothesis (1921, 111). Yet to see Maunder as a standard-bearer for modern science demolishing the pseudo-scientific reasoning of an amateur obscures both the limitations of his critique and his rationale for attacking the canal thesis. As Michael Crowe notes, Maunder was the son of a minister and a "very active member of a small pentecostal and adventist denomination"; he argued in several articles and books against the plurality of worlds thesis on explicitly theological grounds, and resorted to quoting psalms and paraphrasing the Bible in his scientific works (1986, 490–91, 542). Science, for Maunder, is not science if it challenges the God-given, anthropocentric order of the universe. No less than Lowell's theory, his refutation exists within a matrix of scientific, philosophical, and theological assumptions.

While Lowell never explicitly castigated Maunder or his other critics for their religious beliefs, he responded to the illusionist argument with characteristic irony. In his lectures he claimed that it is "as easy from a preconception not to see what exists as to see what does not" (1905a, 29), and noted that Schiaparelli had considered and rejected a version of Maunder's argument in the 1880s. To counter Maunder and Evans, Lowell conducted his own experiments to demonstrate that a wire .0726 inches in diameter was visible at 1,800 feet, the equivalent of .69 seconds of arc. Extrapolating these results to the distance between Mars and Earth, he maintained that he had proved that fifty-mile-wide strips of vegetation could be seen, by an eagle-eyed observer, on the Martian surface (1909, 270–79). In *Mars and Its Canals,* he noted that Flammarion had repeated the Maunder-Evans experiment with French schoolboys, but his subjects did not draw lines between dots. Paraphrasing Sir Norman Lockyer, Lowell concluded that "it looks as if some leading questions had unconsciously been put to [Maunder and Evans's subjects]"; and he characterizes their experiment as "one of those deceptive half-truths which is so much more deleterious than an unmitigated mistake" (1906b, 202; see

Hoyt 1976, 165). This rebuttal produced an exchange that lasted for several years. By 1905, however, Lowell believed that he had proved the reality of the canals once and for all when the expedition he financed to the Andes returned with C. O. Lampland's photographs of Mars. Published side by side with Lowell's drawings, these striking photographs seemed to many observers to prove conclusively the existence of the canals (Lowell 1906a, 132–36).

After Lowell published his photographs of Mars (a topic I treat in depth in the following section) the illusionists were more muted than Maunder and Evans had been. The American astronomer Simon Newcomb responded to Lowell's counterattack by conducting a different version of the illusionist experiment. Newcomb had well-known astronomers—S. I. Bailey, W. H. Pickering, Barnard, and Philip Fox—stand 96 feet from a 38 centimeter circle filled with dotted lines and sketch what they saw. The size of the circle, Newcomb argued, approximated that of Mars during recent oppositions, and the four astronomers drew continuous lines where there were only dashes (Newcomb 1907, 12). Although Newcomb acknowledged "the unequaled continuity of [Lowell's] observations, the care with which the minutest details were looked for, and the generally critical character of their entire discussion," this experiment reinforced his skepticism about the existence of Martian canals, and he advocated additional tests on "visual inference" (13, 17). In replying to Newcomb, Lowell described his own tests that revealed straight lines on paper could be discerned with a telescope at 585 feet, and these results, he maintained, countered Newcomb's initial assumptions about the size of a disk needed to approximate the appearance of Mars and the optical effects that could be measured at that distance (Lowell 1907, 131–40). As in the case with the Maunder exchange four years earlier, both sides claimed to have decisive experiments to support their views, both claimed to have bloodied their opponents, and both felt they had the better of the exchange.

Before 1907, however, the anticanalists had developed another line of argument. In a series of articles in *Popular Astronomy* in 1904, Lowell's one-time assistant William H. Pickering argued that the "canals" were natural formations, analogous to the lines radiating from the dark areas on the moon. These lunar features were commonly supposed to be geological faults or cracks somehow connected to what were taken to be extinct ocean basins. To support his contention that there were natural explanations for the largest canals, Pickering drew a series of comparisons to smaller volcanic cracks on the Earth as well as to features on the lunar surface (1921, 77, 104). Although he doubted the existence of Schia-

The anonymous *Politics and Life on Mars* (1883) includes what Darko Suvin terms "some remarkable utopian-socialist traits" (1983, 25), and in 1891 two women in Iowa published the most interesting of the early Mars novels. In *Unveiling a Parallel: A Romance,* Alice Ilgenfritz Jones and Ella Merchant depict a feminist utopia in which a thick-witted American narrator visits two Martian societies, Thursia and Caskia. In the first, he is confronted by women who vote, operate businesses, hold political office, pay taxes, take lovers, drink intoxicating beverages, "vaporize" (smoke) a potentially harmful substance, and have their own secret societies. The gender equality he observes—the heroine and her brother "respected and honored each other equally" (Jones and Merchant 1991, 19)—both baffles and excites him: he falls in love with the pleasure-seeking and hard-headed businesswoman Elodia, who is as beautiful and charismatic as she is unbound by Victorian conceptions of womanhood. He is in love, she treats him as a fling. In the second half of the novel, the narrator travels to Caskia where a classless society shares the planet's abundant resources equitably so that its members can concentrate on spiritual attainment. The hero once again falls in love, this time with Ariadne, both heiress and schoolteacher, who encourages his spiritual growth toward a kind of religious holism. By the end of the novel, Christian values have been redefined to counter the repressive and pleasure-denying aspects of nineteenth-century theology. The Mars on which women are equal, significantly, is pre-Lowellian: there are no canals, no horrifying deserts spreading across the planet, and no environmental threats to the continued vitality of this gender-neutral society.

For Jones and Merchant, then, Mars retains many of the traits of the idealized utopia: its landscape is edenic, its resources are abundant, and physical labor has been rendered almost effortless by electricity. Paradoxically, though, the authors' vision of a feminist and socialist paradise shares the same set of ecological assumptions about Mars as Greg's conservative fantasy. As a genre, utopias require a natural world that has been rendered infinitely productive, that fulfills absolutely its inhabitants' desires. On Mars, "nature's" ability to satisfy these desires both structures and is structured by the interanimating fantasies of unalienated labor and the abundance of pollution-free power. In both *Across the Zodiac* and *Unveiling a Parallel,* electricity is imagined as an infinite resource that is generated without negative consequences for the environment; mining gets only passing mention as a characteristically antiseptic enterprise, and coal mines, air pollution, and disgruntled workers are banished to a remote past. The Mars of Greg, Jones, and Merchant, in short, is not a dying planet but a vision of what terrestrial society should or might become in the future. The authors spend little time or energy in making the perpetual spring on Mars scientifically plausible; the golden-age landscape is a mythic symbol for the dream of technoscientific transcendence of the problems and inefficiencies of industrialism. Even before Lowell, Martian climatology defines its inhabitants' social destiny.

MARS AND THE POLITICAL IMAGINATION

The effect of Lowell's *Mars* on science fiction was almost immediate, although some minor novelists, such as Edwin Arnold in *Lieutenant Gulliver Jones* (1905), continued the tradition of earlier novels by reducing comparative planetology to "the similarity of many details of [human] existence" (Arnold 1965, 150) on Earth and the red planet. Between 1897 and 1908, three remarkable novels brought home the prospect that encounters with alien civilizations could be more dangerous, compelling, and illuminating than utopian travelogues or satiric parallels unveiled. In *The War of the Worlds,* H. G. Wells crafted the first great novel of interplanetary invasion, a work that exploits the dark underside of the evolutionary worldview by imagining humanity as meat on the hoof for Martian invaders. Less well known to English readers, Kurd Lasswitz's *On Two Planets* (1897) is a moral, political, and philosophical investigation of the complexities and consequences of colonization. An advanced society on Mars eyes the resources—water and sunlight—of Earth and gradually assumes the interplanetary white man's burden of civilizing the reluctant inhabitants of late-nineteenth-century Europe. In *Red Star* (1908) and its prequel, *Engineer Menni* (1913), the Russian revolutionary and physician Alexander Bogdanov reinvigorates the tradition of Martian utopias by depicting the struggles within the mind of his Bolshevik hero to acclimate himself to the socialist paradise that exists on the fourth planet.

In different ways, these novels redefine the generic possibilities of science fiction by offering three versions of an apocalyptic colonialism that haunts the European imagination (Alkon 1994, 47–49). Significantly, Wells, Lasswitz, and Bogdanov were all educated as scientists, and seized on the specifics of Lowell's dying planet to think through the evolutionary implications of an inhabited Mars.[1] All three also were politically left of center: Wells was a socialist, a supporter of women's rights, and a critic of British imperialism; Lasswitz a liberal philosopher and historian whose politics may have cost him a university position; and Bogdanov a committed revolutionary who worked closely with (and, at times, bankrolled)

Lenin. Mars offered these novelists the chance to imagine possible futures for turn-of-the-century European civilization: a complacent and exploitative society sleepwalking to the brink of holocaust (Wells), a civilization eager to transcend the material constraints on progress (Lasswitz), and a marxian utopia (Bogdanov). Yet these differing visions of Mars share a conceptual basis in Lowell's vision of a dying world. In different ways, all three novelists project humankind's ravaging of the Earth's natural resources onto the implacable course of the nebular hypothesis. Even as *War of the Worlds, On Two Planets,* and *Red Star* confront the consequences of pollution, class antagonism, and economic imperialism, they register the anxieties of a rapidly urbanizing and resource-hungry European civilization. In these novels, the fate of Earth already has been inscribed on the face of Mars, and the Martian civilizations register the moral as well as ecological destinies of a dying planet.

In *War of the Worlds,* Wells offers a classic dystopian inversion of European imperialism, projecting onto the Martians a radical alterity, an obscene voraciousness that shocks readers by its very indifference to the moral and social codes that define "civilized" behavior. The nonhumanoid Martians are the implacable instruments of an indifferent universe; they embody a Darwinian absolutism that operates without regard to the moral strictures of human society or convenient reassurances that evolution is part of God's plan.[2] In contrast, Lasswitz, a committed Kantian, holds out a humanist hope for progress through a process of mutual education. Across the gulf of interplanetary space, humans and Martians share desires, beliefs, and conceptions of authority and order; even during the Martian occupation of Earth, the heroes and heroines on both planets embody the faith that cultural differences can be reconciled and antagonisms transcended. Ultimately, humans can end the Martian occupation by capturing a few of the invaders' aircraft only because the inhabitants of the two planets are socioculturally as well as ontogenetically compatible. This liberal hope for interplanetary understanding across the Darwinian divide is both radicalized and undercut by Bogdanov. The utopian socialism that the hero, Leonid, experiences on Mars literally drives him insane, and the debate at the center of the novel between those Martians who favor annihilating humanity and grabbing the planet's resources for their "higher" civilization and those who favor mutual accommodation and interplanetary enlightenment makes clear the brutal costs of maintaining an ideal society. In this respect, what distinguishes Bogdanov from other nineteenth- and twentieth-century utopianists is his insistence that a socialist "paradise" can be gained only by the unceasing war of technologically resourceful "humans" against a hostile nature. Lasswitz's Martians synthesize food from rock; Bogdanov's fictional race confronts the exhaustion of its coal, iron, aluminum, and foodstuffs. The alternatives that Bogdanov poses for his socialist Martians—invasion or accommodation, communism or a Hobbes-eat-Hobbes descent into selfishness and self-destruction—define the generic limits imposed by Lowellian Mars.

WELLS'S *WAR OF THE WORLDS*: APOCALYPTIC DISINTEGRATION

A year before *The War of the Worlds* appeared in serial form, Wells speculated about life on Mars in the *Saturday Review* on April 4, 1896. Scientifically trained and committed to the Darwinism that figures prominently in his fiction, Wells maintains that the question of intelligent life on the red planet "remains unanswered, probably unanswerable." Although spectroscopic analysis indicates that "the chemical elements familiar to us [on Earth] exist on Mars," charting the course of evolution on another planet defies scientific prediction and the powers of journalistic description: "There is every reason to think that the creatures on Mars would be different from the creatures of Earth, in form and function, in structure and habit, different beyond the most bizarre imaginings of nightmare" (1975, 175, 177). *The War of the Worlds* depicts such a radical difference through its nightmarish extrapolation of evolutionary theory. The blood-drinking Martians pose a horrific challenge to bourgeois complacency, even as they give shape to late Victorian culture's masochistic fascination with its own demise (Clarke 2001). While many critics have called attention to the contrast between the narrator's bourgeois lifestyle and the novel's brutal allegory of colonialist exploitation, the "nightmare" of the invasion stems from a biological as well as conceptual decentering of humankind in the universe (Hillegas 1970, 150–77; McConnell 1981; Huntington 1982, 78–84; Hume 1983, 279–92; Fitting 2001, 127–45). If, as Richard Costa suggests, "*The War of the Worlds* is the archetype of all B-grade films that present giant creatures from another world who invade the Earth" (1985, 21), the Martians call into question the ideological equation of intelligence and morality, civilization and sympathy. In Wells's novel, evolution serves the amoral appetites of a master race, and his Martians differ radically from those imagined by his contemporaries.

The humanoid Martians depicted by Greg, Jones and Merchant, Lasswitz, and Bogdanov are fictional constructs, techno-Europeans projected into a future in which the imagined accomplishments of scientific progress fulfill turn-of-the-century consumerist desires for leisure, health, and social order without overt conflict, repression, or labor strife. In their novels, the complexities of political economy—the struggles to produce, profit from, and allocate scarce resources—are subsumed into or displaced by fantasies of antigravity devices, superabundant electricity, and effortless agriculture. Technological and social progress are natural, even foreordained, extrapolations of the laws of biological evolution. The Martians in these utopian or quasi-utopian novels take great trouble to explain and justify the workings of their civilizations because they are offering roadmaps for the hypothesized future of humankind. Although these works offer competing views of the (utopian) future, they all fictionalize the values and assumptions of evolutionary theory from within an anthropocentric, and Eurocentric, framework.

In contrast, the terror induced by Wells's Martians in 1898 and their offshoots (such as Orson Welles's invaders in 1938) stems from their absolute alienness, their lack of a recognizable psychology, sociology, or politics. These invaders give no indication of desires that humankind can understand and therefore resist; they have no imperial strategies (as Lasswitz's Martians do) that can be appropriated and turned back against them. Physiologically, they represent a leap forward in evolution that the narrator can describe but not explain. The Martians apparently have left behind the ability or need to speak: they do not negotiate, dictate, justify their conquest of Earth, declare that resistance is futile, enforce laws, or even subject humankind to bizarre experiments. Consequently, they offer no explanations, no justifications, no meaningful contrasts to or models for nineteenth-century society. "Intellects vast and cool and unsympathetic" (1993, 51) are unreadable; they offer nothing to interpret except the technological trials (the test flights of their flying machines) that the narrator witnesses and their vampiric feeding. Martian anatomy is "simple"; the invaders are "heads—merely heads" that contain a "mere selfish intelligence, without any of the emotional substratum of the human being" (149, 151). The Martians reduce human bodies to foodstuffs—replacements for their indigenous food source, a primitive, if vaguely human-looking creature. The very superiority of the Martians raises the possibility that evolution and intelligence are devoid of any moral purpose. *The War of the Worlds,* in short, seizes Darwin for an anti-teleological materialism.

Although the Martians seem to have a battle plan as they sweep across southern England, and the artillery man conjures up a nightmare vision of these colonial overlords breeding humans for food, their conquest unfolds as a series of vignettes of relentless destruction that exceeds human comprehension. Soon after the invasion, a curate goes mad, asking the questions that the narrator cannot answer: " 'What do these things mean? . . . What are these Martians?" (103). Relentless as a force of nature, they remain inscrutable—presumed to be all-powerful until they succumb to the "putrefactive and disease bacteria," the "humblest things that God, in his wisdom, has put upon this Earth" (184). In one sense, they represent the Lacanian horror of a Symbolic Order—that is, meaning itself—revealed to be impersonal and amoral (Žižek 1989). The "monotonous crying" of the dying Martian—"Ulla, ulla, ulla, ulla" (183)—admits of no interpretation because the "scandal" of evolutionary development and technological progress lies in the narrator's recognition that the only "meaning" that can be derived from the invasion is that intelligence exists merely to serve its own appetites: the Martians' insatiable lust to consume human blood. Humanity's encounter with an unimaginable evolutionary future suggests that the Martians have perfected what the European colonial powers practice with more malice but less efficiency: the technologies to overspread planets, exhaust available resources, reshape what is left of the environment to their own ends, and then seek new territories, new worlds, to invade.

Wells's narrative unfolds with the precision of a computer simulation, and the invasion mimics the supposedly irresistible force of evolutionary laws. Once the initial postulates are programmed—the life-forms on an older, dying planet need new resources to survive—the simulation operates on a self-perpetuating logic over which humans have no control. Wells's novel thus presents alien invaders as relentless scavengers, sardonically displaced reflections of Britain's imperialistic appetites and ambitions, stripped of their ideological justifications. Wells inaugurates a tradition of invasion novels in which aliens do to Earth precisely what humans have been doing to it for millennia—ravaging its resources to supply expanding populations and fuel increasingly sophisticated and resource-hungry technologies. For Wells, imperialist aggrandizement threatens the very conditions that allow humankind and its ecologically compatible species to exist. "Before we judge of [the Martians] too harshly," Wells's narrator says at the beginning of the novel, "we must remember what ruthless and utter destruction our own species has wrought, not only upon animals, such as the vanished bison and the dodo, but upon its own inferior species" (52). For a superintelligent race capable of building thousands of miles of canals to compensate for an irrevocable loss of water and dependent life-forms, invading Earth might seem a cost-effective solution to the problem of what to eat. Rather than synthesize protein, like Lasswitz's humanoid colonizers, or harvest other terrestrial species for blood, Wells's Martians confront his readers with the implications of racism and colonialism by dehumanizing their victims, allowing only the artillery man's fantasies of resistance to their seemingly absolute power.

This projection both provokes and forestalls the curate's question, which might be rephrased as "What do aliens want?"[3] In the simplest terms, the Martians want humans for food, and the curate's theology has all the relevance for the invaders of a steer's philosophical ruminations in a slaughterhouse. Cannibalism has been described by some cultural anthropologists as the ultimate form of consumption, an indication that other sources of protein have been exhausted and therefore a rational, if grim, response to severe ecological distress.[4] In this respect, the fear of being eaten, of bodily violation and ecological devastation, can be read both as an introjection by Victorian readers and their descendents of imperialist guilt and a projection into the future of the consequences of environmental devastation: the end of civilization's aggrandizement threatens to return to the most primitive of appetites—what to eat. In this regard, the fascination with cannibalism in the late nineteenth and twentieth century suggests that the fear of being consumed never can be divorced from the specter of a world stripped of all sources of protein except one's fellow humans. In science fiction, cannibalism is frequently, even obsessively, projected onto Mars as a solution—both rational and horrific—to existence on a dying world. The inhuman Martians of *The War of the Worlds* are the archetypes of the cannibalistic races that prowl the red planet in the novels of Edgar Rice Burroughs and his successors.

The invasion itself, then, destroys Victorian complacency about biological, cultural, and psychological integrity. Wells's novel abounds with images of bodily disintegration and social dissolution. Repeatedly, the narrator comes across scenes that defy rational explanation; the novel exploits the conventions of a literary realism that ironically proves inadequate to describe the horror of the invasion. As people flee London, the institutions of metropolitan government—the police and railway transport—that underwrite civilized order "were losing coherency, losing shape and efficiency, guttering, softening, running at last in that swift liquefaction of the social body" (121). This image is at once visceral and shocking in disclosing the imaginative connections between an identity forged on the reflexive integrity of mind and body and the disorientation produced by the disincorporation of the "social body." Such nightmares, however realistically described, gesture toward a terror that exceeds any language one might use to describe it. Hearing the artillery man's tale of a Martian "fighting machine" that had "picked up nearly a hundred" drunken men and women who had been "dancing and shouting" throughout the night, the narrator exclaims, "Grotesque gleam of a time no history will ever fully describe!" (178). Paradoxically, the rhetorical effect of such attempts at description intensify the unthinkable by making scenes of devastation commonplace. Walking the streets of "Dead London" (the title of chapter 8), the narrator suggests that "it was curiously like a Sunday in the City, with the closed shops, the houses locked up and the blinds drawn, the desertion, and the stillness" (180). This image renders the horrors of devastation as a hyperreal nightmare of biological extinction. The utopian premise that science fiction can offer a simulated reality, what Kim Stanley Robinson terms "*the history that we cannot know*" (cited in James 1994, 113), is inverted by Wells to suggest that history itself exceeds human knowledge, intention, and desire. The narrator's return to an uninhabited cityscape, in a sense, foreshadows the recurring nightmares of the wars against civilians in the twentieth century: the divorce of technological expertise from moral responsibility perpetuates seemingly unimaginable horrors in the name of reason.

Even after the death of the Martians, the inhabitants of London are haunted by their longing for a time before the invasion, a golden-age dream of a past before ecological devastation. As the narrator gazes at the tripods of the last Martians to succumb to Earth's biohazards, he remarks that "those who have only seen London veiled in her sombre robes of smoke can scarcely imagine the naked clearness and beauty of the silent wilderness of houses" (185). Such images of silence, desertion, and devastation eerily evoke a pristine natural world unspoiled by human pollution, London's notorious smogs.[5] "The naked clearness and beauty" that Wells describes gestures toward a world at once presumed to be both "natural" and yet irrevocably lost. There can be no return to a "silent wilderness" except through the unspeakable horrors that "intelligence" perpetuates on itself. Rather than depicting Darwinian theory as an ontology—a bedrock structure that can be analyzed and then used to predict the consequences of dynamic, historical processes—*The War of the Worlds* suggests that evolution discloses the hauntology of anthropocentrism—the originary unreality of the fictions and phantasms that underlie our systems of thought, belief, and self-identity.[6] In this regard, London revivified remains an unreal city, an apparition of itself. At the end of the novel, the narrator tells the reader, "I go to London and see the busy multitudes in Fleet Street and the Strand, and it comes across my mind that they are but the ghosts of the past, haunting the streets that I have seen silent and wretched, going to and fro, phantasms in a dead city, the mocking of life in a galvanised body" (193). This haunting resists a reductive allegorization: it suggests that the complacent bourgeoisie (like Bruce Willis in *The Sixth Sense*) are dead without knowing it. They haunt themselves, and their ghostly social identities mark the irrevocable alienation of "man" from "nature" and of individuals from themselves. These identities, like the curate's, liquefy when confronted by the human inability to understand the consequences of a vampiric imperialism. Although the end of *The War of the Worlds* has been hailed as the ultimate literary precognition of the devastation of the coming world wars, the spectral existence of London is a function of the ways in which the relentless appetites of a dying world—Mars—become the structuring fictions of a civilization that threatens ultimately to be consumed by its own rapaciousness. Wells's London, in other words, is haunted by the ghosts of its (Martian) future.

LASSWITZ'S *ON TWO PLANETS*: PROGRESS, COLONIALISM, AND INTERPLANETARY CIVILIZATION

At first glance, Kurd Lasswitz (1848–1910) seems an unlikely figure to have written the other classic invasion novel of 1897–1898. To a greater extent than Wells, Lasswitz was a figure of the scientific establishment. He received a doctorate at Breslau in physics in 1873, and then spent most of his life as a professor of mathematics, physics, and philosophy at the Gymnasium Ernestium in Gotha where he wrote a popular introduction to Kant (1900) and a multivolume history of atomic theory (1889–1900). Despite the popularity of these works, he never received a university appointment, possibly because of his liberal politics (Fischer 1984, 57–60). *On Two Planets* appeared in 1897 and sold seventy thousand copies in Germany alone by 1933, when it was banned by the Nazis (Fischer 1984, 126). Although translated quickly into several European languages, the novel did not appear in English until 1961. Nevertheless, its influence on continental science fiction as well as on German visions of spaceflight was profound. The 1971 English translation of the novel is prefaced by an epigraph from Wernher von Braun, who declares: "I shall never forget how I devoured this novel with curiosity and excitement as a young man. . . . From this book the reader can obtain an inkling of the richness of ideas at the twilight of the nineteenth century upon which the technological and scientific progress of the twentieth is based" (Lasswitz 1971, 7). As von Braun's encomium suggests, Lasswitz's novel charts an ennobling future for humanity and offers a sustained philosophical reflection on the ways in which the "progress" of civilizations unfolds. Like his scholarly work, *On Two Planets* offers a neo-Kantian view of the development of science as a means to understand both the physical world and the human soul. Yet Lasswitz emphasizes that this progress comes with costs, unforeseen consequences, and inevitable complications, and his depiction of the ways in which both humans and Martians struggle through colonization and conflict testifies to his literary skills as well as "the richness of [his] ideas." A firm believer in the plurality of worlds, Lasswitz argued in 1910 that "the rational order of the universe demands that there should necessarily even be infinite gradations of intelligent beings inhabiting [other] worlds"; and he declared that science fiction could draw plausible "inferences about the future . . . from the historical course of civilization and the present state of science" (quoted in Fischer 1984, 63, 65). In his novel, these "inferences" assume an evolutionary model of civilization and allow Lasswitz to explore the effects on both humans and Martians of social and scientific progress. *On Two Planets* maps an inferential route to the future through the trials of late-nineteenth-century imperialism and sociopolitical unrest.

As William Fischer suggests, Lasswitz's critique of imperialism stems from a self-critical examination of a widespread faith in progress (1984, 155). While the novel ultimately moves toward a reconciliation between Earth and Mars, it offers less a utopian vision of an advanced Martian society than a political fable about the consequences of imperialism for both the colonizers and the colonized. In contrast to *The War of the Worlds*, *On Two Planets* negotiates the problem of cultural difference by imagining the Nume (as the Martians call themselves) as an idealized embodiment of human potential for progress—a paternalistic race that must nurture and suffer through humankind's technological and political adolescence. It is very much an effort to chart in emotional and intellectual terms the future "historical course of civilization."

The novel begins with three German balloonists, lost on their way to the North Pole, being rescued by the benevolent Martians who have a geostationary space station hovering over the pole.[7] The Nume are a classic Aryan race of both physical beauty and advanced intellect. Although Lasswitz's Martians are the antithesis of Wells's vampiric race, their benevolent intentions are qualified by a language fraught with images of colonial conquest that both races employ: a Martian commander declares that the Nume are the "carriers of the culture of the solar system [whose] sacred duty [is] to bring the benefits of Numedom, the quintessence of our many-thousand-year-old civilization, to man" (199). This Kurtz-like statement of official idealism is tempered by one of the German explorers, Grunthe, who pointedly compares the Martians to "Cortez and Pizarro" and argues that, despite their technological superiority, "they also know guilt, sin, and fate" (105). These clashing perceptions of the colonial encounter set in motion a seemingly irrevocable conflict between different ideologies of knowledge and progress: the Martians' "sacred duty" versus the resentment and resistance of an indigenous race. The Nume make no secret that they have come to Earth to siphon off its excess solar energy and water, a project that does not necessarily require "the participation of the humans" (199). But in exchange for access to remote areas where they can collect solar energy without infringing on existing human settlements and rights, the Martians promise an education that will accelerate humankind's technological and social progress. Inevitably, however, this paternalistic philosophy shades into a Martian version of manifest destiny, and the novel offers a trenchant, if thinly veiled, analysis of the consequences of late-nineteenth-century colonialism on both the rulers and the ruled.

Although Lasswitz emphasizes the shared theological "reverence" for a Judeo-Christian "Father" by the "powerful inhabitants of Mars and the weak creatures of Earth" (93), the basis for the ultimate reconciliation of the two races is a common attitude toward the exploitation of resources essential for the progress of both civilizations. Power and weakness are defined in terms of the efficiency of the two civilizations in extracting resources from their respective planets. The idealized unity of Martian society is premised on its unceasing efforts to overcome the limitations of a dying world. Even before the three German explorers are transported to Mars to see for themselves how an advanced society copes with a senescent world, they learn that the "more difficult the aging Mars had made the conditions of existence for the Martians, the more magnificent had been their efforts to develop a technology to dominate nature" (41). This domination of the natural world is cast in economic metaphors that apply to Earth as well as Mars—a language of universal capitalistic exploitation. Gazing down on their world from the Martian spacecraft, the explorers try to explain the problems of coal-fired pollution to a race that has little conception of or need for fossil fuels.

> "What causes these fogs over your large cities?" one of the Martians asked.
>
> "Mainly the burning of coal," Grunthe replied.
>
> "But why don't you take energy directly from the sunrays? You should not be living on the capital but on the interest instead."
>
> "Unfortunately we do not know yet how to do that. By the way, the coal is only accumulated interest, which our predecessors were not able to make use of" (98).

The images of "capital" and "interest" convert the Earth into a storehouse of use-value waiting to be, or in the process of being, exploited. Grunthe's final remark is revealing because it counters the Martians' impression that the Earth is being wasted or reduced by the burning of coal.[8] In effect, Grunthe implies that the Earth has resources that remain undiminished by fossil fuel extraction, the "accumulated interest" that can be used without affecting the overall "capital," that is, solar energy, which can be exploited indefinitely. On Mars, the environment has been depleted by evolutionary processes, and sunlight and water are scarce; on Earth, fossil fuels may pollute the air but they are also seemingly inexhaustible. Burning coal is part of a learning process, a step in the "historical course of civilization" that will lead to the endless exploitation of solar power. In this respect, a seemingly offhand remark by the Martian heroine, La,—"Those clouds are horrible; one cannot see anything clearly" (98)—points to a gap between an ideal of infinite production without pollution (a Martian civilization that synthesizes food from rock and derives all its power from the sun) and the environmental consequences of industrialization. The smog over European cities marks humans as socially, technologically, and morally primitive, bound to an economy of scarcity, pollution, international tension, and social unrest. As they reveal their bemused indulgence of a backward species, the Martians embody the idealized end point of technological progress—a future in which infinite productivity without pollution becomes the measure of social and moral as well as technological superiority.

In keeping with his faith in progress as the motor of civilization, Lasswitz depicts the domination of nature on Mars in idealized, ahistorical terms: the planet is a pastoral antithesis of the nineteenth-century industrial landscapes of Europe. When the Earthmen and the half-Martian, Ell (the offspring of a stranded Nume explorer and a terran mother), first tour the red planet, they are struck by "gigantic trees" that serve, in lieu of a dense, earthlike atmosphere, to moderate "the contrast between incoming radiation during the day and the loss of heat during the night" (172). These "living domes" create a kind of greenhouse effect that allows Lasswitz to explain why the visitors from Earth have little problem breathing on Mars. Despite a population twice that of Earth's, the Martians preserve an idyllic balance between their advanced, solar-powered culture and "the solitude of completely undisturbed nature." Densely populated regions in each district (one thousand Martians per square kilometer) are interlaced with "gardens and parks, . . . flower beds and small ponds," and three-fifths of the land is reserved for "the wilderness of the forest . . . [legally] protected from exploitation and settlement" (173). This vision of civilization and nature in harmony depends on superhuman technologies that transcend problems of labor, scarcity, and the environmental consequences of exploiting the planet's dwindling resources.

Instead of Lowell's dying planet, the Nume inhabit a postscarcity dreamworld that translates desire to accomplishment: "Stones to bread! Protein and carbohydrates from rocks and soil, from air and water without the photosynthesis of the plant cell! This was the progress by which the Martians had emancipated themselves from the early cultural stage of farming and how they had become direct sons of the sun" (175). Technological progress is cast in the language of New Testament miracles—"Stones to bread!" and "sons of the sun"—to mark the transcendence of labor and scarcity. Yet the technoscientific and social histories that have led to this utopia are not described. There is no discussion of a sequence of discoveries, events, conflicts, negotiations, and resolutions, only a brief, romanticized reference to workers, a "culturally underdeveloped groups of Martians," who mine their beloved deserts for raw materials and willingly forego a life of leisure and cultural enrichment along the verdant canals (174). At a time when the industrial heartland of Germany, the Ruhr Valley, was among the most polluted places on Earth, Lasswitz's depiction of technoscientific Mars offers the alluring prospect of transcending the dire consequences of environmental degradation. Lasswitz is careful to emphasize that the Martians are not exhausting or wasting their planet's resources. With its limitless reserves of water and its proximity to the sun, however, the Earth provides the Nume with greater opportunities to sustain the expansion of their civilization and bring the benefits of their technology to a backward, but resource-rich planet. The Nume represent both the idealized conditions to which humankind can aspire and the telos of sociocultural evolution.

The bargain that the Martians offer is peace and prosperity in exchange for coexistence: by covering the Earth with "factories," the Martins will use "the inexhaustible resources of Earth"—solar energy, "air, water, and rocks"—to "produce and distribute food which hardly costs anything" (198–99). By abolishing scarcity, the Nume are confident that they can end war and social unrest "without any utopian leveling of possessions, without egalitarianism" (198). At first glance, it may seem as though Lasswitz himself endorses such an idealized vision of progress; in large measure, the first half of the novel follows a familiar tradition of utopian writing. By the time the Earthmen and Ell return from Mars with a proposal for the peaceful exploitation of the Earth's coveted resources, however, the generic assumptions of the technotopian romance have come undone. The Nume insist on the superiority of their enlightened social order, but their lack of interest in describing how their utopian society evolved foreshadows their inability to understand the humans' widespread suspicion of their motives. If the benevolent colonizers present themselves as a race that has transcended history, their experience on Earth reinforces Grunthe's analogy of the Nume to the conquistadors: the Martians, too, are products of "guilt, sin, and fate."

The responses of the peoples of Earth to the Martian offer vary, but when the British Navy fires on a Martian air machine, the Nume counterattack. The subsequent breakup of the British Empire escalates into worldwide chaos and, to preserve order, the Martians turn the entire planet into an occupied "protectorate." While Wells depicts a nightmare of social disintegration, Lasswitz painstakingly depicts the misunderstandings, errors, and fears that lead idealistic Martians and forward-thinking humans—against their better judgments—to assume the intransigent roles of imperialist occupiers and restive colonial peoples. Despite his "humiliating conviction that [humans] somehow do not deserve a better fate," Grunthe joins a rebellion against the Martians, even as he realizes the ironies of resisting the ideals and progress that the Nume offer: "It is our own tragic fate," he declares, "to stand up against the good! And it is the ill-fortune of the Nume that they have to become bad for the good's sake" (268). For their part, the colonizing Martians split into factions, even as they levy taxes on the Earth's population to pay the cost of their occupation and contemplate genocide: "There were a large number of Martians whose values had been corrupted by the egoistic demand for power—that obsession to rule and that enjoyment of power, which had awakened in them as a result of the conquest of Earth. They maintained that humanity was not at all capable of civilization in the sense of the Nume, that humans had no souls and that they were pitiable, and that it therefore would be better to destroy the inhabitants of the Earth so that their planet could serve the real bearers of culture as an inexhaustible source of energy" (367). The Martians's arguments echo those of the European colonial powers at the end of the century. In this regard, Lasswitz anticipates Joseph Conrad's brutal assessment in *The Heart of Darkness* of the effects of power on the colonial mind: "Exterminate all the brutes." Rather than projecting onto his Martians the obscene desires that characterize Wells's inhuman invaders, Lasswitz charts the corruption and moral incoherence that paradoxically structure idealized visions of (super)human perfection. Because the Nume are disfigured by "guilt, sin, and fate," their efforts to impose "the good" can be experienced only as insensate evil. With dream of a "voluntary adaptation of the humans to the cultural world of Mars" (343) dead, the second half of the novel must resolve the tragic ironies of the colonizers' having to violate the very values that they claim to uphold. *On Two Planets*, then, becomes a tale of the reacquisition of moral responsibility by beings confronted by conflicts between duty and ethical behavior, between liberal ideals and the necessities of empire.

The accommodation that is finally achieved between Martian civilization and the potential for change that exists with a younger and more dynamic race ultimately rests on reinscribing traditional gender roles and gendered categories. Although Lasswitz provides a generic ending suitable for an adventure novel (humans learn to fly Martian spaceships, defeat the conquerors, and then forge a lasting peace between the two planets), the transcendence of interplanetary conflict centers on the love story of the Martian La and one of the original balloonists, Saltner. Acculturation is a sexualized process. Explaining to her Martian friend, Se, her love for Saltner, La characterizes the Earth as a "young planet," "raging like a youth who is shaken by life, fitful and presumptuous, not caring about the creatures it is supposed to protect. Our Nu [Mars] is an old man, who has spoiled us with his secure calmness. How fresh and new and adventurous and beautiful is this Earth, with its weather nonsense and its blind foolishness!" (316). La invests the Earth with an anthropomorphic sexual energy, and Mars is relegated to the stereotypical role of the aged and impotent cuckold. The Martian women, who end up aiding the climactic rebellion of the humans, are seduced by the sublime "power of nameless forces" that both tests their wills and invests their political decision with a destiny—"reason within timeless will"—that transcends space and time. "Love," La declares, "encompasses an insolvable riddle: we are to be this one being consisting of two creatures. . . . To follow destiny is freedom; to satisfy it is dignity" (315, 317). Her rhetoric translates the problems of political economy—the control of Earth's resources—into a sexualized dialectic: a transcendent love that overcomes the technocultural gulf that separates primitive male from civilized female. By having La choose her love for Saltner over loyalty to Numean ideals of civilization, Lasswitz inverts the familiar gendering of colonial exploitation of a "virgin" territory, symbolized by the native woman's willing surrender of her body to her white conqueror. Instead, La and Saltner act out a late-nineteenth-century fable of middle-class domestication: the male's sexual energy captivates and invigorates a languishing woman; her superior knowledge civilizes him for a social order superior to the one into which he was born.

In Greg's *Across the Zodiac,* the technological superiority of the Martians is balanced by the spiritual and cultural values of a younger but more energetic human race; in Lasswitz's novel, the love story of a Martian superwoman falling for a noble terrestrial "savage" validates the erotic and heroic potential of a nineteenth-century, masculinist civilization, boldly exploring the wastes of the Arctic and, if it could, the reaches of space. La's desire for Saltner redeems humanity by suggesting that, as a species, humans deserve a better fate than servile dependence. In another sense, the sexual power that Saltner embodies allows the novelist to convert the threatening image of the powerful female into a lovesick heroine. The end of the novel, then, symbolizes a philosophical synthesis—earthly energy and Martian wisdom—by channeling the superior power of the invaders into a "destined" love that fulfills both "freedom" and "dignity." Lasswitz thus reinforces his belief that science fiction is a means to imagine a future. The reconciliation of Earth and Mars resolves as future history what the bitter conflicts and ethical contradictions of late-nineteenth-century colonialism cannot achieve. In this regard, *On Two Planets* is less an allegory of its present than a simulation that explores the conditions under which one might imagine transcending the evils of sin, scarcity, and oppression that define human history.

BOGDANOV'S *RED STAR* AND *ENGINEER MENNI:* UTOPIAN COMMUNISM ON MARS

In 1920, H. G. Wells visited Moscow and interviewed Lenin at the Kremlin (Wells 1920). Lenin told Wells that if life were discovered on other planets, there would be no need for revolutionary violence. Lenin explicitly tied progress to what he termed the "earthly limit," that is, the material constraints on human progress. “Human ideas,” he reasoned, “are based on the scale of the planet we live in [and] on the assumption that the technical potentialities, as they develop, will never overstep ‘the earthly limit.’ If we succeed in making contact with the other planets, all our philosophical, social, and moral ideas will have to be revised, and in this event those potentialities will become limitless and will put an end to violence as a necessary means of progress” (cited in Stites 1989, 42). As Richard Stites argues, utopian science fiction played a significant role in shaping Russian and then Soviet conceptions of historical and political progress, and behind Lenin’s comments about the effects that extraterrestrial civilizations might have on “human ideas” lie the Mars novels of the physician, revolutionary, and social theorist Alexander Bogdanov (Stites 1989). Bogdanov’s science fiction addresses precisely those questions that Lenin raised with Wells—the dependence of concepts of social, personal, and political identity on economic models of scarcity and competition, and the consequences of confronting a race that had developed the technological and political means to overcome environmental limits on progress.

In *Red Star* and *Engineer Menni,* Bogdanov (1873–1928) challenges the values and assumptions that inform both the nightmare vision of Wells and the liberal humanism of Lasswitz. Born into a middle-class family, Bogdanov studied science and psychology in Moscow before receiving a medical degree in 1899.[9] Appalled by the living conditions of the workers he treated, he became an early member of the Russian Social Democratic Party, then sided with the revolutionary Bolsheviks when they split from the more moderate Mensheviks in 1903. During the 1905 revolution, he worked tirelessly as a propagandist and agitator, and unlike his close associate, Lenin, stayed in Russia after the failed uprising to continue efforts to organize resistance to the Czarist regime. After the brutal suppression of the Russian Left in 1907, Bogdanov, still believing in the necessity of armed revolution, broke with Lenin and withdrew from Bolshevik agitation by 1911 to devote himself to developing his version of systems theory, “tectology,” a term he borrowed from Lowell’s admirer, the biologist Ernst Haeckel.[10] Tectology sought to unify the sciences of social, economic, and biological organization in order to ensure the health of both the individual and society as a whole (Stites 1984, 5). Although a fervent believer in collective labor and the abolition of private control of the production and distribution of goods, Bogdanov was convinced that marxian economic theory had to be brought into line with advances in organizational and physical science. Stites suggests that “Bogdanov did not abandon science for revolution: rather, he deepened and extended his study of physiology, technology, and natural science and combined them with his own version of Marxian sociology" (1984, 2). By 1919, Bogdanov had redefined industrial production in terms of his systems theory: "The proletarian collective," he argued, "is distinguished . . . by a special organizational bond, known as *comradely cooperation.* This is a kind of cooperation in which the roles of organizing and fulfilling are not divided but are combined among the general mass of workers, so that there is no authority by force or unreasoning subordination but a common will which decides, and a participation of each in the fulfillment of the common task."[11] This "common will" is crucial to Bogdanov's political philosophy and fiction, and his resistance to top-down bureaucracies made him a political outsider in the Soviet Union after 1917.

Having served as a military physician during World War I, Bogdanov devoted himself after the Revolution to scientific research and (until 1921 when it was dismantled) to an independent proletarian culture movement. Although he held academic appointments after 1918, he never joined the post-Revolution Communist Party. Fascinated by blood transfusions, Bogdanov conducted numerous medical experiments in the 1920s. The transfusion of blood seems an apt symbol for an idealized collective, a brotherhood of coworkers helping one another battle disease. In 1928, well aware of the consequences, he exchanged his blood with that of a student suffering from both malaria and tuberculosis. Recording his observations on his own condition until he lost consciousness, Bogdanov died on April 7. The student recovered and, in 1983, was reported still to be living in the Soviet Union (Graham 1984, 252). Rather than a martyr to a political cause, Bogdanov died an experimentalist, subordinating the ethics of individualism to the collective good represented by science. In his life as in his fiction, science itself became his utopian politics.

The violence of the revolution of 1905 serves as both the backdrop and impetus for *Red Star.* Drawing on a tradition of Russian utopian literature, such as Vladimir Taneev's *The Communist State of the Future* (1879), and left-wing science fiction, notably Nicholas Chernyshevsky's *What Is to Be Done?* (1862), Bogdanov appropriates the generic conventions of interplanetary travel to envision humankind's communist future (Suvin 1979, 252). A self-consciously political and didactic novel, *Red Star* fictionalizes the elaborate feedback mechanisms and information exchanges that define his vision of tectology.[12] Like Lasswitz, Bogdanov endows Mars with a humanoid superculture which has overcome social divisions to build a collectivist utopia, but *Red Star* offers a very different take on the planet itself. Where Lasswitz's Nume turn "stones to bread," Bogdanov's vanguard socialists inhabit a planet in the full throes of Lowellian decline. His Martians are in dire need of the food and minerals that they are exhausting on their home world, and they come to Earth to see if the inhabitants of the third planet are ready to cast off the chains of socio-political inequality, individualism, and greed—a prerequisite, in their eyes, for interplanetary commerce and cultural interaction.

On Bogdanov's Mars, class conflict has been displaced into what Bogdanov sees as a never-ending struggle of a unified humanity against a hostile nature (Graham 1984, 243). The Martians' heroic efforts to overcome the loss of water and the decline in food production on their dying planet is the subject of Bogdanov's second Martian novel, *Engineer Menni* (1913), which describes the struggles of earlier generations to overcome political, economic, and social divisions in order to build the vast system of canals. For Bogdanov, the canals are not a static backdrop for interplanetary romance but a crucial symbol of a heroic, collectivist undertaking. They represent the socialist future, a working hypothesis for his systems theory that integrates economics, labor, resource extraction, and political will. Like the terraforming projects that dominate late-twentieth-century science fiction, the canals are socialism's epic response to conditions on a dying world.

Yet even as Bogdanov crafts his utopian vision in *Red Star,* he creates a hero, Leonid, who is unable to adapt psychologically to Martian communism. A Russian revolutionary selected by the Martians for his progressive social, economic, and sexual views, Leonid comes to Mars to serve "as a living link between the human races of Earth and Mars" (34). But this experiment at mutual education falters: when the hero cannot adjust to the cultural and psychological norms of a superior order, he becomes disorientated and depressed; finally, in a fit of rage, he commits a murder. In one sense, Leonid's failure measures the gap between utopian desires for political and psychological ideality and the chaotic, retrograde forces that oppress the body and persecute the mind from within. Madness becomes both the cause and effect of a repressive ideology, and Bogdanov, the ardent revolutionary, paradoxically offers one of the most compelling and nuanced depictions of psychological turmoil in science fiction. Political philosophy proves insufficient to overcome the disorientation and self-loathing that the hero feels when he is confronted by the lived experience of his revolutionary ideals. As an allegory of dreams deferred, *Red Star* registers the difficulty of reconciling political idealism and the experience of defeat in the wake of the failed 1905 revolution.

From the start, Bogdanov challenges the bourgeois tradition of science fiction—those miraculous voyages and Faustian bargains that characterize the novels of Verne and Greg. The Martians' discovery of "minus-matter," their mode of interplanetary propulsion, was "not made by any one individual," Leonid quickly learns; instead, it "is the achievement of an entire scientific society" (28). Throughout the novel, individualism is identified consistently with the evils of selfishness and self-aggrandizement that hinder socioeconomic development. Rather than the "ballast of names from the past," the socialist utopia on Mars is founded on the recognition that individuals contribute to a work that is "impersonal," or, one suspects, transpersonal: "science" and "art . . . preserve impersonally the collective accomplishments of all" (44). In contrast to later dystopian novels such as Zamyatin's *We* and Orwell's *1984*, that treat collectivism as a cover for totalitarianism, *Red Star* depicts freedom as the shared fulfillment of a revolutionary future, a sociocultural transcendence of bourgeois individualism, capitalist exploitation, and the false sciences of compartmentalization. This decentering of the bourgeois scientist-hero dovetails with the novelist's refiguring of Mars as a planet approaching the final stages of exhausting its resources. "Collective accomplishments" are the only possible means to survive on a dying world.

In a universe governed by an ironclad evolutionary logic, Martians and terrans share a biological heritage that ennobles them as the apex of life-forms; but the harsh conditions on the red planet have encouraged a cooperative mode of existence, a "course of history . . . gentler and simpler than that on Earth" (53): fewer wars, less-violent class struggle, and a comparatively painless transition from capitalism to communism. On Mars, the history of collectivism takes symbolic, scientific, and economic form in the network of "the famous canals [that] served as a powerful stimulus to economic development at the same time as they firmly reinforced the political unity of all mankind" (55). In contrast to Lowell's interpretation of the canals as the product of a paternalist hierarchy, Bogdanov sees this planetwide system as evidence of what a socialist society can accomplish in its unceasing efforts to combat an irrevocably hostile nature. In an important sense, the canals are both cause and effect of Martian collectivism.

Unlike most utopian writers, however, Bogdanov is as interested in the historical process of revolutionary change as he is with providing a guided tour of its accomplishments—the institutions, workplaces, schools, and living quarters that are a staple of utopian fiction. While *Red Star* describes the voluntary labor and technological complexities of the Martian factory system, the novel devotes more time and energy to depicting the relationships among individuals within a communist society. If the false consciousness of individualism is a baleful effect of capitalism, then, Bogdanov implies, a socialist utopia will produce individuals who embody the evolutionary development of both "species" and "society" (70). The most significant embodiment of this evolutionary belief is Netti, one of Leonid's most trustworthy guides, who, he discovers late in the novel, is a woman. Earlier, the hero had come to recognize that seemingly essential gender differences on Earth are the effects of the capitalist division of labor: "The enslavement of women in the home and the feverish struggle for survival on the part of the men . . . ultimately account[s] for the physical discrepancies between [the sexes]" (76). Netti is the antithesis of the scantily clad Martian princesses who appear in the novels of Bogdanov's successors such as Edgar Rice Burroughs. Unlike Lasswitz's lovers, La and Saltner, whose romance promotes a synthesis of passion and knowledge, Netti and Leonid fall in love as a consequence of their mutual education; because he is on Mars to "*know* [the] future" of revolutionary struggle, the hero must find a love that is based not on masculinist domination but on his heightened consciousness. The heart, too, is a revolutionary organ, and love itself must be accommodated to tectology in order to free the lovers to promote revolution on Earth and responsibility on Mars.

The vanguard socialism of the Martians is carved into the landscape of their planet, founded on their unwavering commitment to an always escalating intensification. They insist repeatedly that "there cannot be peace with the natural elements" (79). Enno, a father figure for Leonid, describes a history of planetary development that highlights the engineering and technological conquests, rather than the politicial conflicts, of his civilization: "[Recently,] we have intensified the exploitation of the planet tenfold, our population is growing, and our needs are increasing even faster. The danger of exhausting our natural resources and energy has repeatedly confronted various branches of our industry . . . [and] at this very moment the struggle has become particularly acute" (79). This struggle—the evolutionary telos of socialist development—is a profoundly antiecological effort to stay one step ahead of a burgeoning and rapacious population. More Martians use—and use up—more of the planet's resources so that the ironclad logic of intensification drives their socioeconomic development. Enno's history, in this regard, emphasizes the environmental consequences of imperfect transitional phases in technological and social progress.

Rather than dodging the climactic and ecological effects of intensifying resource extraction, Bogdanov makes them a crucial impetus for his depiction of both scientific progress and demographic crisis. From the opposite end of the political spectrum, Bogdanov seconds Lowell's views of biological and sociocultural evolution: intelligence—civilization itself—is measured by its responses to a hostile environment. "Seventy years ago, when our coal reserves were exhausted," Enno tells Leonid, the Martians "were forced to destroy a considerable portion of our beloved forests in order to give us time to redesign our machines. This disfigured the planet and worsened our climate for decades. Then, when we had recovered from that crisis, about twenty years ago it was discovered that our deposits of iron ore were nearly depleted. Intense research was begun on hard aluminum alloys, and a huge portion of our available technical resources was diverted to obtaining aluminum from the soil. Now our statisticians reckon that unless we succeed in developing synthetic proteins from inorganic matter, in thirty years we will be faced with a food shortage" (79). This passage offers a hard-headed assessment of the consequences of deforestation, and Bogdanov seems clearly to have in mind the scarred landscapes of industrialized Europe (McNeill 2000). Mars "disfigured" both mirrors and prophesies a future on Earth when social progress—even after Bogdanov's longed-for revolution—will confront the limits of environmental exploitation. Because the Martians refuse to stem their birthrate, their anthropocentric faith in the exploitation of the natural world is figured explicitly in terms of war. To practice birth control is to concede defeat. "We can triumph" over nature, Enno declares, only "as long as we are on the offensive, but if we do not permit our army to grow, we will be besieged on all sides by the elements, and that will in turn weaken faith in our collective strength, in our great common life" (80). By translating the exploitation of nature into these militaristic images, Bogdanov pointedly excludes "nature" from his tectology. Nature is not a system but a storehouse that must be conquered and plundered. Victory, in short, entails the destruction of a planetary ecology, and, by implication, tectology becomes an extreme form of anthropocentrism. Conservative conservationists like Lowell no doubt would shudder at this marxian recentering of humankind in the cosmos.

Bogdanov's logic reveals the grim ironies that underlie neoclassical and Marxian notions of a labor theory of value. In order to make labor itself the source of all value, as it is for Huygens's "Planetarians," the natural world must be blackboxed; it becomes a reservoir of potential value awaiting capital and labor, or a collective will, to extract it. Political economy, in brief, is the patchwork of strategies and practices that tries to make rough sense of the recognition that no resources are infinitely sustainable. In the Lockean tradition, social and economic stratification ultimately must be blamed either on backward societies or, ultimately, as in Lowell's works, on an indifferent universe and the inexorable laws of entropic decline. *Red Star*'s idealized socialist civilization reproduces, as part of its foundational value system, the essential paradox of intensification: although social progress must be sustained by imagining natural resources as indefinitely exploitable, the work of the revolution is to forestall the consequences of depleting these resources. If the inevitable shortfalls of food, land, clothing, shelter, and healthcare in Russia in 1905, for Bogdanov, are the consequences of capitalist exploitation, the ecological strains on his utopian Mars ironically produce a fervent antienvironmental ethos.

Inevitably in *Red Star,* the logic of scarcity and competition is both projected onto a demonized natural world, in Enno's call to an endless struggle against nature, and introjected, in Leonid's case, in his sense of alienation from collectivist ideals. In this respect, the hero's psychological deterioration can be read as an implicit recognition of the incoherence of the utopian ideal—the belief that if one is "good" enough, one should be able to create a society that can escape entropy and intensification, that can solve ecological and economic problems solely by sociopolitical will. An abundance of resources to exploit, according to the theocentric logic that Huygens and Locke employ, should be sufficient for a just society to banish insecurities, greed, shame, and guilt. The good life on Bolshevik Mars relies on a socialist appropriation of the values and assumptions of such a logic. Bogdanov therefore must seek ways to distinguish his tectology from the antiecological and exploitative system—capitalism—that paradoxically underlies his depiction of vanguard socialism.

The question that *Red Star* poses in the wake of the failed 1905 revolution is whether humankind is ready or even capable of ascending to the level of a socialist utopia. Although love, murder, and redemption define Leonid's efforts to explore his own (un)worthiness for this heroic enterprise, the second half of the novel centers on a debate between those Martians who want to colonize Earth and claim its resources for themselves and those who want to exploit uninhabited Venus. Sterni advocates the conquest of Earth and the destruction of its population to make the solar system safe for Martian socialism; Netti favors the colonization of Venus, despite the greater dangers of this enterprise, because she believes that humanity can be elevated to communist brotherhood. This debate allegorizes the political differences among early-twentieth-century revolutionaries; it reimagines the speeches and pamphlet wars of Russian socialism as a set-to between pessimists and optimists, between those who seek a new order through violence and those who favor ongoing efforts to educate and motivate the masses to revolutionary action. In this respect, the principals in Bogdanov's debate offer two forms of colonialist exploitation—the amoral destruction of indigenous cultures and the moral assault on a "virgin" wilderness. For Sterni, the violence and irrationality of Earth's inhabitants, who would project "their spiteful distrust of all other peoples and races" onto Martian colonists, necessitates that the "*colonization of Earth requires the utter annihilation of its population*" (111, 113). Given the Earth's history of violence, repression, and the misery of the masses "under the sway of capital," he reasons, even a socialist vanguard would remain "surrounded by embittered and merciless enemies" (113, 115). In effect, his call to exterminate the terran brutes reflects a loss of faith in the evolutionary processes that have determined the course of progress on Mars. The history of Earth—"national division, a mutual lack of understanding, and brutal, bloody struggle" (115)—condemns its inhabitants to relive the cycles of violence that prevent true progress from taking place.

In contrast, Netti reasserts the potential of dialectical materialism to transcend the conflicts that have defined Earth's past. The planet's inhabitants carry within them "the history of a different natural environment and a different struggle . . . they conceal a different play of spontaneous forces, other contradictions, other possibilities of development" (116). Although such struggles have been more violent on Earth than on Mars, social and biological evolution ensures the triumph of the proletariat. "In the plant and animal kingdoms," she asserts, "millions of species struggled violently and quickly crowded each other out, contributing through their life and death to the development of new, more perfect and harmonious, more synthetic species. The same is true of humans" (117). Her metasystemic theory (a fictional anticipation of Bogdanov's tectology) recognizes that local differences, even violent struggles, among humans contribute to social, moral, and scientific progress. In this process, the individual herself will be reconfigured: the suspicion and hatred that is both projected onto others and internalized as a fundamental condition of bourgeois psychology will give way to a collectivist sense of "impersonal" achievement. The resolution of the debate demonstrates the cogency of Netti's reasoning: Menni decides in favor of her plan to extend Martian civilization to the inner planets by colonizing Venus. The Martians, as good colonialists, will conquer nature rather than exterminate a retrograde bourgeois civilization.

Tellingly, it is Netti who argues for the perfectibility of humanity because she emerges, at the end of the novel, as the means to redeem Leonid after he descends to depression, madness, and ultimately the murder of Sterni. The downward spiral of the hero drives home Bogdanov's insistence that capitalism is a form of internalized bondage as well as political and economic oppression. Because Leonid is the product of a vicious and self-destructive culture, his failure to live up to his socialist ideals drives him to assume the blame for Sterni's voted-down plan to destroy Earth. Unable to assimilate successfully to Martian society, he is anatomized symptom by symptom: lack of concentration, sleeplessness, nightmares, hallucinations, thoughts of suicide, and finally a loss of psychological control: "This was delirium—agonizing, unceasing, endless raving. I did not see any apparitions around me—there was only one black apparition now, but it was inside me and it was *everything*. And it would never go away, for time had stopped. . . . I could not believe in suicide, because I did not believe in my own existence. Sorrow, cold, this hateful *everything*—they existed, but my 'I' was lost among them, where it seemed infinitely small, imperceptible, insignificant. There was no 'me' " (122). In other utopian or dystopian fiction, such suffering measures a satiric distance between a human hero, such as Gulliver in the land of the Houynhmns, and an inhuman ideal: the talking horses of absolute rationality. In *Red Star,* however, Leonid is a victim of a latent individualism that can be figured only as a form of self-hatred, an alienation from the collectivist ideal of camaraderie. He may reject the repressive structures of Earth, but his revolutionary identity can be imagined only in opposition to such a coercive bourgeois ideology. Because no such ideological mechanisms exist on Bogdanov's Mars, the hero has nothing to define himself against, no means to transform himself into an embodiment of the progress that can come about only by the historical incorporation of the "contradictions" and "play of spontaneous forces" that define sociocultural, political, and biological evolution. "Overwhelmed by [the Martian civilization's] loftiness, by the profundity of its social ties, and the purity of its interpersonal relationships" (135), Leonid loses himself when he is confronted by the ideals that previously he could only imagine. Even in a communist utopia, there are no short-cuts in either evolution or revolutionary struggle. In effect, Leonid embodies a marxian analytic that presupposes an originary alienation of "man" from "nature"; bourgeois consciousness is the sad effect of having to fight over scarce resources in a oppressive and technologically backward society. The paradox of individual freedom through collective acculturation represents Bogdanov's recognition that a utopian future can be imagined only as a negation of the "black apparition" of repression.

Leonid's murder of Sterni externalizes the violence that is ingrained in him as part of his evolutionary (that is, genetic) heritage, and his subsequent banishment back to Earth marks the distance between an idealized (Martian) socialist future and the Bolsheviks' experience of defeat in 1905. As in *On Two Planets,* the interspecies romance symbolizes the hope that eventually humankind will find emotional fulfillment and psychological stability in a just society. The Martian women with whom Bogdanov's and Lasswitz's space-faring heroes fall in love combine two idealized and distinct visions of the feminine: the nurturing mother who is wiser than the hero and guides him as a sort of extraterrestrial Beatrice (La, Netti), and the perfect companion who falls in love with the human despite his backwardness—the irrationality that is, at once, the bane of his existence, the "essence" of his sexual attractiveness, and an irresistible incentive to educate him and his kind. Given the symbolic tendency to describe the natural world in the same antithetical terms—a benevolent and generous Mother Nature versus a pristine object of desire waiting to be despoiled—this impossible sexual desire for Martian women discloses the complex role that Mars plays in an ideological economy. Mars is both the redeemed natural world of utopian speculation and a world that already has been exploited. Used-up Mars, in turn, is both the consequence of colonialism and an incentive for exploring new technologies and exploiting new territories. In this regard, the final reconciliation between Netti and Leonid—she journeys to revolution-torn Russia to nurse him back to health—demands the readers' (and novelist's) renewed commitment to a collective revolutionary faith. While it is tempting to read this conclusion as an allegory based on familiar conventions of interplanetary romance, Bogdanov insists that his novel is a call to action—part scientific speculation, part prophecy, and part a reiteration of faith. Even more than Lasswitz, Bogdanov offers a radical nostalgia for the future rather than an escape to a golden-age past.

Engineer Menni extends Bogdanov's fictive exploration of the processes by which a socialist state can arise from the conflicts and contradictions within capitalist society. The novel is a pared-down historical epic that describes the hero's efforts to build the canal system, a "wonder of labor and human will . . . dictated by historical necessity" (Bogdanov 1984, 147). This prequel to *Red Star* is set in the Martian past, at a time when Earth was in the throes of the religious wars of the seventeenth century, and it provides some compelling insights into the widespread fascination with Lowell's canals. Bogdanov projects onto Mars a heroic vision of a technological future, a fictive blueprint for the forms of socio-economic organization necessary for the pharaonic reengineering of the environment. Although the novelist surveys the political debates that attend this project, *Engineer Menni* centers on the psychological and moral transition from atavistic capitalism to a collective realization of the "will" necessary to reshape Mars physically and politically. To reclaim the planet from drought and scarcity, Menni must battle and eventually kill the villain, Maro, and contend with the revolutionary arguments of his son, Netti, over the necessity for labor unions.[13] The father-son conflict stages, in fictionalized form, the ongoing debate between liberal sympathy and revolutionary action. The political struggle of Mars is repeated ontogenetically in the consciousness of the hero, the engineering genius who conceives and supervises the construction of the canals and who struggles to resist his son's insight that the canals represent "a triumph for the united efforts of all mankind, for all-conquering labor" (201). If Menni represents a traditional vision of intellectual labor that seeks to reshape the ecology of vast desert regions according to an aesthetic preconception of a world transformed, Netti insists on the unity of thought and labor, the need to recapture science for the collective good of humankind.

In one respect, Netti's efforts to make his father understand that true progress demands subsuming individual aspirations into a collective will serve as Bogdanov's metacritical commentary on the genre of science fiction. Netti argues that "Utopias are an expression of aspirations that cannot be realized, of efforts that are not equal to the resistance they encounter. Now these aspirations have grown and assumed the form of systematic labor, which is able to overcome such resistance; for this to happen it was necessary that they fuse into a single idea. This is why to me the victory of unified labor and the victory of the idea are one and the same thing" (204). This unity of "idea" and "labor" describes the novelist's self-justification for seeing science fiction as a form of revolutionary action. This idealized unity—given material shape by the canals of Mars—is a fictionalized version of Bogdanov's own tectology, a systems theory capable of the "transformation of the entire planet" (165). The Mars that emerges in *Engineer Menni,* then, represents a future conquest of the forces of nature. Such "victories" over "the elements" (164) are essential, Bogdanov implies, to the development of a socialist utopia spanning Russia from the Baltic to Siberia.

The comparisons of Mars to Siberia in the scientific works of twentieth-century astronomers such as E. M. Antoniadi and Hubertus Strughold suggest one reason why Lowell's canals become so crucial to Bogdanov's political vision: only in the planetwide transformation of a hostile nature to a life-bearing system of collectivist integration can a socialist future be glimpsed. Only on Mars can the vastness of transforming Russia into a communist paradise be rendered imaginable, and only through the telescopes of Schiaparelli and Lowell can the seemingly incontrovertible signs of evolutionary intelligence and social progress be read on the surface of a neighboring world. Ironically, however, Bogdanov must indulge in the kind of idealization that he suggests Menni must outgrow—the imposition of the imagination on a recalcitrant material reality—to envision a planetwide transformation on Earth. The conclusion of the novel—when Leonid's conviction that his love for (the female) Netti means "that our two worlds really are close and that one day they will be able to unite into a single, unprecedentedly beautiful and harmonious order" (152)—stands as a testament to a revolutionary faith that soon would seem as far-fetched as a world of canal-building communists. But *Engineer Menni*'s vision of the canals as the triumph of collective intelligence over a dying world enjoyed a widespread appeal beyond socialist didacticism. In science fiction, at least, the canals represented for two generations of readers a faith that, in the future, humankind will be able to boldly engineer where no man has engineered before.

THE ROMANCE OF THE RED PLANET

In the Soviet Union, *Red Star* was superseded in the public imagination by Alexei Tolstoy's wildly popular *Aelita* (1922), a restaging of the 1905 revolution on Mars that owes more to Western science fiction than it does to utopian socialism.[14] Intensely romantic and politically unthreatening to the Communist regime, Tolstoy's novel exploits the familiar theme of an Earthman falling in love with a beautiful Martian woman; it was made into a film by the Russian director Yakov Protazanov in 1924 that solidified Tolstoy's position within the Bolshevik regime.[15] A nobleman and distant relative of the famous novelist, Alexei Nikolayevich Tolstoy (1883–1945) had gone into exile in France after the revolution in 1917, where he wrote several novels, including *Aelita.* In part, this science-fiction novel celebrating the heroism and sacrifice of his Bolshevik heroes seems to have been an olive branch to signal his desire to return to his homeland under conditions laid down by the new regime. The gesture succeeded. After returning to the Soviet Union in 1923, he became a stalwart defender of Josef Stalin and published his immensely popular historical novel, *Peter the First* (1929–43), to defend the cult of personality that the dictator used to legitimize his rule. *Aelita* appropriates aspects of Russian utopianism but recasts them as an interplanetary romance of doomed love; it is as close to *Flash Gordon* as it is to *Red Star.*

Unlike Wells, Lasswitz, and Bogdanov, Tolstoy makes little effort at scientific plausibility: the hero, Engineer Los, funded by "the republic" (Tolstoy 1981, 8), builds a spacecraft in a garage and journeys to Mars with his proletariat compatriot, Gusev, in less than a day. There they discover an ancient civilization descended, in part, from terran invaders from the lost continent of Atlantis, are introduced to Martian culture by the princess Aelita, and get caught up in a revolt of the workers against a sclerotic and self-destructive political order. The contrasts between Lowellian deserts and civilized regions "intersected by full canals, covered with orange groves of vegetation and canary meadows" (53)—are filtered through the novelist's experience of Russia during World War I. Although Tolstoy depicts shrinking seas, cold temperatures, and expanding deserts, his heroes also encounter the "dead landscape of a destroyed planet" disfigured by reminders of a devastating war: "remains of huge wings, broken shells, protruding ribs" make it seem as though "demons had attacked" (52) the regions outside the still-fertile kingdom of Azora. If such descriptions elicit a visceral response in readers who lived through the Great War, they also throw into sharp relief passages of a hallucinatory surrealism, reminiscent of Greg or Cromie. Mars is, at once, a dying planet and an opiated paradise that "resemble[s] those spring meadows that you remember in a dream from your distant childhood" (53).

This nostalgic evocation of a lost innocence is emblematic of the social as well as the natural world of Mars. On a doomed planet that is freezing into senescence, the upper classes, represented by Aelita's father, Tuskub, embrace the notion that "the history of Mars is over" in order to justify their efforts to destroy the workers' movement and to sanctify their own death wish: if they " 'are powerless to halt the dying out,' " Tuskub argues, then they " 'must use harsh and wise measures to make the last days comfortable and happy' " for the elite (108). The decadence of the ruling classes, their selfishness and irresponsibility, becomes a metaphorical reflection of the death of the planet. While they plan to destroy Azora and retreat to the Martian equivalent of country dachas to wait for the end, the workers call on Los and Gusev to lead them to revolutionary and, apparently, planetary rejuvenation. Gor, the Martian hero of the revolt, knows that " 'Mars is dying out,' " but tells his assembled troops, " 'We will be saved by Earth, by people from Earth, a healthy fresh race with hot blood' " (109). Bolsheviks, it seems, possess the "hot blood" that can overthrow the unjust Martian social order and revivify the dying planet. The situation exploited by Lasswitz and Bogdanov—a superior race willing to midwife a moral, political, and socioeconomic transformation on Earth—is reversed: Soviet heroes of the revolution become the only hope for the masses on Mars who live and work underground and remain drugged and docile on an indigenous narcotic.

Against the backdrop of this ultimately unsuccessful revolt, Los and Aelita fall in love, marry, and spend one night together before she swallows poison to escape the fate of being thrown into a labyrinth for violating her oath to remain celibate. This love story depoliticizes the role of the Martian woman: Aelita is less a representative of a superior civilization than a narcissistic projection of Los's masculinist desires and fears. Even before the lovers are separated, first by the revolt and then by death, the hero's love is defined by a tragic inevitability that renders him the victim of a hostile, and feminized, universe: "Never before had Los sensed the hopeless thirst for love with such clarity, never before had he understood the deceit of love, the terrible exchange of himself for a woman: the curse of the male animal. You open your arms wide from star to star and wait to accept a woman. And she will take everything and will live on. And you, the lover, the father, will be an empty shadow, arms spread from star to star" (112). The "deceit of love" displaces the planetary tragedy of a dying world and the political horrors of a doomed revolution onto a nightmare image of men "curse[d]" by vampiric women. Because the Soviet revolution on Earth already has brought about the utopia envisioned in *On Two Planets* and *Red Star,* Aelita has no heuristic role and reverts to a stereotype of masculinist fantasy: although she introduces the Earthmen to Martian language and scientific knowledge, she is also the voice of an absolute and inaccessible love. If Los becomes both lover and father, Aelita is both siren and daughter, craving protection and yet always in the process of betraying the hero's love. In this respect, she becomes the necessary sacrifice for the hero's revolutionary commitment, the mark of the failure of both true love and, though Tolstoy cannot posit it in this manner, of the possibility of a true revolution.

At the end of the novel, the heartsick Los has returned to Earth and is summoned when radio signals from Mars reach the Soviet Union. The Martian signal carries only Aelita's voice:

> Like quiet lightning, the distant voice struck his heart, repeating sadly in an unearthly tongue: "Where are you, where are you, where are you, Son of the Sky?"
>
> The voice stopped. Los was staring straight ahead with dilated eyes. . . . The voice of Aelita, of love, eternity, the voice of longing, flew through the universe, calling beckoning, entreating—where are you, where are you, love[?]. (167)

Either dead or lost across the vast reaches of interplanetary space, Aelita haunts the hero as a voice of unrealized possibilities. The endings of *On Two Planets* and *Red Star* are again transmuted: the unions of Earth men and Martian woman that signal the hope for the progress of the human race are replaced by a tragic sense of loss—a loss not only of romantic love but of faith in science and the dialectical movement of history. *Aelita* both posits a successful revolution and depicts the impossibility of the kind of change that Bogdanov envisioned. Revolution ends not with a new order and endless progress but with a return to romantic stereotypes: Los and Aelita as interplanetary versions of Hero and Leander. Tolstoy's lack of interest in the scientific and philosophical issues that fascinated Lasswitz and Bogdanov is suggestive, in this regard, of the romanticizing of the red planet in the wake of Edgar Rice Burroughs and the advent of the interplanetary adventure novel. Although there is no evidence that Tolstoy, while in exile, read Burroughs's Barsoom novels, millions of English-speaking readers by 1922 certainly had.

FOUR

Lichens on Mars: Planetary Science and the Limits of Knowledge

In the years after Lowell's death in 1916, the canal thesis led a curious twilight existence. The term "canals" reacquired the ambiguous, semimetaphorical connotations of Schiaparelli's canali and was deployed regularly as a neutral descriptive label that survived even the first Mariner missions. As late as 1968, Samuel Glasstone, Carl Sagan, and James B. Pollock suggested that the Mariner 4 photographs might confirm the existence of two-hundred-mile-wide "canals"—topographical features that retained something of the linear characteristics that Lowell and Schiaparelli had observed (Glasstone 1968, 125–31; Sagan and Pollock 1966, 117–21). From the 1920s to the 1950s, invoking the canals became a convenient shorthand for hedging one's bets about the existence of Martian life-forms or the nature of the planet's atmosphere and surface: astronomers could accept the existence of linear markings yet question or dispense altogether with the "chain of reasoning" on which Lowell had insisted. Although a half century of efforts to detect oxygen and water vapor in the Martian atmosphere gave increasingly discouraging results, it was difficult for anti-Lowellians to prove a negative: that life—particularly low forms of animal life—did not exist on the planet. As the canal builders retreated into science fiction, the idea of "primitive" life on Mars persisted, even in the works of those planetary astronomers, such as Antoniadi, who criticized Lowell's theory. Until the Mariner 4 mission, most astronomers believed that the dark areas of Mars were covered with some form of exotic vegetation or lichen that might "extract the moisture [it] need[s] from the air, having no root system, [and be] able to tolerate almost any conditions" (Ley and von Braun 1956, 98). Yet even as Martian exobiology became a subspecies of speculative ecology, the vision of canals and the possibility of intelligent life on Mars remained ghostly presences on the fringes of scientific explanation, nostalgic traces of humankind's fascination with all things Martian.[1]

Most historians of Mars pass quickly over the half century between the death of Lowell and the Mariner missions. The canal theory, they insist, was quickly discredited, and planetary astronomers, for the most part, had few regrets about relinquishing hope for primitive life once the cratered surface had been photographed. While there is some validity to this account, it simplifies several complex debates in planetary astronomy between 1915 and 1965; overlooks at least one significant comeback by the canal theory in the 1920s; and underestimates the ways in which Lowell's paradigm of a dying planet influenced scientific speculation about the composition of the Martian atmosphere, the character of its surface, and the nature of its putative life-forms. In this chapter, I look closely at three midcentury scientific studies of Mars by E. M. Antoniadi, Earl C. Slipher, and Hubertus Strughold; the continuing controversies about spectrographic analyses of the Martian atmosphere; and the ways in which such scientific debates were framed in popular media. The scientific literature about Mars between 1916 and 1964 is more varied and ambiguous than most historians suggest, and the narrative that Lowell had crafted in the 1890s remained a significant part of midcentury accounts of the red planet. Between the photographs by Lampland published in 1905 and the flyby of Mariner 4 sixty years later, no single technological breakthrough drove a stake through the heart of Lowellian Mars. After 1933, measurements of surface temperatures, atmospheric pressure, and spectral analyses made Lowell's thesis less and less likely. Yet a good deal of the scientific literature—no less than the science fiction of the period—discloses a nostalgia for the possibility of intelligent life on Mars and a hunger for hard data that might answer questions about the planet once and for all. The relative lack of new data prior to 1964 kept some aspects of the canal theory alive yet also created a rhetorical and conceptual need to fit areography into a progressivist narrative of increasingly accurate knowledge. Such progress, though, was defined less by the accumulation of data than by the discrediting of Lowell's thesis.

ANTONIADI AND THE ATTACK ON THE CANAL THEORY

While Martians dominated science fiction during the middle of the century, the scientific study of Mars, like solar system astronomy more generally, was at best an intermittent enterprise. As Ronald E. Doel notes, until the Russians launched *Sputnik* in 1957, "U.S. astronomy remained one of the smallest branches of the physical sciences" (1996, 2). Between 1921 and 1960 only twenty-three dissertations on planetary astronomy were submitted in the United States, and between 1918 and 1960 articles on solar system astronomy never constituted more than 25 percent of all publications in astronomical journals, and often considerably less (236–37). While it has become fashionable to attribute the dearth of planetary astronomers to a sort of institutional horror at the canal debacle (Sagan indicated as much), the problems of trying to interpret the limited data about Mars attracted little research money and promised little in the way of payoffs either in resolving questions about Martian biota or in professional recognition. In contrast, stellar astronomy solidified its position as the dominant area of telescopic research. The development of time-exposure photography revealed stars with magnitudes fainter than those seen by naked-eye observations and allowed astronomers to study distant nebulae; in contrast, planets remained difficult to photograph, and even the best images, often published by the Lowell Observatory, were blurry. The dominance of reflecting telescopes at large American observatories meant that time, energy, and money were devoted to stellar astronomy; while refractors might be better for night-by-night planetary observation, reflectors were superior for stellar photography. By the 1930s, spectroscopy "transformed both the practice and institutional structure of American astronomy" (Doel 1996, 12), yet spectral lines in planetary atmospheres were difficult to interpret accurately because astronomers had to try to distinguish between off-planet chemical signatures and those of gases in the Earth's atmosphere. In contrast, the problems in stellar observation—were nebulae gas clouds within the Milky Way or distant spiral galaxies of their own?—could be answered by new observations and data in a way that debates about Martian canals could not.

During this period, the Lowell Observatory was the only institution that devoted significant time and resources to planetary study (Rothenberg 1981, 305–25; Brush 1978, 771–87). After Lowell's death, the observatory went through a period of uncertainty caused, in part, by a fight over his will, and the canal controversy fostered suspicions between Lowell's staff and the stellar astronomers at other observatories.[2] While the continued study of Mars at the observatory by the Sliphers, C. O. Lampland, and others led to important advances in photography, spectroscopy, polarimetry, and radiometry and generated significant data on planetary atmospheres, in-fighting at the observatory between 1916 and 1927 and the insistence of the Sliphers on promoting the canal thesis fed suspicions within the astronomical community that "Martian science" was an oxymoron (Tatarewicz 1990, 5–12; Crowe 1986, 545–46). While it would be difficult to overestimate the dedication of key figures at the Lowell Observatory both before and after Lowell's death in 1916, internal troubles demoralized the staff and detracted from the task of defending Lowell's legacy against its critics.[3]

The roots of the mid-twentieth-century delegitimation of the canal theory went back to the late nineteenth century and the criticisms directed at Lowell by E. W. Maunder, W. W. Campbell, and E. E. Barnard. Probably the key anticanalist, though, was Eugene Michael Antoniadi (1870–1944), who bridges the gap between the years when the canal debate was at its height and the growing skepticism after 1910 with Lowellian Mars (Sheehan 1996, 134–45; Crowe 1986, 534–36; McKim 1993, 164–70, 219–27). Born in Constantinople of Greek descent and educated in France, Antoniadi became Flammarion's assistant at the Meudon Observatory in 1893 and three years later was appointed director of the Mars section for the British Astronomical Association (BAA). This organization had been founded by Maunder and, in marked contrast to the Royal Astronomical Society, became a venue for scientific studies that disputed Lowell's findings. Antoniadi's early publications were firmly in the tradition of Schiaparelli and Flammarion, but by 1902 he had begun to express reservations about the existence of the canals and broke with Flammarion. During the opposition of 1909, he made a series of observations at Meudon that led him to challenge Lowell's views in a series of papers—many published as reports to the Mars section of the BAA—and to enlist many of the prominent astronomers of his time in his campaign: Maunder, Barnard, Campbell, George Ellery Hale, and Simon Newcomb among them. Antoniadi corresponded with both Schiaparelli and Lowell in efforts to dissuade them from their view of a planet incised with geometrical patterns, but the Italian astronomer died in 1910 still defending his views and Lowell responded with his customary defenses of his thesis.[4] While Sheehan, Crowe, and other historians have emphasized the fundamental disagreements between the canalists and anticanalists about telescopes and techniques for planetary observation, the interpretation of visual evidence, and the composition of the Martian atmosphere, they downplay the fact that many of Lowell's critics still operated within a paradigm that accepted both the evolutionist view of Mars as a dying planet and the possibility of various forms of life. Ironically, then, although the canal controversy went to the heart of differing conceptions of scientific methodology, the lack of canals did not in itself prove that the planet was bereft of higher forms of life. Antoniadi argued that "the true appearance of [Mars] is . . . comparable to that of the Earth and of the Moon" (1975, 23), but if all heavenly bodies are marked by irregular topography, then the comparison of Mars to both the stone-dead Moon and the teeming biosphere of Earth does not invalidate Lowell's vision of the planet so much as it offers an alternative explanation of its appearance.

Antoniadi's *Planet Mars*, published in 1929, summed up twenty-five years of his objections to the canal theory, though his work was not as decisive as later spectrographic studies in banishing Lowell's canal builders from scientific journals to the pulps. In his study, Antoniadi draws on observers in the anti-Lowellian camp, including Barnard and Hale who, using the large refractors at Meudon and Mt. Wilson, had seen masses of irregular detail rather than straight lines on the surface of Mars (Antoniadi 1975, 33–37). While Lowell insisted on using a diaphragm to sharpen the image by reducing its size and filtering out some of the atmospheric distortion, Antoniadi employed the full aperture on the Meudon telescope and waited patiently for conditions of good seeing. While conceding that conditions at Meudon usually are not as good as those at Flagstaff, he maintained that the comparatively few moments of visual clarity allow for much better images of Mars than those seen by Lowell and collaborators. Seen through Lowell's twenty-four-inch telescope, Mars appears twice as far away as it does in the Meudon telescope when the full aperture is employed. During conditions of excellent seeing, this technological advantage, he argues, becomes "decisive," and allows him to achieve an accuracy that the smaller images of the planet drawn in Arizona cannot match: "The photographs of Mars taken by Lowell at Flagstaff are not in accord with his drawings," Antoniadi claims, "but agree well with my drawings made from Meudon!" He goes on to note that "it is ironical that the very worst planetary drawings have been made at the best observing station!" (18). As I shall argue below, this is a claim that E. C. Slipher vigorously disputed as late as the 1960s.

Antoniadi's experience at Meudon convinced him that "nobody has ever seen a genuine canal on Mars." He attributed the "geometrical patterns" observed by Schiaparelli to his older colleague's tendency to draw a straight line every place that the surface "shows an irregular streak, more or less continuous and spotted, or else a broken, greyish border or an isolated complex lake" (40). As this description suggests, Antoniadi's

Mars is structured by terrestrial analogues, and, in some respects, his vocabulary owes as much to nineteenth-century astronomers, such as Flammarion, as it does to either the canalists or Pickering. Yet even as he rejects the canals, he takes pain to integrate Schiaparelli's observations into his own anti-Lowellian polemic. Significantly, Antoniadi accords the Italian astronomer a respect that he denies Lowell, and, as late as 1929, praises Schiaparelli's powers of observation: "Since, for rare instants of about ⅛ second each, I myself have seen the single or double optical lines, the accuracy of Schiaparelli's *observations* is not in doubt. The error here has been in interpretation, not observation; the Milan astronomer wrongly believed in the reality of the geometrical patterns on Mars, and excluded all possibility of illusion. [The canals'] miraculous character defied all explanation. As he wrote to me on 30 April 1900: 'Truly the planet has become a frightening and almost a distasteful subject to me. The more I study it, the less can I understand its phenomena'" (44). Linear patterns emerge only as a trick of the eye and mind, and Antoniadi quotes the letter from Schiaparelli to demonstrate the "miraculous character" of the canals and thereby dismiss Lowell's "explanation."

Paradoxically, however, Antoniadi's insistence on the irregularity of the Martian surface suggests why the canal narrative proved so hard to dislodge. When Antoniadi describes almost epiphanic moments of "perfect" seeing, Mars reveals a surface "infinitely irregular and natural," and his terrestrial analogies, like Flammarion's decades earlier, drift toward something akin to the literary sublime: "The best images of Mars I have had . . . were those of 20 September 1909; they remained perfect for over two hours. . . . in the middle of a crowd of irregular patches I saw, on the site of one of Schiaparelli's linear, single or double 'canals,' an extremely broken-up lake; and the region to the south of the Syrtis Major presented the aspect of a field of fresh grass and woods of darker green, the whole area being dotted with tiny white points. There was no geometrical regularity in the innumerable details seen on this unique occasion!" (15–16). In his efforts to discredit the canal theory, Antoniadi projects onto Mars an earthlike ecology. While Lowell compares the color of the Martian surface to the landscape of the Arizona desert, he refuses to speculate on the planet's biota. In contrast, Antoniadi could be describing the meadows and forests of the French Alps. His image of "a field of fresh grass" renders the planet less alien and less inhospitable than Lowell's dying world, and consequently Antoniadi, like other anticanalists, faces a significant narrative dilemma: the more he insists on the nonlinear character of the "innumerable details" he has observed, the more complicated and uncertain his own account of the planet becomes. As Schiaparelli implied in his letter of 1900, rejecting Lowell's theory offers little help in suggesting alternative explanations for the albedo changes on the planet's surface. Lowell's supporters, like Morse (as I suggested in chapter 2), frequently invoked Occam's razor to highlight the elegance and simplicity of his "chain of reasoning"; his opponents did not, instead emphasizing the complicated "detail" they observed that could not be resolved into either straight lines or straightforward narratives. The absence of the canals did not in itself prove that Lowell was wrong about the character of the Martian surface, and in zeroing in on its "innumerable details" Antoniadi ironically reinscribed many of the key aspects of the American astronomer's description of the planet.

Like many of Lowell's critics, Antoniadi operates within the paradigm of the nebular hypothesis; Mars is a dying world, "undoubtedly in a state of desiccation incomparably more advanced than that of the Earth" (1975, 30). In describing the changing appearance of the surface during the opposition of 1924, Antoniadi imagines a hypothesized "xerophytic vegetation, similar, though not necessarily identical, with ours," exhibiting seasonal changes analogous to terrestrial foliage: "Not only the green areas, but also the greyish or blue surface, turned to brown, lilac-brown or even carmine, while other green or bluish regions remained unaffected. It was almost exactly the color of leaves which fall from trees in summer and autumn" (26, 29). The final comparison seems intended to be (almost) literal rather than metaphoric; the analogy cloaks the alien surface in terrestrial and autumnal colors. This sympathetic resonance between Earth and Mars implies that the advanced "desiccation" of the latter still can be understood in terms of a centuries-old analogy: the red planet embodies almost nostalgically the human experience of changing seasons and intimations of planetary mortality. Antoniadi's Mars, in this respect, is no less a product of early-twentieth-century cultural values and associations than Lowell's.

Acknowledging that it is "very probable" that Mars is covered with vegetation, Antoniadi moves on to consider the ecological arguments advanced for animal life on Mars. His view is closer to Lowell's than to Wallace's:

> The melting of the polar snows and the absence of ice in the temperate and equatorial regions proves that the temperature is not too low to support living things, since animals well known to Man exist

> quite well in the cold of up to -72° in Eastern Siberia. Moreover, the observations of the clouds at considerable heights shows that though the atmosphere of Mars is very thin, it is still able to hold ice-needles or water-droplets in suspension—or even particles of dust. . . . In this connection, we must bear in mind that the people of Tibet have no difficulty in breathing an atmosphere twice as rarefied as ours. If, then, we consider also life's marvelous power of adaptation, one of the aims of the Creator of the Universe, we can see that the presence of animals or even human beings on Mars is far from improbable. This was the view held by Flammarion, Schiaparelli, Moulton and others. However, it seems that advanced life must have been confined to the past, when there was more water on Mars than there is now; today we can expect nothing more than vegetation around the vast red wilderness of the planet. (67)

Following astronomers since Herschel, Antoniadi deduces the existence of ice, water, and an atmosphere (comparable to Earth's at 14,000 meters above sea level) to explain cloud formations and the existence of vegetative life. These conditions allow him to contemplate the possibility of Martian fauna. Terrocentric analogies between hypothetical Martians and terrestrial animals that can survive the Siberian cold, however, can take him only so far, and he resorts to a latter-day version of the argument from design to speculate what the "marvelous power of adaptation" might have achieved on Mars, given the "aims of the Creator of the Universe." Like Maunder and Wallace, then, Antoniadi's view of Mars is cast in the theocentric terms that Lowell and Morse sought to banish. Because these "aims" include the possibility of intelligent life—"far from improbable"—the difference between Antoniadi and Lowell boils down to a disagreement about where Mars stands in its evolutionary descent from a life-bearing world to a dead orb. Their suppositions about Mars are similar, but surface "detail" is interpreted differently: Did these beings build canals or watch their quasi-pastoral world gradually die? Antoniadi banishes his Martians to an unspecified "past"; no advanced civilization struggles to maintain its existence. By projecting intelligent life into the dark backward and abysm of Martian history, Antoniadi also relegates to the past the politico-ecological questions that fascinated millions of science-fiction readers. His variegated Mars is a more somber, and less compelling, version of the Lowellian dying world. Without canals, there is no linchpin to explain the response of intelligent beings to a dying planet, no tragic struggle to point to as a harbinger of the fate of humankind.

Antoniadi's offensive in the first decade of the century did not signal the death of the canals. Although Antoniadi, E. C. Slipher, and a handful of other astronomers engaged in a long battle over the interpretation of photographs and sketches of the Martian surface, many of their colleagues viewed the controversy as one that could not be resolved by appeals to subjective interpretations of lines, streaks, or splotches. Consequently, any new data or approaches that offered the hope of resolving questions about Mars were greeted enthusiastically by the astronomical community. In 1924, as Mars approached its closest opposition to Earth in two hundred years, Swiss astronomers mounted a heliograph in the Alps to flash signals to Mars. The U.S. Navy maintained radio silence for three days to listen to messages from the Martians (Bent 1924). Both efforts, of course, failed to provide the proof of canal-building Martians that some observers had expected. Yet even as efforts to signal Mars failed, the oppositions of 1924 and 1926 witnessed important developments in scientific approaches to Mars brought about by new technologies, notably the vacuum thermocouple and refined spectroscopic equipment (Dick 1996, 107–15). Attaching vacuum thermocouples to telescopes to measure the temperature on the surface of Mars, two teams of researchers—Edison Pettit and Seth Nicholson at Mt. Wilson Observatory and William W. Coblentz and C. O. Lampland at the Lowell Observatory—reached similar conclusions: the temperature rose to as high as 20 degrees centigrade for the dark areas of Mars and ranged from -10 to 5 degrees centigrade for the light areas at the equator. These readings suggested that the dark areas were indeed vegetation, a conclusion seconded by Robert Trumpler at the Lick Observatory, who photographed canals but argued that they were natural formations that harbored vegetation (Dick 1996, 109–14). With these results in hand, Coblentz speculated in articles in *Popular Astronomy* that his readings confirmed the opinions of Lowell and Pickering, who long had held that Mars was warmer than their antagonists claimed. He then went on to offer extended comparisons between a hypothesized Martian ecology and conditions in Siberia (Coblentz 1925a, 310–16, 363–82; see also Coblentz 1925b, 439–41). Like Antoniadi, Coblentz had little difficulty in imagining earthly analogues for a post- but not anti-Lowellian Mars.

Five years later, however, the search for life on Mars hit something close to a dead stop. Since the height of the canal controversy, spectrographic readings of the presence of water vapor in the Martian atmosphere had produced contradictory results; as early as 1894 W. W. Campbell had failed to find any evidence of water vapor, but his findings, as we have seen, were disputed by Lowell and V. M. Slipher (DeVorkin 1977, 37–53; Schorn 1971, 223–36). In 1933, Walter S. Adams and Theodore Dunham, working with a spectroscope attached to a telescope at Mt. Wilson, took into account the Doppler effect in their analyses of the Martian atmosphere. The light from objects approaching Earth shifts slightly to the violet end of the spectrum; objects moving away produce a shift toward the red end. Rather than comparing their readings to those from the Moon, as astronomers since Jules Janssen in the nineteenth century had done, Adams and Dunham tried to separate the blue-shifted lines from Mars from the spectrum produced by the Earth's atmosphere. They found neither oxygen nor water vapor and concluded that Mars was an arid and inhospitable world (Adams and Dunham 1934, 308–16, 1937, 209–11; Sheehan 1996, 150; Dick 1996, 112–15; Moore 1999, 70–71). While astronomers continued to take spectrographic readings during subsequent oppositions, Adams and Dunham's conclusions generally were accepted as definitive, and the hunt for life devolved into speculation about what kind of vegetation could survive the planet's frigid and desiccated conditions. Dunham himself suggested that whatever water and oxygen existed on Mars were locked in the polar caps. But he offered no explanation for what happened to either when the caps melted during the spring and summer. Mars presented a strange, unearthly paradox: a planet with visible seasons but no hydrological cycle.

The volatility of the debate about life on Mars between 1924 and 1933 should not be underestimated. Rather than a sober progression of carefully assessed new data, the controversies about Martian life reflected a variety of opinions and a tendency to accept as semidefinitive the newest astronomical measurements that could be replicated. In this regard, the readings of the temperatures on Mars by the Pettit and Coblentz teams and the spectrographic results obtained by Adams and Dunham highlight a crucial problem in the history and philosophy of science and technology. Any science that views itself progressing asymptotically toward an objective truth is prone to regard the latest results as superior to earlier measurements. Astronomy, like other experimental sciences, is costly and technologically dependent. Investments in telescopes, photographic devices, new equipment, and skilled labor become part of the complex matrix in which "new" data is produced. Institutionally as well as conceptually, new expenditures demand on some level new, more accurate, or more reliable data. New findings by colleagues one knows (if only by reputation) and trusts using state-of-the-art instrumentation reinforce a sense of the reliability of data.[5] While such data typically remain open to interpretation, the default assumption that science always forges ahead is a powerful inducement to credit their reliability as a benchmark against which future results may be judged, even if other researchers offer alternative explanations of their significance. The thermocouple measurements of Coblentz, Lampland, Pettit, and Nicholson breathed new life into Lowell's vision of the planet and, for a few years, redirected the focus of debates about Mars.

On December 9, 1928, the cover story for the Sunday *New York Times Magazine*, written by H. Gordon Garbedian, began with the statement that "science has dropped its traditional theory that Mars is dead" (1928, 1).[6] For this article, titled "Mars Poses Its Riddle of Life," Garbedian interviewed a who's who of astronomers and planetary scientists: Henry Norris Russell (Princeton); Clyde Fisher (American Museum of Natural History); Robert G. Aitken (director of the Lick Observatory); Harlow Shapley (director of the Harvard Observatory); William Pickering; Edward Bryant Frost (director of Yerkes Observatory); William Coblentz (U.S. Bureau of Standards); C. G. Abbot (director of the Astrophysical Observatory, Smithsonian Institution); Wallace Sydney Adams (director of the Mt. Wilson Observatory); George Ellery Hale (Mt. Wilson); and V. M. Slipher. Significantly, this list included stalwart opponents of the canal thesis, including some, like Aitken and Hale, who had sided with Newcomb during the illusionist controversy in 1907. The prospect of a warmer Mars, however, apparently reinvigorated the canal enthusiasts and muted some of the criticism directed at them. Coblentz's data on surface temperatures had produced enough of a consensus that Garbedian could write that "there is life on Mars, the majority of our astronomers, many of whom several years ago scoffed at the idea as 'impossible,' now agree" (1). This substantial article (about five thousand words) includes extensive quotations from all eleven of the astronomers, and it provides the most detailed midcentury account of what scientists believed, or at least what they were willing to say on the record, about Mars. Although the majority of the astronomers hedge their bets about what kind of life might exist on Mars, the overall effect of their rhetorical circumlocutions is that the "riddle" of life on Mars shades into a validation of the canal thesis as a plausible, if hardly the only, interpretation of Coblentz's data. The seriousness with which these men discuss Lowell should serve as a cautionary tale for those who treat the canal thesis as dead, buried, and discredited by 1909 or 1916.

The temperature data from the thermocouple is treated by Garbedian and the men he interviewed as a significant milestone in planetary science. "Contrary to the opinion that the temperature on Mars never rises above the freezing point," Garbedian reported, "the thermocouple gave readings denoting that it rises to 60 degrees Fahrenheit, and even higher, which is not unlike conditions registered in New York, Philadelphia or Washington on a bright sunny day in March or April" (1928, 1). The comparison of equatorial Mars to springtime conditions in major cities on the East Coast and the weak double negative ("not unlike") describe a planet familiar to readers of Edgar Rice Burroughs. In passages such as this one, Garbedian's seemingly balanced and well-written article tilts subtly toward the views offered by Lowellian apologists such as V. M. Slipher and Pickering. In framing the revived debate, the article describes a continuum of scientific opinion ranging from seeing Mars as a planet inhabited by intelligent life to an inhospitable world that supports only primitive vegetation. None of the astronomers rejects or even seriously questions the Lowellian paradigm of a dying world:

> While these astronomers agree that recent discoveries have demonstrated that life is to be found on Mars, they disagree as to the extent that life is there developed. Dr. Pickering takes the advanced view that 'it is almost certain that Mars is inhabited by intelligent beings,' and suggests that the Martians are signaling to us. Dr. Abbot takes the other extreme position, saying that life on Mars, because of the unfavorable conditions, is confined to low types of vegetable life. Others, including Dr. Russell, Dr. Aitken, and Dr. Fisher, hold that the existence of a highly organized animal life, including a civilization of intelligent beings, on our neighbor planet is not 'impossible or even improbable.' But, in their opinion, the present testimony points only to the existence of vegetable and low types of animal life. (1)

Garbedian's rhetoric is revealing: Pickering's view is described as "advanced"; the "other extreme," represented by Abbot's skepticism, is undercut implicitly by a version of a Lowellian ecological argument—the presence of vegetation makes it likely that some form of animal life exists on Mars. According to Garbedian, Aitken, Russell, Shapley, Coblentz, Adams, Frost, Slipher, and Pickering accept this view. Citing spectrographic findings by Adams and Charles St. John at Mt. Wilson (1925), Russell acknowledges that the " 'presence of oxygen and water vapor . . . has been established. There can now be little doubt that, owing to atmospheric influences, the surface of Mars is a good deal warmer than once supposed, and there appears to be no further difficulty on the score of regarding the planet as habitable' " (quoted in Garbedian 1928, 2). Ironically, Russell describes the planet in terms that are more hospitable to terrestrial life than those merciless deserts envisioned by Lowell.

The debates that Garbedian chronicles, then, are not about the likelihood of life on Mars but about the mechanisms by which Martian biota have adapted to a still-forbidding environment. The consensus view, according to Garbedian, is summed up by Russell: " 'Recent observations show only that there is probably vegetable life and low forms of animal life on Mars. There the evidence stops, and the realm of speculation opens. The arguments of the late Dr. Lowell . . . can neither be proved nor disproved at the present time. They rest essentially upon visual observations so delicate and difficult that the results obtained by sincere and impartial observers are in disagreement. No method of solving this difficulty can be found with present telescopes and other apparatus, wonderful as they are. The question therefore must be left open' " (22). By returning Lowell to the fold of "sincere and impartial observers," Russell distances himself from the illusionist camp and reopens the question of the relationship between ambiguous visual evidence and the "established" facts of oxygen and water in the Martian atmosphere. Once opened, however, the Pandora's box of exobiological speculation allows, at one end of the continuum, Pickering and Slipher to champion Lowell's theory. If Pickering by 1928 had long since been isolated from the astronomical community, as Plotkin contends (1993, 101–22), his marginal position is not communicated to readers of the *Times*. The former skeptic who offered a geological explanation for the canals has become a vocal supporter of Lowell's thesis: "Dr. William Pickering, who [has] devoted years to the study of Mars, argues that 'some of the surface markings are so straight and regular that they cannot be accounted for as accidental occurrences of nature and we can only explain them as the result of intelligent beings' " (22). While such assertions by Pickering obviously make good copy, Garbedian's article brings such sensationalist claims within the rhetorical and conceptual bounds of legitimate science. That Pickering may have been regarded as a crackpot by some of his younger colleagues at major observatories is less significant—at least for readers of the *Times*—than his being represented as an acknowledged authority by a disinterested journalist. When Garbedian revised his article for his book *Major Mysteries of Science* (1933), he added additional commentary by Pickering but concluded that "the present observational evidence does not support the enthusiastic speculations of a few men of science, who with Dr. Pickering, see on Mars an extensive intelligent civilization" (1933, 228).

Although Russell gets the final word in the original article, Slipher is quoted more often than any of the other astronomers.[7] His rhetoric is even more polemical than Pickering's: " 'Observations continue to amass more evidence and to furnish further objective proof of the results and conclusions secured by Dr. Lowell twenty years ago, but which were then refused general acceptance. Such results as shown today lead to the conviction that Mars is not a dead world, as most investigators used to think, but an ever-changing planet whose physical conditions differ in degree only from those of our own earth' " (22). The vehemence of a true believer evident in this quotation may offer some insight into Russell's efforts a year earlier to persuade Roger Lowell Putnam, the sole trustee of the Lowell Observatory, to block publication of Slipher's book manuscript that defended Lowell's canal theory (Doel 1996, 36; DeVorkin 2000, 292–308). Slipher obviously considers the case closed and his mentor vindicated by Coblentz's and Lampland's data; Russell leaves matters open. Slipher's stout defense of Lowell, in this regard, may reflect his frustration at having had to knuckle under to Putnam's pressure as well as his commitment to the canal theory. In his remarks to Garbedian, Slipher clearly enjoys having his adversaries on the defensive, and he takes pains to emphasize that his claims are a plausible extension of the facts that other astronomers concede. Rhetorically, then, he can press home the ecological argument—the interdependence of flora and fauna—and conclude with the kind of speculation that made more circumspect scientists wince: " 'With vegetable life fairly assured, animal life is almost certain. Furthermore, if intelligent animal life exists there—as the straight and narrow canals seem to imply—then it seems probable that Mars is a world much more like Earth than we have heretofore been led to surmise. I am now inclined to consider it possible that even human life, if transported to Mars, might exist and perhaps flourish there' " (22). While Russell and the "majority of astronomers" acknowledge that "the observations of science do not preclude the existence of intelligent life on Mars" (22), Slipher insists on tying this half-hearted acknowledgment to the canal thesis. Because the temperature data seemingly remove one of the key objections to Lowell's views, the anti-canalists are put on the defensive, having to concede a key aspect of his "chain of reasoning"—water could melt and flow through canals, possibly covered to prevent evaporation. Russell's objection to Slipher's manuscript in 1927, then, does not indicate either his outright rejection of Lowell's thesis or his support for the views of Maunder and Antoniadi;

instead his reticence stems from a reluctance to endorse Slipher's "speculation" about what is possible—canal-building Martians—as "probable."

Russell's responses to Coblentz's temperature data and then the spectroscopic readings of Adams and Dunham are emblematic of the shifts in scientific opinion between 1928 and 1935. Russell may have been the most influential astronomer of his era and was committed both to theoretical physics and to his Presbyterian heritage and beliefs (DeVorkin 2000, 3–19). A frequent visitor to the Lowell Observatory, he "enthusiastically backed" the observatory's efforts to search for the trans-Neptunian Planet X that had occupied Lowell during the last years of his life and helped to secure much-needed funding from the American Philosophical Society (Doel 1996, 36). In this context, his efforts to block Slipher's canal treatise suggest his sense of the effects that such a work might have in professional circles indifferent to or suspicious of the canal thesis. In Garbedian's article, Russell carefully stakes out a middle ground between canalists and anticanalists, but shies away from the illusionist argument. He concedes that " 'it is generally recognized that there exists an objective basis for the canals in the form of fine detail on the surface of Mars, and it is widely believed that these details have, in a general way, the streaky character of the canals.' " Garbedian notes that this comment "probably expresses the view of most astronomers." These streaks, however, do not seem definitive proof for " 'a large majority of astronomers [and] it is therefore necessary,' " Russell concludes, " 'to render a verdict [of] not proved with regard to [the canal] theory' " (2, 22). Yet Russell, like many of his contemporaries, has conceded the substance of Lowell's view of Mars: a planet warm and wet enough to support life, "streaky" lines on the surface, and the presumption that the presumed vegetation probably supports some form of animal life.

By crediting the most recent—and presumably the "best"—data from the Lowell and Mt. Wilson Observatories, Russell indicates how fragile the anticanalist consensus of the early 1920s—the "traditional theory" that Garbedian invokes at the beginning of his article—seems to have been. The new temperature and atmospheric measurements push the objections of Campbell, Maunder, and Antoniadi to the side, and Russell endorses a biologically robust, if nonanthropocentric, Mars. In his 1933 book, Garbedian echoes Russell's language in concluding his chapter about Mars by acknowledging only that the planet harbors "probably vegetable life and low animal forms" (228). An advanced Martian civilization may not be signaling Earth or building new canals, but there is nothing in Russell's comments to preclude the possibility of intelligent life on the fourth planet. His verdict, in effect, asks Slipher to prove what he himself cannot disprove.

Six years later, after Adams and Dunham's findings, Russell has become far more skeptical, and the canals have begun to retreat into myth. He now accepts the view that Mars lacks significant atmospheric oxygen or water vapor, and, consequently, sees it as a classic example of a planet in a "later stage" than Earth of its evolution (1934, 296). He explains the red surface of the planet as an effect of oxidation and suggests that while the canal theory may be dead, the dying planet struggles on: "May it not be that rock weathering has done its deadly work on Mars and depleted its atmosphere of oxygen? If the oxygen was of vegetable origin, there was once life on the planet. We cannot be too sure that it is not there still, for the exhaustion would be so slow that evolutionary processes might well be able to follow it, producing plants which conserve oxygen as terrestrial desert plants store up water" (297). By 1934, Lowell's canal builders have been banished but exobiology preserved: the hypothesized "evolutionary processes" allow life-forms to adapt to a desiccated environment.

Russell's response suggests something of the ways in which Lowellian Mars faded into its twilight. In popular histories of astronomy written between 1930 and 1964, Lowell's theory often received a fair hearing, even though it was always hedged in by qualifications. As Wily Ley and Werner von Braun wrote in 1956, "It is not easy to say in a sentence why most professional astronomers did not agree with Lowell and why his reasoning is now generally considered wrong in spite of the apparently logical development of his hypothesis." They conclude that it was "probably less a case of [Lowell's] drawing wrong conclusions than one of operating with mistaken assumptions" about the planet's atmospheric conditions and hydrological cycle: overly generous estimates of the percentage of oxygen and water vapor in the Martian atmosphere and the size and depth of the polar caps (1956, 77). A generation earlier, such skepticism was more muted.

Many scientists in the period between the world wars hoped that new data could resolve the paradoxes that Mars presented. In 1929 Mary Proctor, the daughter of Richard Proctor and a fellow of the Royal Meteorological and the Royal Astronomical societies, devoted several pages in her popular study of the planets to the canal controversy. While admitting that "very satisfactory photographs of Mars have been obtained," she declares that "they have given no decisive evidence regarding the vexed question of the canals," and pins her hopes for a resolution of the controversy on a new generation of "giant" telescopes (1929, 81). Proctor treats the canals in a series of carefully crafted rhetorical questions, distancing herself from Lowellian enthusiasm but holding open the possibility that Lowell's vision of the planet could be confirmed: "Should photographs be obtained with the long infra-red waves at a period on the planet corresponding to springtime on ours, when the canals are gradually filled with water from the melting snows caused by the unlocking of the polar caps, will they resemble in appearance and intricacy the hundreds of drawings made by Dr. Lowell and other observers?" (83). Whatever skepticism Proctor may voice, however, she still holds firmly to the nebular hypothesis, much as Russell does: "Mars is an object lesson for us, illustrating the fate awaiting our planet at some remote period, when it will have reached the stage of decrepitude now so plainly visible on our neighbour world" (85). This acceptance of a Lowellian history of planetary evolution underlies her refusal to reject the canal thesis out of hand. Like the American astronomers interviewed by Garbedian, Proctor is alert to the implications of the nebular hypothesis; if one accepts the "decrepitude" of Mars, one may reject the canals without dismissing the vision of a dying planet. Mars may be farther along its evolutionary path than Lowell had assumed, and whatever intelligent life-forms it may have harbored may be extinct.

A decade later, the discouraging spectrographic evidence of the 1930s colors the case for life on Mars. In his 1940 study *Life on Other Worlds,* H. Spencer-Jones, Britain's astronomer royal and a fellow of the Royal Society, summarized Lowell's theory as "attractive, ingenious, and logical, provided that the observational basis can be accepted" (1940, 220). But this "observational basis" is open to widely differing interpretations because "the detail [on the planet's surface] is so intricate that much of it is finer than the grain of the photographic plate"; for astronomers trying to sketch what they see, "the detail . . . is at the limits of vision and only to be glimpsed momentarily at rare intervals" (202, 215). His emphasis on the difficulty of observing Mars allows Spencer-Jones to stake out a middle ground between Lowell and his critics. Like Pickering thirty-five years earlier, he accepts the reality of the canals—"there can be no question about the existence of at least the most conspicuous of them"—but considers them "natural formations" (222, 227). In rejecting Lowell's description of the canals, however, he is careful to distance himself from both "extremes" and from the incendiary rhetoric of the debate:

> It must not be thought that there is any question of dishonesty or bad faith on the part of the observers to account for such extremes as the delineations of Lowell and Barnard. These two observers

> studied Mars for many years under favourable conditions and both were trained and experienced observers; each of them has honestly recorded the appearance of the planet as he saw it. The only possible explanation of the difference is that the observation of these faint elusive details is subject to complex personal differences. . . . The truth may lie somewhere between the two extremes, though it appears probable that Barnard's delineation is the nearer approach to the truth. (224–25)

The logic of this argument is suggestive of the continued persuasiveness of the nebular hypothesis. Like his contemporaries, Spencer-Jones does not discount Lowell's training, care, or scientific ability; the "complex personal differences" he invokes may glance at Lowell's desire to believe in the canals, but the Flagstaff astronomer remains firmly within the scientific community. By taking the fail-safe approach of maintaining that the "truth may lie somewhere between the two extremes" (244) represented by Lowell and Barnard, Spencer-Jones reproduces a weak version of the consensus of 1928. Although life on Mars is "almost certain," the logic of Lowell's theory can be rejected by redefining the canals as "natural formations." Like Ley and von Braun, Spencer-Jones does not have a surefire means to explain the eclipse of the canal theory; he can only voice his opinion that Lowell's assumptions about the planet have been severely qualified, even as his narrative of a dying world retains scientific currency.

POSTWAR RESEARCH

By the outbreak of World War II, Martians had become crypto-Nazis in much science fiction; but, in scientific studies, for the most part, they had been reduced to extraterrestrial lichen. The more attenuated the Martian atmosphere appeared to scientists, the more restricted views of the planet's ecology became. Consequently, the values and assumptions that astronomers brought to the study of Mars verged on those of what we now might call an extremophile exobiology—the study (or speculation) about how life-forms might adapt to conditions far more hostile than any encountered on Earth. And the more extreme the Martian environment appeared, the more the philosophical questions that had characterized the canal debate—geocentrism, biocentrism, and anthropocentrism—were projected outward beyond the solar system and into the galaxy. As the nebular hypothesis began to lose favor in the mid-twentieth century, the red planet seemed less certainly a harbinger of the fate of the Earth than a planet perpetually locked into conditions that permitted the existence of only primitive life.

After World War II, the U.S. Air Force initiated a series of research projects on planetary atmospheres in order to learn more about the global circulation of weather patterns on Earth. Mars was the beneficiary of renewed scientific attention. Efforts were also directed toward spectrographic analysis in the infrared spectrum; these too were characterized by the "overwhelming importance of military sponsorship" (Smith 1997, 56; Doel 1996, 45–50). The Dutch-born astronomer Gerard Kuiper turned to planetary research in the 1940s and, in 1947, detected what he argued were small amounts of carbon dioxide in the Martian atmosphere (Tatarewicz 1990, 19–24). During this period, the Lowell Observatory became the center for the study of planetary atmospheres and the site of Air Force funded conferences; these projects "reinvigorated the observatory's planetary research" (Tatarewicz 1990, 8). Conferences in Flagstaff in 1950 and 1951 were attended by Kuiper, Henry Giclas, V. M. and Earl Slipher, and Lampland, among others; a study by Harry Wexler, chief of the Scientific Services Division of the U.S. Weather Bureau, used meteorological principles to derive an estimate of 80-millibars surface pressure on Mars (Tatarewicz 1990, 9–11; Doel 1996, 56–63). Earl Slipher served as project supervisor for a report on planetary atmospheres, funded by the air force (Slipher et al., 1952). European efforts, disrupted by the war, resumed; in France, an important overview of what was known about Mars by Gerard de Vaucouleurs (1954) became the standard scientific work on the planet. Still, despite the upsurge in military funding, there was little consensus about the significance of space research in the overall military defense strategy until the launch of *Sputnik* in 1957.

As a result of allied military efforts during World War II, advances in radiometry and the study of the infrared spectrum gave astronomers new tools with which to study Mars. Better understanding of meteorology and geology allowed astronomers to refine the questions they posed about the fourth planet (Tatarewicz 1990, 6–7). However, fundamental uncertainties remained about the composition of the Martian atmosphere. The persistence of the belief in vegetation reinforced terrestrial analogies; consequently, some astronomers suspected that the difficulty in determining the composition of the Martian atmosphere was probably the result of a nitrogen-rich atmosphere. Nitrogen is inert and difficult to find in the visible spectrum, and because it comprises 78 percent of the Earth's atmosphere, many astronomers suspected that the thin envelope of the Martian atmosphere would have a similar composition. In 1954, Vaucouleurs estimated that nitrogen comprised 98.5 percent of the atmosphere of Mars (1954, 16), a figure that was contested by a 1963 spectrographic reading from the Mt. Wilson Observatory that showed low atmospheric pressure and the presence of carbon dioxide (Kaplan, Munch, and Spinrad 1964, 1–15; Wilford 1990, 55–56). Consequently, scientists before 1965 were inclined to accept vegetative life on Mars as fact and speculated that Martian plant life had evolved to secure sufficient moisture to survive despite the harsh conditions (Salisbury 1962, 17–26). Aldouin Dollfus conducted extensive studies of the polarization of light reflected from the dark areas on the planet's surface and concluded that seasonal variations were consistent with results obtained from terrestrial algae and lichen (Dollfus, in Kuiper and Middlehurst 1961, 380–81). Nonbiological explanations for the wave of darkening during the planet's spring and summer, such as volcanic ash blowing across the surface of the planet, evoked too many unknowns and made too many assumptions about Martian volcanism to displace the visual evidence of vegetation coming to life as the polar caps melt. The consensus view of Martian "biota" was voiced by the Nobel laureate Joshua Lederberg in 1960. "There seems little doubt," he wrote, "that many simpler earthly organisms could thrive [on Mars]." While discouraging the use of terrocentric terms such as "vegetation" that might imply "that the Martian biota will necessarily fall into the taxonomic divisions that we know on earth," Lederberg maintained that "the most plausible explanation of the astronomical data is that Mars is a life-bearing planet" (1966, 133). With the advent of v-2 rockets at the end of World War II, such statements suggested to some within the scientific community that Mars deserved careful study as a possible goal for interplanetary exploration and colonization of the sort envisioned by Arthur C. Clarke, Ray Bradbury, and others. With the publication of Werner von Braun's *The Mars Project* (1948; translated into English in 1953), the science-fiction vision of humans on Mars was given a veneer of probability.

On May 2, 1945, von Braun surrendered to American troops in Austria and brought with him to the United States 115 members of his rocket team, 100 of the v-2 rockets that had devastated London, and crates of data on rocketry. Even as he had directed German efforts at Peenemünde, von Braun had advocated developing interplanetary spaceflight and had spent two weeks in prison in 1944 for arguing that spaceflight was more important than military weaponry (von Braun 1992, vii-viii). In the preface to the 1962 edition of *The Mars Project*, von Braun declared that his vision of a fleet of spaceships heading to Mars would be "possible at a cost which will be only a minute fraction of our yearly national defense budget" (von Braun 1992, xv). This "minute fraction" would buy a "flotilla of ten space vessels manned by not less than 70 men" (3). Using guesswork calculations to solve the problems of launching thousands of tons of men and equipment into Earth orbit and from there to Mars, von Braun envisioned solving the engineering problems of human spaceflight by brute force: intercontinental ballistic missiles would be converted into launch vehicles to heave three-stage spacecraft on a 260-day, one-way trip to Mars, and "nuclear-powered ion ships" would be developed to send humans on interplanetary flights (xvi-xvii). The crew could be housed in seven of the ten passenger vessels; the others then could serve as "boats" to ferry astronauts to the Martian surface.[8] The landing party of fifty, an interdisciplinary crew of scientists, would be able to remain on Mars for 400 days. Operating on the assumption that the planet's atmosphere has a density of about one-twelfth that of Earth's, von Braun envisions that his first "landing boat[s] will make contact with the Martian surface on a snow-covered polar area, and on skis or runners" to minimize the risk of a crash (66). The return trip in seven spacecraft would require another 260 days. Acknowledging the "many gaps" involved in theorizing spaceflight on this scale, von Braun states that his "objective" is in stimulating scientists, engineers, economists, diplomats, and businessmen to begin to fit together the pieces that might bring his vision to reality (von Braun 1992, 3; Ordway 1996, 69–95; Turner 2004, 35–57).

If *The Mars Project* seems oddly familiar, it is in part because it looks both backward to the science fiction of Kurd Lasswitz and forward to the Mars mission novels of the second half of the twentieth century. Having "devoured [Lasswitz's] novel with curiosity and excitement as a young man" (Lasswitz 1971, 7), von Braun recasts his future astronauts in the image of the fictional Martians of 1898—explorers dedicated to bringing civilization to a new world. As a thought experiment, von Braun's study gives scientists a host of baseline calculations to fine tune or confirm: pages of speculative data on escape velocities and orbital insertions, flight paths to and from the red planet, payload capacities, pounds of thrust, radio communications across interplanetary space, and living conditions onboard the ships in the flotilla. In its vision of a large-scale expedition, *The Mars Project* both feeds off and reinvigorates science fiction as humankind nears the era of spaceflight. In an important sense, the novels of Ludek Pesek, Gordon Dickson, Stephen Baxter, Ben Bova, Geoffrey

Landis, and others that imagine various scenarios for human missions to Mars are the stepchildren of *The Mars Project*: they envision the costs and consequences in human as well as technoscientific terms of exploring a mysterious world. Scientific writers, particularly proponents of nuclear-powered rockets to put humans on Mars, like Martin J. L. Turner, emphasize the seminal role that von Braun has played in all subsequent mission scenarios (Turner 2004, 38–43). In turn, novels such as Pesek's—a searing vision of a mission succumbing to failures of both technology and spirit—illuminate the sociopolitical, physiological, and psychological problems that haunt the scientific literature of near-future exploration.

The Mars Project represses—or, more accurately, bulldozes—these problems through imagined engineering triumphs that owe as much to Tom Swift as they do to hard science. In one respect, it seems a compensatory fantasy for the realities of military-sponsored science in the 1940s and 1950s. Von Braun's vision of researchers spending 400 days on Mars imagines a utopian depoliticizing of space travel that severs its costs from the demands of both military and nonmilitary expenditures. Elsewhere in his work, von Braun offers a familiar vision of Mars in the 1950s, but one that lacks the overriding incentive—real live Martians—to justify the United States devoting a "small fraction" of its military expenditures to his interplanetary project. In 1956 von Braun and Wily Ley described Mars as "a small planet of which three-quarters is cold desert, with the rest covered with a sort of plant life that our biological knowledge cannot quite encompass." Although "this plant life seems to be doing quite well," the authors admit that the planet cannot "be inhabited by the kind of intelligent beings that many people dreamed of in 1900" (99). Without canals, dying civilizations, or the possibilities of exoarchaeology, the search for life on the fourth planet becomes less compelling, and a "flotilla" of rockets traveling to Mars, more than a half century later, remains the stuff of science fiction.

While movie audiences traipsed vicariously across the surface of Mars in the early 1950s in films such as *Rocketship X-M*, a small cadre of scientists offered serious speculations about the ecology of the red planet. In 1953, Hubertus Strughold published *The Green and Red Planet: A Physiological Study of the Possibility of Life on Mars*, an offshoot of his work on pilots' responses to supersonic flight. Strughold received his Ph.D. from Muenster in 1922 and his M.D. a year later from Wuerzburg. Like von Braun, he had been an important figure in Germany's war effort, having served as director of the Aeronomical Research Institute in

Berlin. In 1947 he was brought to the United States to serve as director of the School of Aviation Medicine at Randolph Air Force Base and then became the first professor of aviation medicine at the Air University. A pioneer in the study of the effects of supersonic flight on human beings, Strughold turned to what he called the "new field [of] planetary ecology" as a physiologist interested in "examining the possibilities for life on other planets [in order to] acquire . . . a better knowledge of the limitations and abundance of life" on Earth (1953, 3, 5). Given his disciplinary training and interests, it is not surprising that he offers a sympathetic assessment of Lowell's work, even as he rejects the canal thesis. "The most impressive, most original, and—so to speak—classical work" on the possibility of life on Mars, he tells his readers, "is that of Percival Lowell" (6). Like Lederberg and other would-be exobiologists of the era, Strughold adheres to a "weak" version of the Lowellian argument, modifying the dying planet into an environment capable of sustaining only "a primitive type of plant life, similar to the lichens that grow on our desert rocks and Arctic tundras" (92). Still citing Coblentz's optimistic 1925 measurements of the surface temperatures, Strughold concludes that "the deficiency of oxygen" definitely eliminates "the possibility of warm-blooded creatures and the higher types of cold-blooded creatures [existing] on Mars" (58).

Yet this scientific assessment is tinged with a nostalgia for a Lowellian vision of the planet and the possibility that humans can adapt to this most earthlike of planetary environments: "It would be rash of any writer to assert categorically, from the inconclusive evidence we now have, that life of an order comparable to ours is impossible on Mars. . . . We can only say—regretfully—that it seems unlikely, in view of the physiological limitations that apply to life on earth . . . . Yet with a few mechanical aids—of a sort that we now supply to pilots soaring into the upper levels of the atmosphere—a man should be able to maintain himself on Mars for a considerable time without discomfort" (97, 89). Strughold's analysis, like those of Vaucouleurs, Ley, and von Braun, concludes that while the dreams of ancient civilizations may be dead, the planet is not irrevocably hostile to exotic biota or even human colonists. His comparison of conditions on Mars to those in the upper reaches of the Earth's atmosphere brings human occupation of the planet within the technological capabilities of standard-issue oxygen masks. A nitrogen-rich atmosphere, temperatures occasionally above freezing, and hardy plant life define the Mars of the 1950s in scientific speculation as well as science fiction. It remained a Lowellian world with important lessons to teach astronomers—and potential explorers—about Earth.

Optimistic views of conditions on Mars received significant scientific support at the end of the 1950s. In 1957 and 1959, William Sinton, an astronomer at the Lowell Observatory, published two articles that focused on the 3.4 micron region of the infrared spectrum, the carbon-hydrogen (or C-H bonds) resonance, to suggest that the dark areas of the Martian surface were indeed covered with plant life. The controversy that followed suggests some of the ways in which views of the red planet continued to depend on technologies of visualization and measurement—and on efforts to explain ambiguous data. Organic molecules have strong absorption bands in 3.4 micron range, and because light reflecting from vegetation penetrates beneath the surface and then re-emerges, it can reveal the absorption bands of plant life. Sinton reasoned that he should be able to compare spectrum analyses taken during the Martian opposition of 1956 to those from Earth lichens. His 1957 article, provocatively titled "Spectroscopic Evidence of Vegetation on Mars," found that the dark regions of Mars—the "seas" and "oases" of earlier observers—had a characteristic signature in the infrared spectrum at 3.45 microns that indicated the existence of lichen-like plant life. In 1958, Sinton used the 200-inch reflecting telescope at Mount Palomar to identify three distinct absorption bands in dark regions of Mars, all of which had analogues to lichen or algae found on Earth (Sinton 1959). In 1961, N. B Colthup suggested in a letter to *Science* that the "Sinton bands," as they were now called, nearly matched the infrared signatures of organic aldehydes, and speculated that acetaldehyde "may be the end product of certain anaerobic metabolic processes" such as "the metabolic fermentation of carbohydrates to acetaldehyde" (Colthup 1961, 529). The Mars of Arthur C. Clarke, if not Burroughs, seemed to have been given strong scientific support.

Sinton responded favorably to Colthup's suggestion, but two years later a team from Berkeley demonstrated that two of the Sinton bands (at 3.58 and 3.65 microns) were evidence for heavy water molecules (HDO) rather than organic compounds. While they speculated that this interpretation might indicate the presence of heavy water in the dark areas of Mars, they were careful to limit their claims: "We unfortunately have no answers for the major questions raised here. At present we know of no satisfactory explanation of the Martian bands" (Rea, Belsky, and Calvin 1963, 927). Two years later, two members of the team—in a paper co-authored with Sinton—suggested that the absorption bands at 3.58 and 3.65 microns were most likely the result of HDO in the Earth's infrared spectra. Once again, the difficulty of distinguishing spectra produced by the Earth's atmosphere had affected astronomers' views of Mars. At the end of the article, Sinton retracted his earlier views about Martian vegetation: "Too much emphasis has been placed on the association of bands with organic molecules as was too strongly suggested in [his 1959 article]" (Sinton 1965, 1287; cf. Horowitz 1986, 89–90; Murray 1989, 33–34). The 3.45 absorption band more or less dropped out of the scientific literature after that date, although there has been no widely accepted account of what it might signify. The problem of the absorption band at 3.45 microns, however, remains in the words of astronomer James Bell "enigmatic" (DiGregorio 1997, 82), and, in the debates about possible bacterial life on Mars the last of the Sinton bands has found its way into discussions about the life-detection experiments aboard the Viking landers. The afterlife of the Sinton bands controversy, in this regard, must be picked up again in the post-Viking era and the debates about what was and was not registered by the three life-detection experiments.

THE CANAL THEORY AND PLANETARY PHOTOGRAPHY

On the eve of the Mariner missions in the mid-1960s, the canals were not quite dead. Carefully tempered versions of the canal thesis could be accommodated to the known facts and extrapolated to guesses about Mars. The official NASA map used to plot the flyby of Mariner 4 in 1965 still had Lowell's canals marked on it.[9] Not surprisingly, this image of Mars was the product of the Lowell Observatory and was derived from the last great monument to the canal thesis, Earl Slipher's *A Photographic History of Mars 1905–1961,* published by the observatory "with the assistance of the Aeronautical Chart and Information Center [of the] United States Air Force" in 1962.[10] The book was produced under research grants from the Air Force in order "to provide the most complete and best available coverage of the Planet Mars as a contribution of the United States Government in the national space effort" (Slipher 1962, n.p.). Slipher reproduced 511 photographs of Mars, culled from the thousands taken at the Lowell Observatory and other institutions in North and South America, Europe, and Africa between 1905 and 1961 (Raeburn 1998, 48). These photos were analyzed and juxtaposed to drawings by Lowell, Sliphers, Trumpler, and other canalists as well as to the sketches of the author's chief antagonist, Antoniadi (figure 3). Arguing that "since identical conditions [of viewing the surface of the planet] recur only after a lapse of at least fifteen years," Slipher asserts that "no single observation can be *exactly* repeated even by

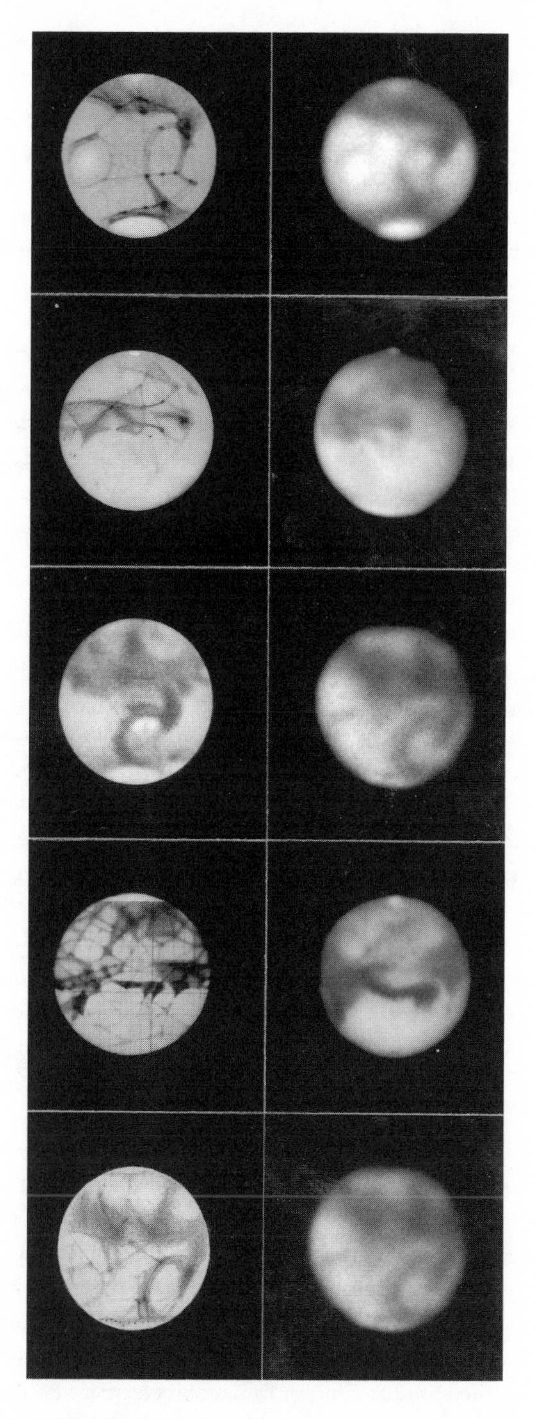

3. SKETCHES (LEFT COLUMN) AND PHOTOGRAPHS OF MARS BY VARIOUS ASTRONOMERS, REPRODUCED FROM SLIPHER 1962. COURTESY OF LOA.

the same observer," and consequently only an aggregation of photographs compared to drawings by multiple observers can reveal "the minute or faint details" (1962, 11) on the planet's surface. While Slipher's careful arrangement and analysis of these photographs now seems about as relevant as the star charts of Ptolemaic astronomers, his study represents a sophisticated attempt to defend what was left of Lowell's thesis by yoking it to a consensus view of Martian biota: the lichen-like organisms that intrigued von Braun, Strughold, Lederberg, and Sinton.

Although he is uncompromising in his defense of the canals, he does not simply endorse Lowell's thesis; his descriptions of the Martian surface echo those of Antoniadi as well as the would-be exobiologists of the fifties. In describing the albedo changes on the Martian surface, Slipher maintains that "nothing we know of but vegetation" can explain these changes: "The blue-green lapses into ochre and revives again to blue-green just as vegetation on Earth does . . . Certain large dark areas . . . turn to chocolate-brown and ochre at times, the color of dead plants and fallow ground" (Slipher 1962, 69–70). Such images, as we have seen, are standard descriptions of the dark regions of the Martian surface. But if this vegetation is the remnant biota of a "dying world," Slipher nonetheless defends vigorously key aspects of Lowell's planetary science. Mars must have enough water vapor in its atmosphere to account for its seasonal changes; if the "cirrus clouds" above the polar caps do not consist of snow and ice particles, he maintains, then "the ever-changing surface features of the planet and its atmosphere have no satisfactory explanation" (70). Like Russell, Slipher attributes the "strong red color of the surface" to oxidation and surmises, like Vaucouleurs, that "the bulk of the Martian atmosphere may well be molecular nitrogen" (69, 30). At best, he concedes, if enough oxygen once had existed "to sustain life, and if the exhaustion by weathering were exceedingly slow, it is conceivable, but not considered probable, that the evolution of life may well have kept pace with it" (70). Although plants may have adapted to the desiccation of the planet, "there is no decisive evidence whatsoever that intelligent animal life exists on Mars." The canal builders envisioned by his mentor have retreated to a realm of speculation that dovetails with the science fiction of the fifties: "It is possible that Earth's human race, at its present level of intelligence, would be able to secure its survival, though in diminishing numbers, in enclosures supplied artificially with oxygen, provided it had the millions of years of warning that changes would undoubtedly give and provided that it took the necessary precautions" (70). With the Martians banished to extinction or to the domed cities of science fiction,

Slipher can then direct the bulk of his study toward a defense of the reality of the canals, whatever their origin and nature.

What distinguishes Slipher's study from the speculations of other midcentury planetologists is the photographic evidence he presents. At several points, he offers detailed comparisons of individual photographs of Mars to Antoniadi's sketches and finds them wanting. Trumpler's drawings of canals, in contrast, are praised for their correspondence to the photographs.[11] To clinch his point that although "the photograph[s] may not show the precise character of extremely minute markings," their demonstration of "the reality of controversial markings . . . is undeniable," each copy of the book includes a photographic plate pasted in rather than an offset reproduction (162). "Plate XLVII [figure 4] is published here as a direct photographic reproduction," Slipher tells his reader, "in order to provide each reader with the best possible opportunity to distinguish the fine lines and to judge their reality for himself" (162). This updating of Lowell's strategy from 1905—displaying photographs of Mars side-by-side with his drawings—remains a shrewd strategic move a half century later. Slipher, like Lowell, asks individuals to judge for themselves the correlations between albedo variations on the planet's surface and the competing sketches by Antoniadi on the one hand and by Slipher, Lowell, and Trumpler on the other. This invitation is made explicit by Slipher's suggestion that, as a kind of experimental control, "a number of discriminating observers independently plot the positions of lines for several of the photographs and then compare their results" (162). This counter to the illusionist thesis—again an updating of Lowell's original rejoinders to his critics—is buttressed by his invoking his mentor's favorite argument: "skilled observers" need to go repeatedly "to the best available site[s]" (162) in order to convince themselves of the reality of the canals. Astronomers in England, Germany, and "northern France" must study Mars during perihelic oppositions at an angle only 15 degrees above the horizon, and "observations of Mars made under such difficult conditions are hardly comparable with those made [by] Lowell [Observatory] astronomers [who] observed Mars in Chile in 1907 and in Bloemfontein, South Africa in 1939, 1954 and 1956" (162). In singling out Antoniadi for his attacks, Slipher retreats to the Lowellian standby: the conditions of good seeing at Flagstaff and at other select locations around the world guarantee the accuracy of the canal sightings by "Lowell astronomers." Citing recent "confirmations" of the canals at the Lick Observatory, he dragoons many astronomers, including Campbell and Barnard, who—at one time or another—saw streaks, fault lines,

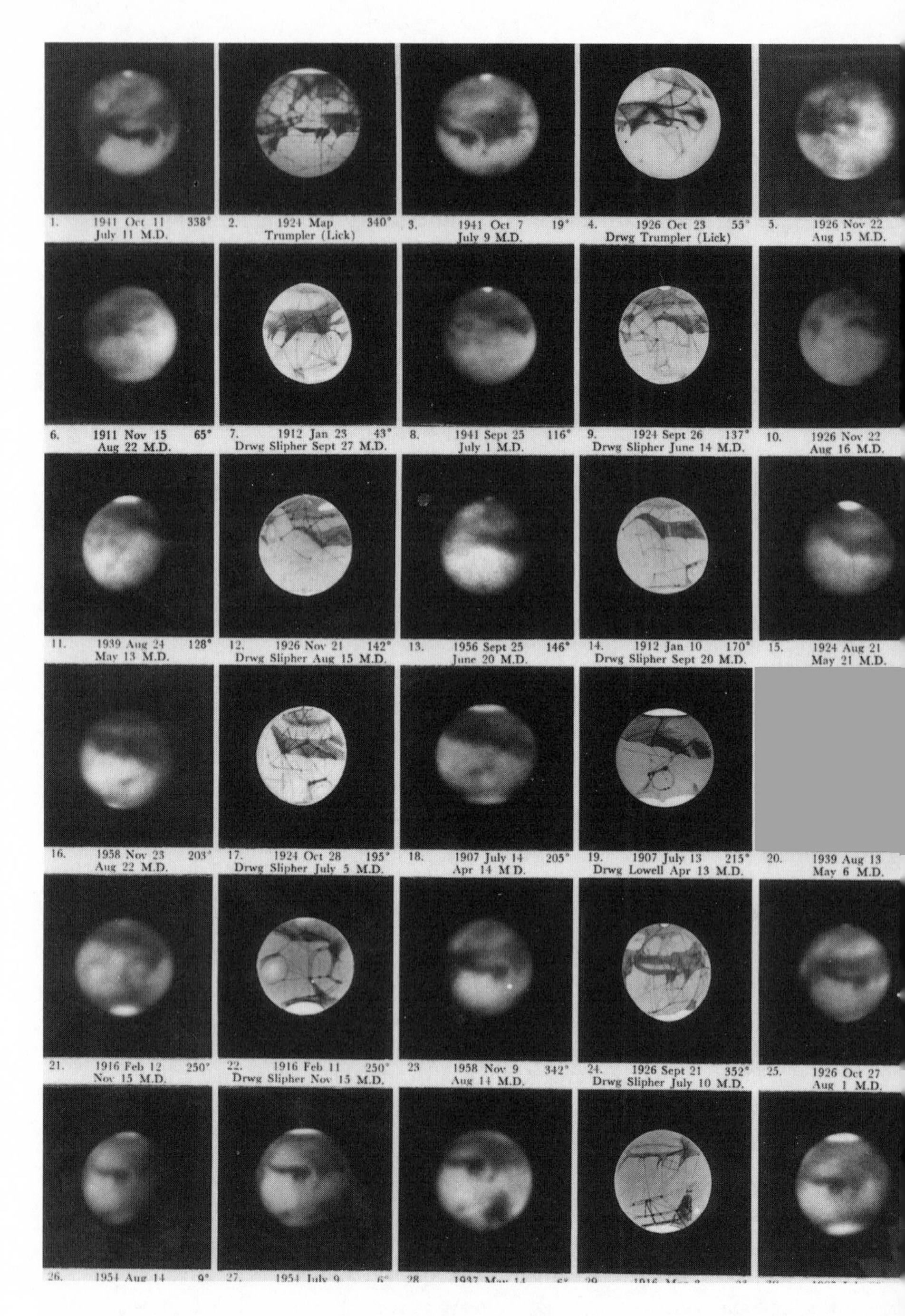

4. PHOTOGRAPHIC PLATE OF SKETCHES AND PHOTOGRAPHS OF MARS, REPRODUCED FROM SLIPHER 1962. COURTESY OF LOA.

or canals, into the Lowellian camp. These skeptics and waverers are set implicitly against Antoniadi, and the reader is redirected to the irresolvable question of what the canals signify: "While there is room for difference of opinion as to the interpretation of the canals," Slipher concludes, "their existence as true markings on the planet has been clearly established" (162). With the imprimatur of the Air Force, Slipher manages to keep alive the remnants of a scientific debate about these "true markings."

The Photographic Story of Mars documents the significant role that the Lowell Observatory continued to play in the early 1960s. Researchers associated with the observatory besides Slipher contributed to a continuing scientific literature on the canals as real surface features. In an article published in 1964 in *Icarus* (then edited by Sagan), F. A. Gifford Jr. suggested that the canals might be long chains of sand dunes formed on a windy desert planet. Adapting terrestrial analogies to explain anomalous features on Mars, he notes the "striking analogies" between chains of sand dunes in southern Saudi Arabia that reach two hundred miles in length and the "Martian canal systems" (132). Citing Lowell and Antoniadi as scientific antagonists in a still-simmering controversy, Gifford, a meteorologist who had participated in the Lowell Observatory conferences on planetary astronomy in the fifties, summed up the opposing positions as he perceived them. Rejecting the critiques of anticanalists, Gifford argues that there is too much disinterested and independent confirmation of the existence of the canals for them to be merely tricks of the eye or brain. The contention that "the canals are entirely illusory" is characterized as an "extreme point of view[,] . . . untenable because many canals have been seen repeatedly, by different observers, in the same, well-established positions." On the other hand, the Lowellian thesis of their "artificial origin" has been "rejected by most disinterested scientists, to a considerable extent on intuitive and philosophical grounds" (130). His compromise position is to suggest that both views have a basis in observable fact. Even as he admits that "the broader canals and canal systems . . . have been resolved telescopically under conditions of excellent seeing, into small, irregular light and dark patches," Gifford nevertheless argues that the best observations do "not destroy [the canals'] essential appearance of linearity" (131). The solution that he proposes—that the linear features are sand dunes that stretch for hundreds, or even thousands, of miles—extrapolates what is known about desert conditions on the Arabian peninsula and projects them onto Mars. The dunes retain their linear character even as they shift, degrade, and reform on what he believes is a relatively flat and featureless surface. Calculating that the winds of Mars would have sufficient force to remold the surface, Gifford offers a "proposal" (131) that fits the known facts claimed by latter-day adherents of both the Lowellian and illusionist arguments. The Mars of waterways, oases, and ecological disaster has vanished; the canals have a reality that can be explained by a new analogy to terrestrial surface features. If the Martian "desert" has been stripped of much of its Lowellian "horror," Gifford's acceptance of Lowell's description of the planet's featureless topography indicates that the canal thesis remained current enough to be cited by reputable scientists. The assumption that "the Martian desert areas are a vast unbroken area for dune formation, free from mountain ranges or sea coasts" (134) is yet another aspect of Lowell's complicated legacy: a plausible hypothesis that could not be rejected out of hand until the era of spaceflight.

CODA: GHOSTLY CANALS

After the first photographs of Mars were returned by Mariner 4 in 1965, the canal debate lingered as astronomers compared photographs of the planet's surface to Slipher's chart. Carl Sagan and James B. Pollock took radar Doppler spectroscopy readings of the planet and argued that "radar evidence and Mariner 4 photography suggest that the Martian canals are ridge systems or associated mountain chains in analogy to similar features in terrestrial ocean basins" (1966, 121). Although these "ridge systems" proved as illusory as Lowell's canals, Sagan and Pollock were among the first to argue for extreme variations in the topographical relief of the Martian surface. Samuel Glasstone, in a NASA-sponsored study, accepted the outlines of Sagan's and Pollock's argument, suggesting that "the canals are relatively long and narrow, almost linear features, having the same nature as the dark areas, that run across the bright regions of Mars" (1968, 127). He lamented the fact that "the [Mariner 4] spacecraft crossed the northern hemisphere, where most of the canals are located, during the local winter when the canals are very faint or invisible against the bright background" and that the resolution of its cameras "was of an area approximately 320 kilometers (200 miles) square. . . . similar to the width of many canals" (127). While the photographs "revealed a number of what appear to be linear topographical features known to geologists as lineaments," these features are "much narrower than the canals. They could not have been detected from Earth by the best telescopes even under ideal viewing conditions" (128). The "canals," for Glasstone, still have a nominalist existence. They remain a descriptive term that has a ghostly life of its own, the remnant of a now-discarded view of Mars that persists almost independently of observed features. As subsequent Mariner missions photographed more of the cratered Martian surface, "canals," sand dunes, and "narrow lineaments" disappeared from the scientific lexicon. In place of Lowellian Mars or the lichen-covered world of the 1950s, the planet that came into view in the 1970s offered new challenges to scientists seeking to resolve the fundamental questions about areography and exobiology.

FIVE

Mars at the Limits of Imagination: The Dying Planet from Burroughs to Dick

When I was ten . . . I came upon the Edgar Rice Burroughs
novels about John Carter and his travels on Lowell's Mars. It was a
world of ruined cities, planet-girding canals, immense pumping stations—a
feudal technological society. The people there were red, green, black, yellow, or
white, and some of them had removable heads, but basically they were *human.*
I didn't realize the chauvinism of making people on another planet like us; I
simply devoured what seemed to me the riches of another planet's biology.
. . . I loved those books. They were full of new ideas. . . . I tried to imagine
my way to Mars, the way Carter did: I would go into a vacant lot,
spread my arms, and wish to be on Mars.–CARL SAGAN
(quoted in Henry Cooper, *The Search for Life on Mars*)

A map representing Edgar Rice Burroughs's vision
of Mars has hung on the hallway wall outside Carl Sagan's
office for more than twenty years.–OWEN B. TOON,
"Environments of Earth and Other Worlds"

• • ● • •

BURROUGHS AND BARSOOM

If no historian or philosopher of science ever has made a career by championing Lowell, no literary critic has made her or his reputation by arguing for the aesthetic significance of the Martian novels of Edgar Rice Burroughs and his imitators. But just as the canals occupied the attention of a generation of scientists, Barsoom and its offshoots dominated the interplanetary fiction of the first half of the century (Locke 1975). Carl Sagan was hardly alone in his fascination with Mars. In science fiction between 1912 and 1964, readers confronted the implications of Lowellian Mars. The dying planet offered powerful images of ecological decay and social disintegration that resisted—or recast—the didacticism that had characterized the science fiction of Lasswitz and Bogdanov. Against this backdrop of cosmic fatalism, displaced Earthmen (and occasional Earthwomen) acted out fantasies of archaic heroism.

Six weeks after Wells's *The War of the Worlds* appeared in serialized form, the journalist and popular science writer Garrett Serviss wrote a thoroughly Americanized "sequel," *Edison's Invasion of Mars,* which appeared in the New York *Evening Journal* in 1898 and in book form in 1947. Wells's novel itself had been bowdlerized for American audiences; the "yellow press," as Suvin notes, "simply suppressed the parts without action" (1979, 214). But Serviss went much farther, and his novel anticipates an important development in twentieth-century American science fiction—the popularity of the Edisonade, a genre characterized by its uncritical faith in narratives of technological progress, power, and cowboy morality (Suvin 1979, 260; Clareson 1985, 213–16; Franklin 1966; Bleiler 1990). Led by Thomas Edison and other scientific giants of the late nineteenth century (Lord Kelvin, for example), Earth forces invade Mars, thanks to Edison's invention of an electrical spaceship and a disintegrator ray. Yankee ingenuity triumphs over an anthropomorphized race of fifteen-foot-tall Martians, frequently compared by the narrator, an American journalist, to American Indians. Wells's savage satire of European imperial ambitions is transformed into a tale of manifest destiny gone interplanetary.

If *Edison's Invasion of Mars* exhibits the hallucinogenic optimism of nineteenth-century American scientific romance, it also describes the limits of efforts to restage the "conquest" of the (American) frontier on the red planet. The technological faith that informs Serviss's yellow novelism, for the most part, is questioned in Burroughs's eleven Martian novels, a series that exerted a profound influence on science fiction and, as Sagan suggests, even planetary astronomy. In contrast to Lasswitz and Bogdanov, who appropriate Mars to probe the political and social problems of European civilization, Burroughs reimagines on Mars—rechristened Barsoom—a heroic version of the vanishing American frontier, but one interlaced with strains of chivalric medievalism, technological romance, social Darwinism, Lowellian planetology, and sword-and-sorcery adventure. This strange brew of heroic values, science fiction, and nativism produced in the figure of John Carter the first of the superheroes who would dominate popular culture for nearly a century. Although Burroughs has been savaged by many literary critics, he was probably the best-selling fiction writer of the first half of the twentieth century (Taliaferro 1999, 13–14) and he exerted a profound influence on the development of popular fiction. While his twenty Tarzan novels spawned an industry of films, comic books, and merchandising spin-offs, his Martian novels fired the imagination of generations of science-fiction readers and writers. As Mariner 9 approached Mars in 1971, Ray Bradbury declared that "[Burroughs] has probably changed more destinies than any other writer in American history," and credited his predecessor with firing the imagination of two generations of scientists as well as dreamers: "Burroughs and his alter ego John Carter, seized off to Mars by impossible dreams, pulled ten million boys after them and changed America's scientific territory forever" (Bradbury et al. 1973, 17, 18). Participating in the same discussion at Caltech, Arthur C. Clarke voiced a sentiment echoed by Sagan: "It was Edgar Rice Burroughs who turned me on [to space adventures], and I think he is a much under-rated writer" (27).

Such praise from science-fiction writers stands in sharp contrast to critical assessments that dismiss Burroughs's "Martian saga and its subliterary descendants" as "fantastic romances" (Philmus 1970, 3; Rovin 1978, 58–61), read them as autobiographical wish fulfillment (Lupoff 1976, 62–63), or pounce on Barsoom as "a perfect world for puerile dreams" (Mullen 1971, 230). This gap between the views of scientists and science-fiction writers on the one hand and literary critics on the other reveals the significant tensions that exist between aesthetic values imported from realistic fiction and the standards invoked by defenders of science fiction. For Bradbury, Clarke, and Sagan, Burroughs's Mars provides the dialectic of cognition and estrangement that calls into question the values and assumptions that underlie both "realistic fiction" and the fantasies of inexhaustible resources and unending productivity that fuel post–World War II political economy.[1] On Barsoom, Burroughs intuited, readers could confront a dying planet and an enervated civilization, the dark underside of the American technotopia.

Burroughs grew up in Chicago, the son of a well-to-do businessman and Civil War veteran. There's a Zelig-like ubiquity to Burroughs's life story: a would-be cavalry officer who failed the entrance exam to West Point; a soldier in the Arizona Territory, fruitlessly chasing Apache renegades; a one-time prospector, stationery store owner, pencil salesmen, night watchman in a Utah train yard, and then head of the stenography department at Sears Roebuck in Chicago; a popular writer who incorporated himself in the 1930s; a rancher; a Southern California real estate impresario who sold off his 550 acre Tarzana ranch in lots; a Hollywood personality; and the oldest war correspondent in the Pacific theater during World War II. In 1912 at age thirty-five he sold the serialized version of *A Princess of Mars* to *All-Story Magazine* for $400; within a few years he had published several Tarzan novels and two subsequent Barsoom novels, *The Gods of Mars* and *The Warlord of Mars*, which made him a wealthy man. He spent money, however, as quickly as he made it, and his efforts to write "serious," highbrow fiction to tap into a wider market did not find a significant readership. In his own mind, he became a prisoner of his success as a pulp novelist. "I write better about places I've never seen than those I have," he decided (quoted in Taliaferro 1999, 84). Although he claimed as early as 1920 to be tired of both Barsoom and Africa, Burroughs continued to churn out Tarzan potboilers, another seven Mars novels, the Carson Napier series set on Venus, the Pellucidor novels that chronicle adventures at the center of the Earth, and two novels, *The War Chief* and *The Apache Devil*, that feature an Apache warrior as their hero. Most of these novels are still in print, and the University of Nebraska Press is reissuing some of the Mars novels in its classics of science fiction series. In many of these novels, Burroughs draws on and transmutes his experiences in the American west to re-create a vision at once nostalgic and surreal. On Barsoom, he fashions a popular and ideologically compelling response to the insecurities and anxieties of white masculine identity at a time of financial turmoil and doubts about whether the self-made entrepreneurs of the nineteenth century could survive in an era dominated by conglomerates (Stecopoulos 1997, 170–91; Bederman 1995; Lears 1981; Saxton 1991).

While Burroughs's critics have decried the racial stereotypes that characterize the Tarzan series, his depictions of race and religion on Barsoom, filtered through the author's shaky grasp of evolutionary theory, are more ambiguous than they may at first appear. The first three Mars novels, in particular, challenge default assumptions about the racism of Burroughs's populist fiction. Writing for a Pocatello, Idaho, newspaper in 1898, Burroughs, the local store owner, savaged the imperialist apologetics of Rudyard Kipling's "The White Man's Burden" in a satire that can be traced, in part, to his family's strong abolitionist views (Porges 1975, 12–15). While Burroughs supported eugenicist programs in the 1920s, "The Black Man's Burden" is a searing indictment of racist ideology.

Take up the white man's burden;
Poor simple folk and free;
Abandon nature's freedom,
Embrace his "Liberty";
The goodness of the white man
Who makes you free in name;
But in her heart your color
Will brand you "slave" the same.

Take up the white man's burden;
And learn by what you've lost
That white men called as counsel
Means black man pays the cost.
Your right to fertile acres,
Their priests will teach you well,
Have gained your fathers only
A desert claim in hell. (Porges 1975, 708)

By inverting racial stereotypes, these stanzas insistently undercut the white race's self-styled "burden" of civilizing the non-Western world. Although Burroughs invokes a romanticized, even patronizing vision of "poor simple folk" living in harmony with nature, his Swiftian ferocity is directed at the ideologies of justification—legal ("counsel") and theological ("priests")—that leave African Americans with only a "desert claim in hell." The loss of "fertile acres" figures their identity, labor, and freedom in explicitly material terms, a reflection perhaps of Burroughs's experience in the American west where crop yields and water rights often had as much currency as hard capital. The poem's final lines sardonically reveal the racism and self-satisfying superiority that informs social Darwinism:

Peruse a work of Darwin—
Thank gods that you're alive—
And learn the reason clearly:—
The fittest alone survive. (Porges 1975, 709)

The ironies of this stanza suggest an ideology of supremacy eroded from within; the very values that Kipling promotes are revealed as a cynical justification for oppression and exploitation. Yet despite its fervor, this early poem is also an indication of a peculiarly American distrust of colonialism that resonates with what Taliaferro calls Burroughs's "selective progressivism" (1999, 216; Porges, 1975, 72–73). Over a decade later, his suspicions about the white man's burden inform his interplanetary populism on Barsoom. If his poetic satire allows for a measure of subversive skepticism, the Mars novels offer a fascinating—if politically incoherent—hodgepodge of views that yoke a nativist obsession with the purities and impurities of racial identity to the little guy's distrust of all forms of social and theological authority. The result is a compensatory fantasy of individualist integrity in a hostile environment where physical prowess and the certitudes of love and honor offer a means for a social misfit to transcend heroically the complications of life in an increasingly bureaucratized and morally ambiguous America.

Burroughs's first Martian novel, *A Princess of Mars* (originally serialized as *Under the Moons of Mars* by "Normal Bean"), strips Eurocentric science fiction of its imperialist legacy. The nightmare invasion of Wells's Martians, the complex politics of Lasswitz's interplanetary relations, and the utopian didacticism of Bogdanov's socialism are replaced by a framing narrative in which the narrator receives the manuscripts of his mysterious great-uncle, John Carter, a Confederate captain turned prospector. Pursued by Apaches and overcome by mysterious fumes in an Arizona cave in 1866, Carter is transported mysteriously to Barsoom. While his biographers cannot pinpoint what science-fiction novels Burroughs may have read, his creation of a hero who has no memory of his youth, never ages, and yet has "the same horror" of "the real death from which there is no resurrection" (1979a, 11) marks a decisive break with the traditions that invest Martian society with didactic religious or politico-philosophical significance. His provisional, term-limited immortality grants Carter a kind of mythic status, stripped of the Christian sentimentality that characterizes Percy Greg's *Across the Zodiac*. Where his predecessors such as Camille Flammarion and Louis Gratacap had made Mars the planet of resurrected souls, Carter's transportation to Barsoom signals a ruthless populist campaign against metaphysical excess. Paradoxically, it is Burroughs's lack of interest in assuming the burden of ideological justification—Christian or marxian—that allows him to dramatize powerfully a Lowellian vision of a dying world. On Barsoom, social organization, personal relations, and economic activity are held thrall by the "stern and unalterable cosmic laws" of planetary senescence (66).[2]

The appeal of Burroughs's Mars novels lies both in their escapist fantasies and in their projecting onto the natural world a cosmic indifference, against which humanoidkind must define its virtues without a metaphysical safety net. When he first awakens on Barsoom, Carter finds himself among the ten-foot-tall, four-armed race of Tharks, nomadic raiders whose tribal structure, stoicism, and indifference to "modern" values identifies them with American Indians. Burroughs depicts this imagined race with a nostalgic respect that anticipates his treatment of his Apache hero in *The War Chief* and *The Apache Devil;* the green-skinned Tharks become apt symbols for his insistence that social and personal identity are shaped by the harsh conditions of a dying world. Their life, says Carter, "is a hard and pitiless struggle for existence upon a dying planet, the natural resources of which have dwindled to a point where the support of each additional life means an added tax upon the community into which it is thrown" (40). For Burroughs, the crucial term is "dwindled": on a dying planet, embattled nomads and now-contracted civilizations find in the ruins of deserted Martian cities a nostalgia for both vanished greatness and for "the gorgeous foliage of the luxuriant Martian vegetation" (66). Descriptions such as these—characteristic of the Barsoom novels—reveal a Mars haunted by its past. Again and again in twentieth-century science fiction, the dying of the indigenous ecology and its civilization produces both an awe inspired by the inconceivable vastness of Martian history and the accomplishments of its imagined civilizations and a terror at the prospect of the "unalterable" desertification of the planet. The idealized Mars of magnificent civilizations and "luxuriant . . . vegetation" is irrevocably haunted by the "real" consequences of ecological collapse, but ironically the "real" Mars, as the ten-year-old Sagan knew, remains haunted by the phantoms of science fiction. Burroughs self-consciously exploits this double haunting to great advantage.

In *Thuvia, Maid of Mars* (1916), Burroughs's fourth Barsoom novel, Carthoris, Carter's son, finds the city of Lothar guarded by phantom bowmen and its streets peopled by "a countless multitude," who "live the old dead past of ancient Lothar" (92). Uncannily, the descendents of some of the ancient Lotharians continue to exist as "permanent materializations" (1979c, 96), ghosts who have a "real" existence. The "etherealists" among the Lotharians believe that they may be "the deathless minds of individuals long dead" (97) who stand in for and "merge into the real" (96). In satirizing the Cartesian division of mind from body in the debates between "etherealists" and "realists," Burroughs suggests the bankruptcy of a philosophy that values abstractions over material existence, notably the belief of some Lotharians that they "could not have existed all these ages without material food and water had [they] been material" (97). On Mars, life is haunted by a vanished past, and the dried seas and vanishing vegetation are blasted images of a golden age of abundance. If the "real" Barsoom is haunted by its lost past, the ideals of love, honor, and nobility that Carthoris embodies reciprocally are haunted by the grim competition for resources symbolized by the princess's body. The Lotharians are consumed by their fears of planetwide catastrophe, of universal processes of entropy and desiccation that dwarf the individual. Ghosts proliferate on Mars in science fiction precisely because the dying world confronts its readers with the specter of extinction.

The Lotharians are suggestive, in this respect, of the ways in which Burroughs conceives of character. If, as Stecopoulos suggests, *A Princess of Mars* offers readers a form of compensation for the feared inadequacies of white masculine identity, it also overcompensates for the inability of politics, property, and negotiation to prevent a Darwinian competition for scarce resources (1997, 182–86). Burroughs uses the implacably hostile environment of Barsoom to externalize the conflicts, insecurities, and moral ambiguities of human identity. The environment becomes, in effect, the real enemy: the various warring races of Mars embody the dangers of a planet that both endows Carter and his surrogates (Carthoris and Ulysses Paxton, another Earthman, are the heroes of some of the later novels) with superhuman powers and yet presents an unending series of tests of will, courage, skill, and honor. Throughout the Barsoom novels, heroism is a continuing process of initiation—kill the monster(s), get the girl, and (after Carter becomes warlord of Mars at the end of the third novel) rule wisely, that is, with an intuitive sense for the just allocation of resources. If this narrative structure recalls those of the dime novels of the American west, Barsoom recasts frontier heroism by setting it within the metanarrative of the downward spiral of civilization and planetary ecology. On a dying world, all triumphs are temporary. The cowboy faces a similar set of obstacles—defeat the Indians, save the white woman, and either become a respectable citizen or, alternately, forestall that domestication to ride again—but the readers know that the frontier is always about to close, that the Indians are about to be herded onto reservations, and that the cowboy both heralds and resists the coming of civilization. But if the dime western represents a dying past, the progressivist settling of the frontier runs backward on Barsoom; the nostalgic appeal of the American frontier is displaced into a vision of the future of the Earth. Carter's swordplay and derring-do may represent a romantic escape from the cold calculus of declining stocks of water, fertile land, and air, but his adventures also suggest that such acts of heroism, friendship, and love are played out in an indifferent universe. Without the metaphysical guarantee of a progressive order to and in history, Barsoom offers only materialist consolations on a world where the material conditions for existence are eroding.

Unlike the archetypal white hero, whose seduction of the native princess justifies the possession of her body and the lands that she symbolizes, Carter is the one who seems born for and to Mars, always and already seduced by the red planet. At the end of *Princess,* after he awakens in the Arizona cave, Carter can think of little besides the fate of his wife and her planet. Twenty years later (on the brink of a sequel) Mars and Dejah Thoris become interlocked symbols of his seduction: "Tonight she seems to be calling to me again as she has not called since that long dead night, and I think I can see, across the awful abyss of space, a beautiful black-haired woman standing in the garden of a palace, and at her side is a little boy who puts his arm around her as she points into the sky toward the planet Earth" (159). The feminization of Mars in this passage is striking: the masculine god of war, destroyer of men and cities, is transformed into a symbol at once sexual and maternal. In this respect, Mars provides a psychological and familial structure for a hero irrevocably alien on Earth—a past for a man who remembers no past, a memory that provides a means to envision a future on a red planet for which he "feel[s] a bond of sympathy and love even greater than for the world that gave [him] birth" (1979b, 32). Forever cut off from the memories that would grant him a stable sense of self, Carter assumes his identity only in an imaginative space in which he can construct roles for himself as husband, father, and eventually ruler. The very dangers that he experiences promise at least the possibility of death—a sense of closure to an otherwise ageless, paratactic existence.

But Mars also haunts what for John Carter is the "lost world" of Earth. Gazing at the dead seas of Barsoom or walking through its ruined cities, he sees the future of his home planet: the retreat to the ramparts by an embattled civilization, the mixing of races, and a descent toward an inevitable doom. The civilized red race is a composite of the white, black, and yellow-red races that have been forced to interbreed in order to defend ever-smaller areas of fertile land against the "green hordes" (1979a, 62). This remnant conserves the ray guns, flying machines, and videoscopes of its technoscientific past, but it is beset by the threat of the Tharks and by conflicts among its own city-states. Politics reverts to a bronze-age Manichaeanism (Carter 1977, 60–67; Holtsmark 1986, 16–32). Helium, the home of Carter's beloved Dejah Thoris and the seat of her father, Tardos Mors, the jeddak or king, is the embodiment of all that is noble on Mars; Zodanga, given to endemic corruption, greed, and lust, is the evil empire that must be defeated. This moral economy obviates the need to make tough choices about who deserves what: evil is manifest in the insatiable lust of villains to monopolize the resources of the dying planet; goodness is linked to an ordered social structure that does its best to maintain a civilized generosity. Carter's appeal for ten million adolescents stems from a moral clarity that privileges adventure over negotiation, action over thought. But this nostalgic, hypermasculinist vision of a heroic past also offers more than its share of intriguing gadgets, alien technologies, and scientific romance. Barsoom, in brief, offers an imaginative escape into both the past and the future—a bronze-age social structure equipped with a futuristic technology. It is both the "origin" of traditional conceptions of honor, propriety, and masculine identity and a storehouse of technological wonders—massive airships, sophisticated telescopes that reveal the smallest details of life on Earth, and the atmosphere plant that maintains the conditions for life on Mars. A vanished heroic age makes the reader's present seem muddied, insignificant, and tawdry; the achievements of Martian technology humble the pride that informs the Edisonades of Serviss and other pulp writers.

In contrast to the know-it-all heroes of the Edisonades, Carter is a techno-primitive, amazed by the semi-lost arts of Martian civilization and relying on his strength in the lower Martian gravity to outfight the avatars of love and honor cultdom at their own game. At once, Carter seems a wandering knight on a quest to be restored to his "rightful" place in the royal hierarchy of Barsoom and an outcast—an American hero who can remember no past, no youth, and consequently must reinvent himself in an ongoing process of initiation into an alien culture. If *A Princess of Mars* reinscribes a white male identity "at once predictably elitist and flamboyantly transgressive" (Stecopoulos 1997, 172), the category of race itself in the novel is riven by the unending competition to exploit dwindling resources. The first three Barsoom novels fictionalize competing claims of racial and evolutionary supremacy. They are less a coherent vindication of white superiority than a repetition, in different keys, of the determinant relationship between nature and cultural and individual identity.

Martians—green, red, black, white, and yellow; four-armed and two-armed—are hatched from eggs after a five-year gestation. In *Princess*, the egg-laying Martians seem objects of condescension and scorn: newborn Martians, of uncertain parentage, "are the common children of the community, and their education devolves upon the females who chance to capture them as they leave the incubator" (40). The social and familial structures of the "five million green Martians who rove the deserted cities and dead sea bottoms of Mars" (38) are fictionalized versions of white perceptions of Indian tribal arrangements. Although the chief of the tribe, Tars Tarkas, soon comes to play the role of the noble savage who befriends the hero and remains unswervingly loyal, the oviparous Martians are racially and biologically alien. Egg laying becomes an obvious symbol for the lack of emotion, love, and intimacy that characterizes the nomadic raiders of the dead sea bottoms. Yet the novel's action hinges on Carter's falling in love with Dejah Thoris, the beautiful daughter of the ruler of Helium, an oviparous humanoid who spends much of the novel naked and with whom Carter eventually fathers a son. How the reproductive systems of the Martians work is never explained, and readers are left to take it on faith that Carter and Dejah Thoris are capable of sexual intercourse. Egg laying aside, the heroine is described as both exotic and familiar: "Her very feature was finely chiseled and exquisite. . . . Her skin was of a light reddish copper color, against which the crimson glow of her cheeks and the ruby of her beautifully molded lips shone with a strangely enhancing effect" (46). In this and later Barsoom novels, Dejah Thoris and other beautiful heroines, such as Thuvia in *Thuvia, Maid of Mars,* are subjected repeatedly to threats of rape and forced marriages to malevolent tyrants of various hues: red, yellow, black, and white. The red woman, in traditional fashion, stands synedochically for the resources of her world—material, political, even spiritual. Like Dejah Thoris, Thuvia is pursued by an evil prince bent on marrying her and thereby monopolizing the riches of Helium, the last bastion of an ancient and noble Martian civilization. In this respect, the life-or-death struggles over Thuvia and Dejah Thoris are ultimately contests for the dwindling resources of a dying Barsoom. To possess, or rape, Dejah Thoris is to gain access to the promise that Helium symbolizes—the idealized city governed by a wise ruler that has "solved" the problems of resource exhaustion by its guardianship of the atmosphere plant. (Helium, unless under attack, never seems to exhaust the material to keep red bodies and heroic souls together.) But, in another sense, she embodies a transcendent seductiveness, and it is Carter's intuitive recognition of the stability she represents that distinguishes his adventures from those of cowboys out to tame the wilderness and civilize "inferior" peoples.

In contrast to his predecessors who depict utopian societies gone to seed (Cromie) or faced with tough choices about the survival of their civilization (Bogdanov), Burroughs is the first novelist to follow Lowell and play up the desperation that the canals represent. At the same time, his novels offer a refracted image of the controversies that surrounded the canals. In this regard, the inconsistencies that Alfred Russel Wallace noted in Lowell's theory find a fictional correlative in the technological black box that keeps Barsoom from dying—a gargantuan "atmosphere plant" that uses the "ninth ray" of the Martian prism (one of two "new and nameless" rays that exist beyond the "seven colors of our Earthly prism") to keep the planet habitable (1979a, 112). Like his contemporaries, Burroughs is fascinated by electricity, and the atmosphere plant allows him to finesse the problems for humanoids trying to survive in the tenuous Martian atmosphere: the ninth ray is separated from sunlight, stored, "treated electronically," and then released to interact with "the ether of space" to create an "atmosphere" (112). The ninth ray, in brief, encodes an antiecological dream: the generation of infinite power that allows the red race on Barsoom to fly in huge airships, power its cities, maintain food production, and replenish the resources that stave off the planet's fate. Yet even as the Barsoom novels exploit a modernist faith in science, they set the values and assumptions of progress in the contexts of looming ecological disaster and mysteries that lie beyond rational knowledge. Although all red Martians understand the principles behind Barsoom's atmosphere generator, only two men know the telepathic secret needed to enter the plant, and their deaths at the end of the novel provide a perfect setup for a sequel: the production of air has ceased, and Carter must rush from Helium to try to enter the plant and save the planet. Staring "bravely into the face of their unalterable doom," Tardos Mors recognizes that the plant's failure means that " 'tomorrow's sun will look down upon a dead world which through all eternity must go swinging through the heavens peopled not even by memories' " (155). The fate awaiting Barsoom can be forestalled, it seems, only by memories that see the present as already past, an artifact created by a vision of—and from—the entropic future. The almost-lost wisdom that the atmosphere plant represents thus inverts narratives of scientific progress. To face the reality of a dying world without the sustenance provided by a past technological knowledge is to confront planetary doom: Carter loses consciousness as he crawls toward the controls in the atmosphere plant, then awakens back in the Arizona cave, unsure whether he has succeeded in saving Mars.

After a failed attempt to write a medieval saga, Burroughs was encouraged by Thomas Newell Metcalf, the managing editor of *All-Story Magazine,* to write a sequel to *Princess,* in order to develop the " 'semi-religious, semi-mystical regions' " alluded to in the novel (Taliaferro 1999, 72). In *The Gods of Mars* (1913), Carter again is transported to the red planet and finds himself caught in a battle between white and black races for supremacy. The racial situation on Barsoom is complicated and the biology bizarre, but Burroughs insistently challenges early-twentieth-century truisms about race and religion. He inverts the racialized expectations of most of his readers: the white Therns are cannibals, servants of the false goddess Issus, an aged, flesh-eating queen who embodies the hypocrisy and superstition of organized religion. In contrast, men of the black race, the "First Born" are "handsome in the extreme," "identical" physically to Carter himself; the "polished ebony" of their skin "adds to rather than detracts from their marvelous beauty" (57). Characterized by their aristocratic pride, these black warriors survive by piracy, descending from their airships to raid the settlements of other races. Xodar, their prince, declares, "We are a non-productive race, priding ourselves upon our non-productiveness. It is criminal for a First Born to labour or invent. That is the work of the lower orders, who live merely that the First Born may enjoy long lives of luxury and idleness" (111). The First Born seem like Barsoomian Vikings, and their rejection of the American values of ingenuity, hard work, and moral probity is not condemned so much as it is used to render them worthy opponents for the sword-wielding hero. Their pride is given an "evolutionary" basis in the bizarre biology that Xodar describes—a twenty-three-million-year descent from a single Tree of Life: plant life begets all animal life, and all animal forms are offshoots of either a sixteen-legged worm, the first white ape, or the first "renegade black man" (69). The white race improbably is descended not from other humanoids but from the white apes, and the red, green, and yellow races are miscegenized offshoots of black and white. As absurd as this evolutionary biology may be, Burroughs is consistent in using interplanetary chivalry to undermine religious and racial orthodoxies. The red race of Helium and the ebony First Born represent the telos of Martian evolution, a mocking challenge to pretensions that if "the fittest alone survive," it will be the Kiplingesque hero who will be the last man standing.

The nightmarish underside of such racialist ideologies is found in the dungeons of the Therns where "the offspring of the prisoners from the outside world" have interbred and produced "odd freaks," the "deformed" and "maimed," who, as "they lay sprawled about the floor . . . suggested instantly to me the grotesque illustrations that I had seen in copies of Dante's *Inferno*" (54). Burroughs exploits this rare literary illusion to ironic effect: the moral order of Dante's Hell that determines suitable punishments for each sinner has been perverted on Barsoom into a cruel, sadistic joke. When Barsoomians of all races end their thousand-year lives by undertaking their final journey down the River Iss toward what they believe is a glorious afterlife, they are thrown into prison or fed to yet another race of monsters: the cannibalistic Therns eat only the flesh of victims whose bodies have been drained of blood by the blue plant men, a race of vampiric, evolutionary dead ends—mindless, almost ancephalitic, feeding machines. Burroughs leaves his reader little interpretive leeway to dodge the implications of this satire on false religion because there are no "true" or unblemished spiritual institutions on a planet where archaic conceptions of personal honor embody the highest good. *The Gods of Mars* is relentless in its debunking of religious superstitions and institutions.

In an important sense, the Therns' cannibalism becomes the ultimate expression of a fundamental antagonism in which all races are pitted against each other. Deploying a different version of the rationale that animated *The War of the Worlds,* Burroughs's second novel depicts a dying planet on which human flesh is a significant source of protein. At the heart of the Martian code of honor lies the bitter knowledge that "each dead man means so much more of the waning resources of this dying planet to be divided amongst those who survive" (65). Those resources, of course, include the flesh of one's enemies.[3] Cannibalism is the ultimate recourse for true believers in social Darwinism. In effect, cannibalism and "honor" are dialectical responses to a hostile universe in which goddesses turn out to be flesh-eating harpies and individual worth is determined only by resisting—often in battle—the malevolence of criminal aggrandizement. Honor and identity are always under assault from the very conditions of a perpetually dying world.

The cliffhanger ending of *The Gods of Mars*—Dejah Thoris is trapped in a prison whose lock can be opened only once a year—is suggestive of the ways in which the logic of Burroughs's narrative must defer closure: the fate of Mars has been determined by the nebular hypothesis, and therefore to contemplate the future is to recognize that all life exists on borrowed time. The planet's fate—like the hero's—is always threatened and always deferred. In the third Barsoom novel, *The Warlord of Mars* (1914) Carter must again rescue Dejah Thoris from yet another evil manifestation of a corrupt order. Confronted by the imminent destruction of the air fleet of Helium by the "yellow hordes" of the far north, Carter springs to action: "The impulse that moves me and the doing of the thing seems simultaneous; for if my mind goes through the tedious formality of reasoning, it must be a subconscious act of which I am not objectively aware. Psychologists tell me that, as the subconscious does not reason, too close a scrutiny of my mental activities might prove anything but flattering; but be that as it may, I have often won success while the thinker would have still been at the endless task of comparing various judgments" (131). This celebration of action over thought lies at the heart of the appeal of the action-adventure saga; Carter embodies a populist distrust of explanation, a preference for chivalric instinct and emotion over the pale cast of thought. The "tedious formality of reasoning" implies a self-knowledge or self-seeking that could only adulterate what Carter calls his "great and wondrous love" for Dejah Thoris, an emotion that transcends "racial distinctions, creed, or religion" (102). If reason is the mark of civilization, its "endless" weighing of alternatives leaves the Achillean hero outside the idealized legal strictures of justice and good government that he ostensibly defends. On Barsoom, the chivalric hero has only to defend communal values against tyrannical desires, not directly assume the responsibilities of governing, managing resources, or adjudicating the competing claims of political economy, all of which remain in the hands of a benevolent jeddak. Nonetheless, the initiation of the hero never truly can end because tyranny, rape, and power mongering are the inevitable byproducts of the struggle for authority on a dying world.

At the end of *The Warlord of Mars,* Burroughs patches up an ending to conclude his trilogy. John Carter's victory over "the hateful tyrant of the north" (154) signals an idealized peace for Barsoom: just rulers such as Xodar in the south and Talu in the north defeat pretenders to their kingdoms and claim their rightful positions; Carter's son, Carthoris, marries the princess Thuvia of Ptarth, thus ensuring peace between her city and Helium; and a golden age descends on a planet where the hero and Dejah Thoris come to embody a future of unending prosperity. In closing off his trilogy, then, Burroughs superimposes two seemingly incommensurate views of Mars: the dying planet that depends on Helium's rationalized management of resources and the timeless world of chivalric honor. "What matter ages," asks Carter rhetorically, "in this world of perpetual youth?" (155). This appeal conveniently invokes a narrative amnesia to encourage readers to forget what they know about conditions of Barsoom with its dead sea bottoms and its iffy atmospheric generator. Burroughs would later return to Barsoom when he needed money, although he wrote in 1919 that "if it wasn't for the lure of filthy lucre I should never write another Tarzan or Martian story" (quoted in Taliaferro 1999, 162–63). His subsequent Barsoom novels become increasingly formulaic, following the adventures of other heroes, Carthoris and, in

The Master Mind of Mars (1927) and *Synthetic Men of Mars* (1939), Vad Varo, the Earthman Ulysses Paxton, who has his legs shot off in World War I and wakes up on Barsoom where he does battle against evil scientists and chophouse androids. With the exception of *Swords of Mars* (1936) in which he infiltrates an assassins' guild, Carter becomes a gray eminence as his avatars defend Helium's honor against incarnations of evil who embody an envious or absolute desire to monopolize Barsoom's scarce resources. Carter, his son, and Vad Varo thus serve functions analogous to the atmosphere plant—they represent radical interventions in a failing system that are nonetheless necessary to forestall the ultimate fate of Barsoom.

MARS IN PULP FICTION

Burroughs's novels exerted a profound influence on science fiction about Mars from the 1920s to the 1960s. Even as he churned out his later Martian novels, Burroughs had imitators who capitalized on the widespread fascination with the dying planet. To explain how humanoid life could survive on Mars, Otis Adelbert Kline in *The Outlaws of Mars* (1933) has a scientist announce that he is "presently in [telepathic] contact with [Martians] who, to our niche in space-time, have been dead for millions of years," so that the hero's adventures take place in the remote past (Kline 1961, 4). The paperback cover of the 1961 reissue of this novel promises "Interplanetary adventure in the best Edgar Rice Burroughs tradition," but, like Michael Moorcock's *Warrior of Mars* trilogy in the 1960s, *Outlaws* is of the Edgar Lite Burroughs school: dying planet, shrinking seas, and a black boxing of mysterious ancient technologies in a culture that has reverted to martial primitivism. Like Sagan and Bradbury, and "ten million" others, Moorcock was an avid reader of Burroughs as a child and dedicates the reissue of his novels to the creator of Barsoom (Moorcock 1981, 9–10). But another response to Burroughs took more seriously the implications of cultures trying to survive the inexorable forces of ecological collapse. Two broad, interanimating traditions of science fiction emerge in extending Burroughs's version of Lowellian Mars: efforts to imagine the life-forms that might exist on a forbidding planet and efforts to think through the problems of human colonization on a world that harbors the remnants of once-great civilizations. Both traditions follow Lowell in assuming that environmental decay on Mars heralds the fate of the Earth. Leigh Brackett, one of the most successful of Burroughs's heirs, makes explicit the ecological concerns that motivated pulp writers in the 1930s, 40s, and 50s: "We were all very high on ecology in those days, but it was all so simple" (quoted in Carter 1977, 261). This simplicity—the moralizing of the problems of environmental pressures and resource extraction—stems, in part, from a conviction that the Earth's future already had been written. Popular redactions of entropy, the nebular hypothesis, and Darwinian evolution fed into anxieties that the Great Depression and the world wars presaged an apocalyptic collapse of civilization. Even as the canal thesis waned in scientific circles, Mars continued to provoke dialectical responses of dread and hope for the fate of the Earth: humankind's worst Hobbesian fears were projected onto the planet and set against an imagined future of spaceflight, colonization, and the manifest destiny of humankind in the cosmos (McCurdy 1997).

After the stock market crash in 1929, the political and social implications of worldwide economic crisis shaped the genre of future history. Given the complex associations that Lowellian Mars held for millions of readers of science fiction, scientific skepticism about Lowell's canals had comparatively little effect in dislodging the red planet from the popular imagination. As a dying planet of canals, warring races, and dead sea bottoms, Mars dramatized sociopolitical and ecological anxieties and funneled them toward an imaginative center—the metanarrative of planetary desiccation meant increasing hardship, violent conflicts, and downward pressures on standards of living. In the 1930s, the great era of pulp fiction, the American Midwest suffered through drought, wind erosion, and mass migrations: nine million acres in Texas, Oklahoma, Kansas, New Mexico, and Colorado were severely affected by wind erosion in 1938 alone; in the same year soil scientists for the Department of Agriculture estimated that 500,000 square miles "had been seriously damaged by erosion" (Worster 1994, 225). In 1935, 80 percent of people in some counties were on relief and, as topsoil washed away, one-quarter of the farm population of Oklahoma—275,000 people—moved from one farm to another, "an aimless wandering" that exacerbated economic hardships that dated back to the nineteenth century (Worster 1994, 224). Lowellian Mars offered fictional analogues, even uncanny anticipations, of the problems that confronted millions of people worldwide: food shortages, massive unemployment, drought, the dust bowl, forced migrations, political turmoil, violence, and belated efforts, such as WPA projects, that sought to preserve the productivity and beauty of the natural world.

The feudal codes of militaristic honor that characterized Barsoom and its offspring may look cartoonish, or dangerous, to critics in the twentieth- and twenty-first century, but for writers in the 1930s fascist dictatorships in Italy and Germany that appropriated the trappings of imperial Rome may have seemed a reasonable extrapolation of the course of terrestrial politics (Mullen 1971, 229–47; Holtsmark 1986; Wright 1996, 24–26). By the 1930s, Mars became a site for anxieties not only about ecological degradation but about its sociopolitical consequences—about the ability of any government to deliver the necessities of life and preserve democratic freedoms; about political unrest; about masculine identity in a world in which millions of men, confronted with unemployment, doubted their ability to provide for themselves and their families; about population growth; about race relations; about militarization; about an American frontier that had disappeared into the unheroic tasks of managing water, urban growth, zoning regulations, and declines in wildlife populations; about the complications of delivering goods and services over rail lines and a road system that were increasingly seen as inadequate; about the distribution of electricity and other technological innovations; and about the gap between "modern" goods and traditional patterns of resource use (Davis 1998; Brown 1993, 129–63; McNeill 2000). If Brackett and others often adopted "the ancient, but comparatively static, culture of prerevolutionary China" as a fictional model "for their older-than-Earth Martian civilizations" (Carter 1977, 68), their inverted orientalist approach was less an escapist fantasy of Western superiority than an implicit threat: in the face of the economic and ecological onslaught of the 1930s, America and western Europe, too, could go the way of political chaos, communism, or fascism.

Even as Martians became interplanetary bogeymen in much popular culture, writers such as Stanley Weinbaum, P. Schuyler Miller, and C. S. Lewis explored the philosophical, ecological, and theological implications of human encounters with alien life. None of them, however, were diehard Lowellians, and their fictional renditions of Mars look more like Antoniadi's planet, a "vast red wilderness" (1975, 67), although one inhabited by enough exotic life-forms to fire readers' imaginations. Weinbaum's "A Martian Odyssey" (*Wonder Stories* 1934) became an instant classic of human efforts to communicate with nonhuman, nonmonstrous, intelligent life forms. Separated from his comrades, the hero, Jarvis, encounters strange Martian creatures on an odyssey across the deserts of the red planet and learns to recognize in Tweel, a vaguely ostrichlike being, a true comrade. At first frustrated by his efforts to talk to or understand the sounds made by his Martian traveling companion, Jarvis finds that Tweel is an adroit mathematician, and they use mathematics to communicate sophisticated ideas, to distinguish "one," intelligent and moral beings, from "not one," sentient but hostile or unthinking creatures. The simple formula, "one plus one," comes to assume the symbolic values that unite Jarvis and Tweel across planetary and species barriers—the essential, "human" qualities of loyalty, friendship, and responsibility. In telling his fellow Earthmen how Tweel put his own life at risk to defend the hero against barrel creatures, Jarvis pays his Martian companion the ultimate tribute, calling his alien friend "a man" (Hipolito and McNelly 1976, 62). This quest for a means to communicate across barriers of time, space, culture, and physiology figures prominently in subsequent science fiction, displacing, to some extent, the interspecies love stories between Martian women and Earthmen that had sexualized the problems of bringing the alien and its planet within the bounds of a familiar sociosexual economy.

If "A Martian Odyssey" is a story about communication, intelligence, and reason, P. Schuyler Miller's "The Cave" (1943) is a short classic of planetary hard times and the life-forms that adapt to them.[4] In a letter to *Astounding Science-Fiction* (October 1941), Miller asserted that "cultures do backslide" and that "there are plenty of precedents for a people hitting bottom and staying there" (cited in Carter 1977, 245). This is a fair description of his Martian novella *The Titan*, set on a dystopian "dying world" (Miller 1954, 16). An enervated master race literally drinks the blood of its servant class, imprisons a visitor from Earth, and attempts to put down a slave revolt; Martian society has collapsed into a darker version of Burroughs's Barsoom, beset by oppression, addiction, and ignorance. This fear of reverting to a primitive existence also animates "The Cave," in which an entire planetary ecology has deteriorated to a point that the "surface had been desert for more millions of years than anyone [on Earth] had yet estimated" (Hipolito and McNelly 1976, 121). To a greater extent than any writer in the previous half century, Miller explores the implications of desertification, bringing them, at a time when the dust bowl and Depression loomed large in public consciousness, to their grim and logical ends. The planet's natural wealth "had been exhausted by a native Martian civilization pursuing its inevitable way to an inevitable end at a time when Adam and Eve probably had tails" (126). Ecological crisis produces an embattled mutual dependence that is both a response and counter to the Darwinian struggle for existence. With resource exhaustion and the dearth of water, the remaining sentient life-forms constrict their civilizations to a few settlements of "Greenlanders" and the inhabitants of the deserts. On Miller's Mars, "all living things

[are] united in the common battle for existence against a cruel and malignant Nature" (125): "Millions of years of unceasing struggle with the forces of an inclement environment on a swiftly maturing and rapidly dying planet have ingrained in the native Martian race, greenlanders and drylanders alike, the fundamental concept that Nature is their undying enemy. . . . You find it in the oldest legends: always the wily native hero is outwitting—there is no other word for it—the evil purposes of the personified, malignant Universe" (131). Such descriptions make explicit with a vengeance the implications of Lowell's "nameless horror": human (or nearhuman) agency, heroism, and intelligence are defined as a stoic resistance to nature's implacable hostility. In such an environment, the essence of every Martian's self-definition, "the very core of his existence," is the realization "that all beasts are brothers" (130). Art, technology, and identity are all shaped by Miller's nightmare vision of planetary desiccation.

Harrigan, the only human in Miller's story, is a miner who stumbles into a cave during a dust storm. Surrounded by indigenous life-forms that he fears and does not understand, he kills a lizardlike predator, the *zek*, that goes for his water. In turn, Harrigan is killed by a *grak*, an intelligent, nonhumanoid being, for violating the basis of Martian law: the sharing of water that binds all living things on the planet—all *grekka*—together in their unending battle against the "eternal foe," "the forces of Nature" (137). Any violations of this mutual alliance brands the transgressor, even an ignorant one like Harrigan, as the agent of a "personified, malignant Universe." In killing the miner, the grak can claim a "victory won for the brotherhood of living things against the Universe" (137). Rather than the romanticized efforts of individual heroes to stave off ecological collapse, Miller depicts the grim consequences of Antoniadian Mars: the environment is past reclamation, and not even the ghosts of seas or canals haunt the imagination of its sentient life-forms.

But the vast red wilderness of Mars could be used for very different purposes, and in *Out of the Silent Planet* (1938), C. S. Lewis contests the philosophical values and assumptions that Burroughs and Miller share. Yoking "science fiction's literary tradition . . . with the mythology of the Bible" (Pierce 1989, 40), Lewis brings the dying planet within the orbit of Christian theology. In this respect, his novel updates the spiritualism that had characterized the fiction of Flammarion and Wicks, and it offers a fictional analogue to the criticisms of the canal thesis by Maunder and Wallace that, in part, were based on theocentric conceptions of the universe. Lewis's hero, Elwin Ransom, a philologist, is drugged and taken aboard a spaceship for Mars by the ironically named villains, Devine and Weston; the latter, a renowned physicist, becomes a mouthpiece for the godless scientism that Lewis satirizes. Arriving on the planet, the three space travelers find that Mars is both a blasted environment, its atmosphere ripped away by the "bent one," Satan, and an unfallen world, Malacandra, on which the surviving species exist in a divinely ordained ecology of low-lying waterways. These life-supporting environments, handramits, the hero learns, are both "gigantic feats of engineering" (Lewis 1997, 143–44) and the miraculous results of divine intervention. The Oyarsa, or planetary intelligence, of Malacandra tells the Earthmen that the "bent one . . . smote your moon with his left hand and with his right he brought the cold death on my *harandra* [planet] before its time; if by my arm Maleldil [God] had not opened the *handramits* and let out the hot springs, my world would have been unpeopled" (120). The specter of entropy that haunts planetary science and science fiction is pressed into the service of a Christian cosmology. The Oyarsa tries to explain to Weston, "the great physicist" (15) who wants to conquer Mars for humanity, that his science is in vain. He and Devine are trying to conquer a world " 'older than your own world and nearer its death. . . . Soon now, very soon, I will end my world and give back my people to Maleldil' " (138). The ecological endgame of Lowellian Mars is superceded by a divine order in which entropy becomes a form of spiritual apotheosis.

The true antagonist, though, for Lewis is not Lowell or Burroughs but H. G. Wells who projects the sins of fallen humanity—rapaciousness, greed, and self-interest—onto his Martian invaders (Pierce 1989, 37–40). Early in *Out of the Silent Planet,* after escaping from his kidnappers, Ransom catches sight of one of the three intelligent life-forms on Malacandra, the sorns, and imagines them as the monstrous embodiments of science fiction come to life: "He had read his H. G. Wells and others. His universe was peopled with horrors such as ancient medieval mythology could hardly rival. . . . He saw in imagination various incompatible monstrosities—bulbous eyes, grinning jaws, horns, stings, mandibles. Loathing of insects, loathing of snakes, loathing of things that squashed and squelched, all played their horrible symphonies over his nerves. But reality would be worse: it would be an extra-terrestrial Otherness—something one had never thought of, never could have thought of" (37). For Lewis, Wells and the science-fiction tradition of "extra-terrestrial Otherness" assumes the irrevocable nature of sin. Monstrous aliens, in effect, embody the fears of Darwinian evolution gone inhuman. Moreover, such nightmares link this alien otherness to fears about environmental decay and the evolutionary horrors it might produce. But on Malacandra Ransom finds three intelligent life-forms distinguished by both biology and a division of labor: the hrossa are beaverlike guardians of the planet's waterways and the poets of their world; abstract knowledge, philosophy, and theory are the province of the mountain-dwelling sorns; and technical knowledge, crafts, and construction are entrusted to the pfifltriggi. Biology is not only destiny, it is labor harmony, social stability, and salvation as well. After being rescued by the hrossa, Ransom quickly finds that his fears of alien otherness have vanished, and that he has entered a realm of true spirituality "as though Paradise had never been lost and earliest dreams were true" of animals with "the charm of speech and reason" (59). If Devine and Weston threaten to turn an unfallen planet into a mining colony, Ransom recognizes that human desire—figured as the insatiable hunger for what one cannot have—is itself a mark of the fall of Thalacandra (Earth)—a planet that, thanks to the bent one, suffers a form of spiritual excommunication. On Lewis's Mars, the premises of "The Cave" are inverted: the unity of all living things marks the sufficiency of resources; the lack of desire registers an acceptance of individual and planetary mortality and an absolute faith in the justness of the universe. Ransom's kidnapping becomes a pilgrimage, and Malacandra offers a glimpse of an Edenic existence before the advent of sin and scarcity.

THE WAR OF THE WORLDS AND THE PANIC OF 1938

In the late 1930s Mars struck back, and in film serials, comic books, pulp magazines, and radio programs resource-hungry Martians invaded the Earth. The most famous and influential of these imaginary invasions was Howard Koch's adaptation of *The War of the Worlds* for Orson Welles and the Mercury Theatre, broadcast on October 30, 1938. Koch's radio script updates Wells's novel and transforms its narrative for a new medium, one that makes different kinds of demands on its audiences. Rather than a fictional narrator recounting the inexorable progress of the Martian invasion, this adaptation cleverly exploits the conventions of radio melodrama to produce an auditory realism: news bulletins interrupt a musical program; reporters interview an astronomer from Princeton; government officials plea for calm; and news flashes cut away to battlefield reports from correspondents. Wells's prose is pared down and recast in 1930s American colloquial speech, and Koch's adaptation mimics the immediacy of radio news reports. An announcer stands on the roof of a building in Manhattan and, with car horns in the background that suggest a massive traffic jam, tells his listeners, "The enemy's now in sight above the Palisades. Five—five great machines. The first one is crossing the river. I can see it from here, wading, wading the Hudson like a man wading through a brook" (Holmsten and Lubertozzi 2001, 47).[5] Colloquialisms and repetition create the illusion of an unscripted encounter. The unworldly invaders are transformed into an "enemy" whose approach invokes the nightmarish memories of World War I. The tripod "wading the Hudson" conjures the visual image of a "man wading a brook," but the sheer size of the invader suggests a terror that exceeds the announcer's homely simile. The Martian machine in its immensity and the speed of its approach gives auditory form to the fears rampant in the late 1930s that the United States could be dragged into another European conflict. When the Martian machines begin spewing "black smoke," the "people in the streets" respond by "dropping in [to the East River] like rats"; Wells's "swift liquefaction of the social body" (Wells 1993, 121) is transformed into the horrors of poison gas and the unknown terrors of new generations of weapons. The Martian cylinders "falling all over the country" (Holmsten and Lubertozzi 2001, 47) seem part bomb, part aircraft, as though the United States is experiencing the horrors of the Japanese invasion of Manchuria. The broadcast, in this regard, gains much of its power by playing on fears about the outbreak of a new war in Europe with the next generation of weapons of mass destruction.

In 1940, Hadley Cantril, Hazel Gaudet, and Herta Herzog for the Princeton Radio Project interviewed 135 people who heard the Koch-Welles broadcast and published a detailed sociological study of listeners' responses to the program in an effort "to determine the underlying psychological causes for [the] widespread panic" (Cantril 1940, vii).[6] Six million people heard the 1938 broadcast; of those interviewed, 29 percent admitted to thinking *The War of the Worlds* was a news bulletin and 63 percent of those who tuned in late and missed the opening disclaimer thought that the invasion was real (58, 73). If the jammed phone lines and calls to radio stations in northern New Jersey that night are any indication, some of the skeptics may have been nervous as well. A nurse who was having a party that Sunday evening reported that "everybody was terribly frightened. Some of the women almost went crazy. The men were a little calmer. Some of the women tried to call their families. Some got down on their knees and prayed. Others were actually trembling. My daughter was terribly frightened and really suffered from shock. A ten year old child who was here was petrified. He looked like marble" (123–24). Such anecdotal evidence suggests that the figure of 29 percent may understate the contagiousness of the invasion-induced panic. The other 70 percent or so of the respondents may have been "a little calmer," but the interviewees' comments indicate that some nervousness and uncertainty remained beneath their skepticism.

Many of the respondents testify both to a faith in authority figures—the "scientists" and "government officials" interviewed early in the broadcast—that overwhelmed any willing suspension of disbelief and to the ability of radio to create a sense of the "reality" of events. "I believed the broadcast," said one listener, "as soon as I heard the professor from Princeton and the officials in Washington" (71). Another declared, "If so many astronomers saw the explosions [on Mars], they must have been real" (71). In such instances, the medium itself reinforced the seeming reality of the broadcast because listeners placed their faith in the sociopolitical function of radio. One individual who admitted to thinking that the invasion was real suggested that "in a crisis [radio] has to reach all the people. That's what radio is here for" (70). Significantly, however, 25 percent of those frightened by the broadcast thought that they were listening to a foreign attack rather than an invasion from Mars. For some, skepticism about Martian invaders did little to compromise the verisimilitude of the broadcast: "I never believed it was anyone from Mars. I thought it was some kind of a new airship and a new method of attack. I kept translating the unbelievable parts into something I could believe" (99). Another listener stated: "I knew it was some Germans trying to gas all of us. When the announcer kept calling them people from Mars, I just thought he was ignorant and didn't know yet that Hitler had sent them all" (100). This translation of the unimaginable invasion into a sneak attack by Nazi airships and poison gas gives human shape to the fear of invasion. In one respect, converting Wells's Martians into Hitler's shock troops renders the horror of technological warfare in terms made familiar by radio and newsreels. As opposed to an enemy "different beyond the most bizarre imaginings of nightmare," German invaders can be resisted, defeated, and forced to pay reparations. In another, Nazis with poison gas and "a new airship" assume many of the characteristics of the terror inspired by Wells's Martians: a ruthlessness without meaning, a penchant for genocide, an utter disregard for human—or at least American—life. Like Nazis, Martians embody evil incarnate.

If Wells's novel is the ur-text of interplanetary invasions, subsequent generations—in the United States in 1938; Santiago, Chile, in 1944; and

Quito, Ecuador in 1949—have confronted fictional Martians and the fears that they induce over airwaves.[7] The semipanic caused by an updated invasion scenario broadcast in Buffalo in 1968—well after the Mariner 4, 6, and 7 missions had photographed a cratered and apparently lifeless planet—indicates that "Martians" continued to signify a radical alterity, particularly in those media such as radio, comic books, and science-fiction novels that can depict aliens without the expense, and limitations, of filmic special effects. As the archetype of all remorseless invaders, "Martians" signify the dark, nightmarish underside of a modernist ideology that places its faith in science, technology, and progress. At the same time, they have served as stand-ins for paranoid fears of malignant intelligence: Nazis, communists, body snatchers, fifth columnists, and all aliens who sap human strength and subvert identity from within. Martians, in short, are both the Other, the embodiment of an absolute intelligence that we can neither understand nor defeat, and those others who rival and mirror our desires and weaknesses and feed off of our own fears. In both cases, Martians mark a return of the repressed: a force of evolutionary terror acting as retribution for humankind's gutting of the Earth's resources or a doomed race that reflects humankind's blindness to the consequences of mindless exploitation. Martians are the skull beneath the skin: they give shape to our fears that civilization is always on the verge of collapse and that we are all "phantoms in a dead city."

The range of reactions to the 1938 version of *The War of the Worlds* reveals an underlying dialectic in midcentury science fiction about Mars: paranoia about invaders from space set against a profound cynicism about the gullibility of those listeners taken in by the scientific improbabilities of a radio play. Although both "paranoia" and "cynicism" may seem to have commonsense definitions, both need to be explored in the contexts of midcentury politics and the uneasy oscillation in attitudes toward life on Mars. In effect, paranoia and cynicism describe two poles on a continuum of individual and national self-definition.

In 1965, Richard Hofstadter described what he called "the paranoid style in American politics," in which a " 'vast' or 'gigantic' conspiracy [becomes] *the motive force* in historical events . . . set in motion by demonic forces of almost transcendent power [and] infallible rationality" (1965, 29, 34, 36). Sequences of cause and effect themselves are construed as evidence of an "infallible rationality" at work. While Hofstadter focuses on the public arena, the narrative of history itself mirrors the individual's desire to validate his or her own worth by uncovering the "demonic forces" that constitute him or her as an intelligible subject—

that is, as a self forged by his or her heroic resistance to a "transcendent power." The logic of paranoia thus creates a double bind. In the clinical literature as well as in popular culture, paranoia is considered both a psychological disorder, often unambiguously delusional, and an extension of the dynamics of "normal" identity (Fenigstein and Vanable 1992, 129–38; Goldwert 1993, 326; Menuck 1992, 140–41; Mirowsky and Ross 1983, 228–39). Characterized by projective thinking, hostility, suspiciousness, centrality, delusions, fear of loss of autonomy, and grandiosity, paranoia in practice becomes difficult to distinguish from "normal" processes of forging logical connections among disparate phenomena or observations or, more simply, negotiating one's way through daily existence.[8] Freud, for example, recognizes that "the delusions of paranoiacs have an unpalatable external similarity and internal kinship to the systems of our philosophers," leading his commentators to suggest that paranoia is a form of hyper-analysis—the belief in a method or system that is indistinguishable from the search for meaning, order, and coherence (Freud 1953–74, 17: 261).

Paranoia creates an analytical paradox: we want to believe in the coherence of the world and in the explanatory power of science and reason, but efforts to render the world coherent always threaten to slip from order into obsession. To resist the tyranny of too much coherence—a tyranny in which we are paradoxically complicit—we must resist the tendency to forge absolute meanings from fragmentary data. The more paranoid we become in constituting our identities in opposition to a tyrannical order, however, the stronger our identification with that order; and the more we resist that identification, the more forcefully introjected our alienation from our supposedly coherent selves becomes. In this regard, paranoia is the effect of what we fear is an originary alienation both within ourselves and from others—the double alienation from a dying world and from his own psychological interiority that John Carter defines as his mode of self-definition. This recognition of our double alienation must be acknowledged and repressed for us to constitute ourselves as both makers of meaning and as coherent subjects. If, in effect, all knowledge is paranoid, then resisting the production of knowledge leaves us desiring to be stable subjects—as we believe or fear everyone else is—but always clinging to the fantasy that our knowledge will fail us, that in not making sense we can find a means to resist the meanings imposed on us.

In a paranoid universe of invading Martians or hostile riders of the dead sea bottoms, cynicism becomes paradoxically a mode of resistance and resignation, an inoculatory abreaction (an effect preceding its cause) that therefore can imagine the future only as an extension of the past. Cynicism ironically validates existing forms of ideological order by rejecting all forms of analysis. In science fiction set on Mars both before and after World War II, the future technohistory of a new frontier is set against the grim evidence of dying or extinct civilizations: the Martians and their human avatars repeat the mistakes of the past and obstinately project these same failed strategies into the future. In the novels and short stories of Leigh Brackett, Lester del Rey, and Philip K. Dick, trips to Mars turn all heroes into cynics. This future tense of cynicism protects us from false hopes that the revelation of the truth is imminent, that Stygian politics will be cleansed by revolution or reform. Abstracted from its dialectical relationship with paranoia, however, cynicism is not a heroic shield against the forces of oppression but a mark of stupidity, comparatively uninteresting to cultural critics and psychologists because it seems either the aftereffects of a romantic disillusionment or a maddening refusal to acknowledge the imperatives that motivate their intellectual endeavors. If in contemporary popular culture, paranoia is a mark of intelligence or perspicuity (*The X-Files*), then cynicism preserves an obstinate, and paradoxically attractive, resistance to tyrannies of imagined coherence (Polan 1986), as in the fiction of Ian Douglas, whose *Semper Mars* (1998) pits noble U.S. Marines against the evil forces of the United Nations, out to monopolize the resources of Mars for themselves. Such contemporary science fiction, however, has deep roots in the postwar period, and fears of Martians or Nazis "trying to gas all of us" are projected later onto the conflicts of the cold war.

MARS AND POLITICAL CRITICISM IN THE 1950S

World War II brought the threat of invasion home to millions of English and American readers of *The War of the Worlds.* In the 1940s and 1950s, the pulps betray a widespread fascination with the technologies—from radar to power generation—that were essential to the prosecution of the war against Germany and Japan. While patriotic fervor was often in evidence, some science fiction depicted thoughtfully the ways in which technology might bridge the temporal and physical gulfs between hardworking American scientists and a long-vanished Martian civilization. George Smith's "Lost Art" (1943) depicts twenty-first-century scientists working to understand the "power beam" that made Martian civilization possible: alternating between a distant past on Mars and an archaeological future, the story depicts the construction and, eons later, reconstruction of a generator that fulfills the dream of perpetual energy—an output of electrical power that always exceeds the resources required to keep it going. Techno-archaeology figures as well in H. Beam Piper's "Omnilingual" (1957) in which Earth scientists try to decode Martian language from a civilization dead for fifty thousand years. Entranced by a mural that depicts the progress of Martian civilization from "skin-clad savages" to an advanced civilization in decay, the scientists recognize a universal narrative of progress and decline: "Seaports on the shrinking oceans; dwindling, half-deserted cities; an abandoned city, with four tiny humanoid figures and a thing like a combat-car in the middle of a brush-grown plaza" (Hipolito and McNelly, 196). The key to the aliens' language is uncovered when the archaeologists find the Martian periodic table of the (then) ninety-two elements. But if chemistry is a universal semiotics, intelligible to all "advanced" beings, "Omnilingual" also naturalizes a metanarrative of progress and ecological dissolution. The warfare symbolized by the "combat-car" is the inevitable end of the evolution of worlds.

The emphasis on the technological accomplishments of long-dead Martian civilizations also could be marshaled to call into question postwar American triumphalism. If an advanced civilization on Mars could degenerate, then the United States and Europe too might "backslide" into political repression. For a vocal group within the science-fiction community, McCarthyism threatened to undo the trappings of technological progress. Paul Carter argues that "the 1950s in science fiction were a time of trenchant social criticism," and Judith Merril, the science-fiction novelist and anthologist, claims bluntly that during the McCarthy era the genre was "virtually the only vehicle of political dissent" (Carter 1977, 140; Merril 1971, 74). In her collaborative novels with C. M. Kornbluth (published under the pseudonym Cyril Judd), *Gunner Cade* (1952) and *Outpost Mars* (1953), Merril offers a sardonic view of capitalism and political corruption, one seldom found in "mainstream" postwar fiction. In some respects, the enthusiasms of pulp writers and readers for a "utopian politics" of science fostered a hope for an enlightened political morality that would match postwar technological capabilities; but such hopes presumed that humanity would learn from its past mistakes, and writers on the Left were achingly aware of the consequences of failure: cynicism, annihilation, and consumerist dystopias. The result, in part, was to invert the premises of earlier utopias and depict nightmarish images of humankind repeating the mistakes of the past.

The pulps were rife with a cantankerousness that often approached intellectual rebellion. In an editorial in *Other Worlds* in 1952, Raymond A. Palmer mocked the scientific certitude that characterized war games and sneered at the logic that underlay nuclear tests: "maybe war is important to soldiers and bombs are top priority, but to just plain people it's peace that's important, and less lying, thieving, immorality and dictatorial governing" (4).[9] Such provocative comments, however, could easily mask feelings of isolation and impotence. The leftist leanings of many writers could slip into cynicism or the kind of resignation voiced by Merril's sometime collaborator, Cyril Kornbluth. Near the end of his short life, he concluded reluctantly that "science fiction is socially impotent," usually turning readers "inward to contemplation" rather than "outward to action" (1959, 75, 55). Having internalized formalist standards of literary evaluation, Kornbluth discusses the exploration of character in more-or-less Freudian terms and dismisses his former work. Ultimately, he concludes that science fiction's political "criticism is massively outweighed by unconscious symbolic material more concerned with the individual's relationship to his family and the raw universe than with the individual's relationship to society" (Kornbluth 1959, 75).

Merril and Kornbluth offer radically different assessments about science fiction's usefulness as a vehicle for social change and political critique. If one expects, as Kornbluth came to, demonstrable political effects arising from works of fiction, one is likely to end up like Lemuel Gulliver, who laments in the preface to *Gulliver's Travels* that six months after his book has been published the world has yet to reform. Judged by a strict logic of cause and effect all satire "fails": Kornbluth even disparages his novel *The Space Merchants* (1953), written with Frederick Pohl, a brilliantly ferocious satire of a future run by rival ad agencies, as without "high literary merit" (73). As Merril implies, science fiction practices a kind of seat-of-the-pants, ad hoc critique of ideology in which present-day tensions are extrapolated either into the future or onto different worlds. Such cognitive estrangement produces a range of dialectical responses—cynicism or outrage, despair or the conceptual groundwork for imagining a different kind of social, political, economic, and technoscientific future. These reactions, however, stem from a common vision of postwar society in which social and political progress lagged far behind technological progress. While Flash Gordon and Buck Rogers pursued pulp fiction's manifest destiny in space, Mars remained a favored setting for critiques of American capitalism and imperialism, and for bitter condemnations of humankind's stupidity and blindness in transposing its self-destructive tendencies onto another world. Rather than celebrating humankind's technoscientific future and the extension of cowboy justice to the final frontier, much of the science fiction set on the red planet betrays the cynicism that characterizes *The Space Merchants* and its satire of ecological and sociopolitical dystopia. The descendants of John Carter on 1950s Mars tend to be either outcasts or disillusioned functionaries struggling against endemic corruption, more like Raymond Chandler's Philip Marlowe than a warrior-hero.

In *Gunner Cade* (1952), Kornbluth and Merril update the anti-totalitarianism of Yevgeny Zamyatin's *We* (1921) by exploring the mindset of their hero, Cade, a member of the elite military police force, who begins to doubt his mission as the enforcer for a totalitarian regime that rigidly controls scientific knowledge. His descent into uncertainty and psychological turmoil ultimately becomes a fall into a recognizable identity—complete with a heretofore forbidden sexual desire for women, the renegade princess Jocelyn in particular. In a post-holocaust world, she is one of the few who recognizes that history existed before the ten-thousand-year reign of the current empire. Time, she tells the hero, now fallen from "gunner" to outcast and rebel, is not an eternal present of subservience and duty but a history of conflict over vanishing energy sources on a stagnant Earth: "There was a time—I know from History—when men powered their machines with the metal uranium. It's gone now. Thorium was used next, and now it's gone too. And now the iron. Earth's iron is gone. When the Mars iron is gone too, what next? There should be ten million men working day and night to find a new power source, but there are none" (143). The corruption of power and the stifling of the human spirit are cast starkly in images of a single-minded, ecological devastation: resource extraction is a one-way vector—minerals are mined to depletion. The climactic rebellion of miners on Mars makes assumptions and reaches conclusions similar to those of Alexander Bogdanov's *Red Star;* planetary exhaustion can be extrapolated from biology to geochemistry. In McCarthyite America, though, there are no reasoned debates or higher order of beings to control civilization's appetites for raw materials and conquest. Projected into a dystopian future, Kornbluth's and Merril's nightmarish capitalism treats humans, minerals, and foodstuffs as commodities that are used brutally, unthinkingly, without regard to consequences.

In *Outpost Mars* (1953), Kornbluth and Merril take a very different tact and introduce narrative and physiological concerns that are picked up and adapted by subsequent writers of Martian science fiction. Rather than repeating the dystopian nightmare of *Gunner Cade,* they anchor their second Mars novel in realistic, even gritty detail to depict a wide-ranging struggle between an idealistic, communal settlement and an evil mining conglomerate. Most of twenty-first-century Mars is devoted to mining and agriculture made possible by the discovery of "the magic pink pellets, containing the so-called 'oxygen enzyme'" (9) that enables humans to breath the Martian atmosphere. This "magic" enzyme allows human beings to colonize Mars, chemically altering indigenous plants to make them palatable for humans or adapting terrestrial crops so that they can grow in Martian soil. These scientific advances, however, lead to a repetition of the mistakes that have led to political, economic, and environmental crises on the "damned, poverty-ridden, swarming Earth! Short of food, short of soil, short of water, short of metals—short of everything except vicious, universal resentments and aggressions bred by other shortages" (76). The villains are the forces of American capitalism projected a century into the future. The novel's hero, Tony Hellman, the doctor of the communal Sun Lake colony, recognizes that "his own high-flown thoughts mocked him. . . . Already the clean air of Mars was thickening with the eructations of Earth's commerce" (13). Shortages, pollution, and environmental degradation are depicted as consequences of the damn-the-torpedoes industrialization and social stratification that the authors see as the twin evils of the postwar era. The Sun Lake colony, in contrast, is an outpost of sanity, communal decision-making, and democratic socialism that harks back to a homegrown tradition of American utopianism.

In *Outpost Mars,* the symbol for the contradictions—political, economic, and somatic—that define the political struggle for the future of colonization is "marcaine," an addictive drug grown and refined on Mars, then exported to Earth more for its black-market value than its legitimate use as a pharmaceutical. The drug functions as a *pharmakon,* the Greek word for both poison and cure. The novel's plot centers on a missing shipment of marcaine that is blamed on Hellman and other colonists at Sun Lake; threatened with an embargo that will ruin their progressive experiment, he discovers that marcaine is essential to the survival of an infant born on Mars who can breath the atmosphere without the oxygen enzyme. Resisting corporate and government efforts to terraform Mars, the colonists discover that the mythological Martian "dwarves"—gremlins blamed for the suspicious breakdowns that befall the pollution-spewing factories—are actually human children who have gone feral, adjusted to breathing on Mars, and who exist on marcaine and indigenous plant life. Rather than pathologizing the physiological differences of the children who have gone native, Kornbluth and Merril offer a parable of biological adaptation as a form of utopian speculation. The children's genetic anomalies—fatal to them on Earth—allow them to survive on Mars: in a sense, they embody the utopian ideals of the colony. Evolution serves a progressive political agenda. The novelists inscribe on the bodies of these children an alternative to a future of continued intensification that depends on exploiting both the fragile Martian environment and the underpaid, drug-addled human labor (refining marcaine is a dangerous and addictive job). In this regard, marcaine serves as a complex symbol for the possibilities of adaptation to a seemingly hostile environment and for the fatal consequences, biological and economic, of nonadaptation. By the end of the novel, the "Mars-viable gene" of these "mutants" (219) offers an economic future for the embattled Sun Lake colonists who can prosper by aiding the adaptation of humans to Mars. The possibility for social regeneration, symbolized by the telepathy that these children share, lies in rejecting the exploitation and profiteering that have brought Earth to the brink of destruction.

Even as the popular western provided a resilient vehicle for representing idealized images of American national identity, the science fiction of Kornbluth and Merril contested the metaphor of the frontier that had been applied to Mars since Burroughs (Canfield 2001). Other Martian novels in the 1950s appropriated the conventions of hard-boiled detective fiction and sent to Mars rugged loners seeking some semblance of justice on a corrupt and violent world. In Lester del Ray's *Police Your Planet* (first serialized in 1953, then published as a novel in 1956 and revised in 1975), Marsport becomes a science-fiction redaction of Chandler's Los Angeles, and del Ray's hero, Bruce Gordon, a disillusioned, futuristic version of Dashiell Hammett's Sam Spade: ex-boxer, ex-gambler, ex-cop, and ex-muckraking journalist. Deported to Mars, Gordon encounters a city rife with corruption: Marsport is a sinkhole of racket bosses, violent gangs, drug addicts, gamblers, prostitutes, corrupt police, brutal protection rackets, utter squalor, and rigged elections. "Nobody except romantic fools," Gordon recognizes, "ever thought frontiers were pretty" (3). As in *Outpost Mars,* humanity's future on the planet is constrained by the scarcity of oxygen and water. Although "air machine[s]" (105) are producing oxygen from the ground, beginning a process of terraformation, and third-generation colonists are breathing the thin air of Mars for longer and longer periods of time, the control of such resources pits warring factions against each other. Confronted by starvation, the ragtag elements in the slums surrounding Marsport rebel and destroy the dome over the city; ultimately, the future of Mars lies in destroying as well the fantasy that the colony can exist only by "trying to be Earth," by adhering to the morally bankrupt hierarchies of "slave and master" (195). Their ringleader, Gordon, part intentionally, part opportunistically, emerges as the representative arm of Solar Security, a shadowy organization, much like Hammett's Continental Agency, that seems the only bulwark against utter chaos. With the dome destroyed and the worst of the crooks dead or on the run, Gordon's next assignment will take him to the mines on Mercury, accompanied by his tough-as-nails wife, Shelia, and armed only with one hundred credits, a knife, and a deck of marked cards. By deploying the conventions of detective fiction, del Ray universalizes the cynicism and amorality of the hard-boiled genre: the crime and corruption of Marsport are inherent in frontier ideology.

As Samuel Delany suggests, the generic impulses of science fiction are never pure, and the 1950s critiques of ideology superimpose competing conventions, narrative strategies, and ideolects (1994, 192–93). The best example may be the Martian fiction of Leigh Brackett (1915–1978), who began her career as a writer for pulp magazines and went on to become a successful mystery and science-fiction novelist and screenwriter.[10] The Ace paperback editions of her novels carry a foreword by her husband, the science-fiction novelist Edmond Hamilton, who notes her debts to Burroughs and comments on her "incurable romanticism" about the fourth planet: "She maintains that when the first astronauts land on Mars they will find dead cities, fierce riders and wicked beautiful queens" (Brackett 1964, 2). Brackett's "romanticism," however, assumes complicated forms. More so than the John Carter tales, her Martian novels strike an elegaic note for lost civilizations and a lost green world. In her time-travel classic, *The Sword of Rhiannon* (1953), the terran hero Matt Carse and the warrior queen Ywain find adventure and lust on the Mars of a million years ago. At the end of the novel, they return to the present but are haunted by the vision of a planet before its desiccation: "The desolate land and the ghosts of the past were all around them. Now, over the bones of Mars, Carse could see the living flesh that had clothed it once in splendor, the tall trees and the rich Earth, and he would never forget" (141). This scene is emblematic of a tradition that collapses Martian history into legend, and then defines the vanished past as a ghostly haunting of present desires. Bracket's prose enshrouds this mythic past with qualities that can be intuited, paradoxically, only by their absence. More than her contemporaries, she is a stylist of the pop sublime, using the repetition of words and phrases to gesture toward an unrepresentable past of both technological grandeur and barbaric heroism. The last descendents of the canal builders, contemporary Martians are the heirs to a "history . . . so vastly long that it fades back into a dimness from which only vague legends have come down—legends of human and half-human races, of forgotten wars, of vanished gods" (11). This nostalgia for long-vanished ecologies and cultures underscores the symbiotic connection between the barren surface of the planet and the fate of its inhabitants: social identity is as much a function of ecology as it is of culture. Brackett's novels also betray a generic nostalgia for oriental or African adventure tales. The senescent planet is part Orient, part Arabia, full of cities such as "Barrakesh" and "Jekarta" and featuring such horrors as a glowing "beast jewel" that sends its devotees tumbling down the evolutionary ladder to physical as well as moral bestiality.

Like other pulp writers, Brackett uses the fictive colonization of Mars to offer a thinly veiled critique of European and American imperialism. In her stories from the 1950s and early 1960s, collected as *The Coming of the Terrans,* the indigenous Martian civilization has succumbed to a planetwide "exhaustion of resources"—metals, water, and crops—and consequently suffered an "inevitable reversion to the primitive" (1967, 204–5). This fantasy of decadent races at the mercy of Earth's managerial elite is used to challenge the liberal ideology of the New Frontier. In "2038: The Road to Sinharat" (originally published in 1963), the Martians, "these so-called primitives," reject Earth's technological aid because they "have been through all this before" (212). "All this" is civilization—hi-tech strategies of intensification to exploit the planet's dwindling resources. As one of the Martians tells us an American ethnologist: "We do not want our wells and water courses rearranged. We do not want our population expanded. We do not want the resources that will last us for thousands of years yet, if they're not tampered with, pumped out and used up in a few centuries. We are in balance with our environment; we want to stay that way" (171). The "primitives" ventriloquize the rhetoric of the independence movements of 1960s Africa, Latin America, and Southeast Asia, and the terran colonists are identified with the worst excesses of postwar capitalism. Significantly, the resistance of the Martians—"we will fight" (171)—is cast in terms of opposition to the kinds of large-scale development projects that were a cornerstone of a postwar American policy in the Third World. Brackett's interest in ecology has developed from her self-confessed "simple" sentiments of the 1930s to a broad-based critique of top-down development, carved paradoxically from the seemingly rigid conventions of planetary romance.

In *The Nemesis from Terra* (1961), the discourses of heroic fantasy and hard-boiled detective fiction are effectively juxtaposed to de-idealize the hero, Rick Urquhart, and to offer a Chandleresque critique of the corruption, brutality, and moral degeneration that accompanies mining operations on Mars by the Terran Exploitations Company. Early in the novel, Ed Fallon, the head of the company, views the exploitation of Martians and the underclass of Earthmen from his office overlooking one of his mining operations, a gigantic raping of the planet for "Fallonite," a mineral that is revolutionizing the plastics industry on Earth:

> Death was out there. Age and cessation. Fallon thought no more of it than he did of last year's worn-out shoes. He watched the life of his Company, the thunder and sweat and surge of machinery and the men who bossed it, and it was his own life, his own blood and sweat and surging energy.
>
> Young, that baby, like Earth's intrusion onto dying Mars, but already stretching out muscular hands to close around a planet. . . . A planet practically untouched by outland hands until the discovery of Fallonite. It was disunited, ingrown, weak, an easy touch for the first strong man who could see wealth and power springing out of its fallow fields. (14)

While the portrayal of corruption may be formulaic, Brackett appropriates the style of Chandleresque fiction to offer a cultural history of the deterioration of Mars. By displacing the rhetoric of Western colonialism onto Mars, she recasts such exploitation in a peculiarly American idiom; her metaphors—"worn-out shoes," the company's "muscular hands" choking the planet—yoke the orientalist descriptions of the dangerous back alleys of Martian cities to both an economic and environmental sterility. The imagery is overtly masculinist. Mars is figuratively impotent, the Earth hypersexual in its rape of the planet. Unlike the "thrusting, aggressive world" of Venus, Mars is "old, passive, faded and worn out. Even the Martian hatred of the Earthmen, the invaders, was a silent thing, festering in barren darkness. The stream of Martian trade flowed . . . like the chilling blood of an old man already three-quarters dead" (37). Brackett inverts and extends the sexualized rhetoric of age and desiccation. If Mars is aged and "worn out," it is also an alien world that is reimagined in the image of oriental societies that, from a twentieth-century perception, may have "hit bottom and stay[ed] there" (Schuyler Miller; quoted in Carter 1977, 245).

In this respect, Brackett's "desert forgotten of God and man," superimposes radically different perceptions of the planet: "In the Martian spring, the gorges ran full with the thaw-water that fed the canals. There were mosses and lichens and a few tough flowers. But the black rock was rotted and split by time, ice, wind and water, and it looked as untouched by humanity as the Moon" (125). Brackett give shape to different areologies—the Martian spring that holds out the hope for (at least) primitive plant life and near-lunar desolation. It is a correlative of the picture painted by von Braun and Ley—a kind of Himalayan Mars of thin air and hardy survivors—filtered through the blasted romanticism of her hybrid form: heroic adventure, orientalist cultural anthropology, and hard-boiled cynicism. As in other 1950s science fiction, the dying planet structures an overarching narrative that links the degeneration of the planet's resources, the enervation of its inhabitants, and the corruption that both produces and is produced by corporatist exploitation.

Serious critiques such as Brackett's, to some degree, were offset by the popularity of Mars as a setting for adolescent fiction and comic-book adventures. Major science-fiction writers, such as Lester del Ray, turned to the juvenile market with teenager-in-space adventures such as *Marooned on Mars* (1962). Yet even in modes that make no great claim to transcendent literary merit, like juvenile fiction, Mars retains its cultural significance. Donald A. Wollheim's *The Secret of the Martian Moons* (1955) was reissued in 1963 by Tempo Books, a series published by Grosset and Dunlap for high school readers, with an "editorial advisory board" of two professors and the supervisor of English for the Denver Public Schools; other series titles included *National Velvet* and *Lassie, Come Home.* Wollheim's novel is dedicated "to the memory of Percival Lowell." In his introduction, Wollheim claims that he has "chosen to depict Mars as presented by the late Professor Lowell, of Flagstaff Observatory, and his followers. It has always been the view most exciting to men's minds and it is still upheld by a substantial section of planetologists" (1963, 10). It is doubtful that a plurality of planetary scientists in the 1950s expected to find space-faring civilizations orbiting Mars in their artificial ships, the erstwhile moons Phobos and Deimos.[11] Yet the dissemination of such an image of Mars to impressionable high school students—was *Secret of the Martian Moons* on Dan Quayle's reading list?—is suggestive of the appeal of this romanticized image of Mars and the educational significance attached to it during the space race. It is significant that one of the lessons the "editorial advisory board" wants students to learn is about value systems. One of the Martians tells the sixteen-year-old hero, "I learn from your mind and language that gold apparently occupied the major role in your development of an exchange medium. On Mars it was water rights from the very first, and our original and oldest currency consisted of pledges and permits for water" (165). On Wollheim's Mars, all questions of political economy are recast in materialist terms: the ultimate standard of value—water—redefines politics as the control of the scarcest and most valuable of resources—again water. With these rights and resources secured—and even twenty-first-century Earth has developed "synthetic atoms" that free future history from finite resources—the idealized American values of the 1950s come to the fore as both the means and the end of our destiny as a species: Wollheim's Martians periodically go off on interstellar vacations, closing up their planet for the holidays.

BRADBURY'S *MARTIAN CHRONICLES*

Although Mars dominated the speculative fiction of the mid-twentieth century, science fiction as a genre continued to be a poor relation in literary circles. Even popular successes such as Lewis's *Out of the Silent Planet* were consigned to the margins of children's fiction, and ambitious novels such as Lasswitz's *On Two Planets* remained unknown to English-speaking readers. The first crossover book of the postwar period—really the first book since Wells's *The War of the Worlds* to challenge the divide between literary art and pulp fiction—was Ray Bradbury's *The Martian Chronicles* (1950), which included stories originally published in science-fiction magazines. The original British title, *The Silver Locusts,* suggests something of Bradbury's view of the plague of Earthlings descending on the fourth planet. Set in a future that runs from January 1999 to October 2026, Bradbury's novel—a series of thematically linked stories and vignettes that trace the rise and fall of human colonization of the red planet—reasserts the values of a liberal humanist establishment, even as it offers a pointed critique of McCarthyite America and the consequences of a frontier ideology transposed into the twenty-first century. In an important sense, *The Martian Chronicles* brings the anti-establishment suspicions of the science-fiction underground into a form that intimates that the evils of colonization betray the tenets of American democracy.

In the opening tale, Bradbury's readers encounter a familiar Mars of "fossil sea[s]," deserts, canals, and naturalized gender roles (2). Mars is introduced in terms of an interspecies desire: Mr. K grows jealous of his wife and her prophetic dream of Earthmen landing in rockets. As fans of Burroughs no doubt recognized, the love of the alien woman for the human hero is a tried-and-true means of legitimizing and mystifying the colonial appropriation of native peoples and their resources. It is played out repeatedly in the pulps of the 1950s, such as in John Wyndham's "Dumb Martian," a quasi-feminist revenge fantasy of a Martian wife who finally outwits her abusive terran husband (1952, 49–74). Bradbury recasts the exotic Barsoom of ancient, debilitated cultures into an evocative re-creation of the tensions and anxieties of Earth-based nostalgia. Mrs. K anticipates the violent encounter between her husband and an Earthman in images that merge Mars and the American Midwest: "It was like those days when you heard a thunderstorm coming and there was waiting silence and then the faintest pressure of the atmosphere as the climate blew over the land in shifts and shadows and vapors. And the change pressed at your ears and you were suspended in the waiting time of the coming storm" (11). The "waiting time" of this extended simile suspends the reader between planets—distance and difference evaporate. By employing a second-person mode of address, Bradbury evokes a nostalgic terrestrial experience, recasting Mars in the emotive images of an imagined, temperate past: on a summer's evening "upon the placid and temperate planet Mars . . . boats as delicate as bronze flowers drifted" on "green wine canals" (14). Stylistically, such elegiac passages evoke a world far different from Burroughs's Barsoom or Brackett's Mars. Rather than a dying planet, Bradbury's landscape is reminiscent of an idealized small-town past. This world, however, is already gone—the Martians inhabit a vanished, prelapsarian past that antedates the coming of Earthmen. In turn, nostalgia for a "Martian" past structures the laments for the loss of a balance—ecological, social, and psychological—in a technocratic society.

In this sense, Bradbury presents fables of human presumption and insensitivity that herald the despoiling of Mars: "To introduce a strange, silly bright thing like a stove . . . would be a kind of imported blasphemy. There'd be time for that later; time to throw condensed-milk cans in the proud Martian canals; time for copies of *The New York Times* to blow and caper and rustle across the lone gray Martian sea bottoms; time for banana peels and picnic papers in the fluted, delicate ruins of the old Martian valley towns. Plenty of time for that" (49). Bradbury's vision of the coming pollution of Mars superimposes time frames—social and ecological—of the past and present, even as this superimposition suggests the incompatibilities of terrestrial and Martian realities. Colonized Mars is defined by the garbage of a consumerist culture; the "delicate ruins" have become tourist stopovers. Bradbury, like Brackett, recasts the colonial past of the Americas and the consequences of colonization on indigenous peoples: four of the five Martian cities the Earthmen encounter "have been empty for thousands of years" (50), and, in the fifth, the Martians are dead of chicken pox. They succumb to disease in a manner reminiscent of those American Indian tribes devastated by European viruses, but they exhibit a Zen-like acceptance of their fate. Their resignation suggests a sensitivity to the spiritual and artistic aspects of existence that the colonizers lack.

The hero of "June 2001:—And the Moon Be Still as Bright," Spender, turns renegade and must be hunted down and shot by other members of his landing party. Before he dies, he tries to explain to his commander why he has been killing his fellow Earthmen. Spender realizes that the vanished Martian civilization "blended religion and art and science because . . . science is no more than an investigation of a miracle that we can never explain, and art is an interpretation of that miracle." He displaces onto the Martians his own epiphany: they "never let science crush the aesthetic and the beautiful" (67). If this putative redemptive power of art is the liberal humanist's means of voicing resistance to "wars and censorship and statism and conscription and government control of this and that, of art and science!" (31), Spender is acutely aware that the ancient culture and unspoiled landscape of Mars will fall prey, irrevocably, to ecological devastation and self-destructive militarism: "No matter how we touch Mars, we'll never touch it. And then we'll get mad at it, and . . . rip it up, rip the skin off, and change it to fit ourselves. . . . They'll be flopping their filthy atomic bombs up here, fighting for bases to have wars. Isn't it enough that they've ruined one planet, without ruining another; do they have to foul someone else's manger?" (54, 64–65). Spender's isolation, culminating in murder, allies him with those outcast heroes of postwar science fiction, as in *Gunner Cade* and *The Nemesis from Terra,* who serve as agents of their creators' protest against the forces of greed and aggrandizement: "It's simply me against the whole crooked grinding greedy setup on Earth," Spender recognizes (54). Yet if his developing moral awareness becomes a form of political self-realization, his violence indicates the limits of such protest. Spender does not stop the process of colonization, but the inevitable despoiling of Mars suggests the hubris of a species intent on conquering and fouling another world. In this regard, Spender's recognition that "no matter how we touch Mars, we'll never touch it" signals the pettiness of even environmental degradation and political control. The Mars that cannot be touched by human will or malevolence persists as an image of planetary, even cosmic, destiny that dwarfs humankind's dreams and desires. Spender's death foreshadows the ultimate failure of the colonists to refashion the red planet in the image of midcentury America.

Unlike his contemporaries, then, Bradbury refuses to turn Mars into the setting for a morality play pitting hard-bitten moralists against a corrupt colonial power. Spender's violence is too little, too random, and too late. Even the eco-idealists are deluded. In "December 2001: The Green Morning," Benjamin Driscoll envisions himself as a Martian Johnny Appleseed bringing life to an enervated world: "He would plant trees and grass. That would be his job, to fight against the very thing that would prevent his staying [on Mars]. There lay the old soil, and the plants of it so ancient they had worn themselves out. But what if new forms were introduced? Earth trees, great mimosas and weeping willows and magnolias and magnificent eucalyptus. What then? There was no guessing what mineral wealth hid in the soil, untapped because the old ferns, flowers, bushes, and trees had tired themselves to death" (75). Driscoll wholeheartedly buys into a myth of ecological restoration, a kind of early terraforming scheme that, in the postwar era, equates making the desert bloom with the idealism of projects ranging from a new Israeli state to the Southern California land boom. In a sense, Driscoll's tale anticipates a crucial direction of science fiction in the seventies and eighties because it suggests that the tradition that yokes spiritual and moral renovation to ecological renewal is only another dream. The dying Driscoll hallucinates "great trees, huge trees"—"nourished by alien and magical soil"—and "oxygen, fresh, pure, green, cold oxygen turning the valley into a river delta," but, like the other humans, he has failed "to beat the strange world into a shape that was familiar to the eye, to bludgeon away all the strangeness" (77, 78). If other settlers are more destructive in their efforts to reshape the planet—"replacing old Martian names of water and air and hills" with "new names: IRON TOWN, STEEL TOWN, ALUMINUM CITY, ELECTRIC VILLAGE, CORN TOWN, GRAIN VILLA, DETROIT II, all the mechanical names and metal names from Earth" (102–3)—this would-be Johnny Appleseed succumbs to the lure of terrocentric ecology. If the litany of "mechanical names and metal names" marks the grim logic of intensification, the failure to transform the red planet into a second Earth suggests the failure of an ecology of nostalgia—a return to a romanticized past. As *The Martian Chronicles* ends with the evacuation of Mars and the subsequent destruction of Earth in a nuclear holocaust, the two planets remain separated by the oppositions foregrounded by liberal humanist culture: spirituality versus materialism, art versus mechanistic science, poetry versus machinic rationalism, wisdom versus know-how, tolerance versus fear and racism, and fluid versus linear time. The nuclear holocaust that engulfs Earth, in one sense, is the ultimate inversion of Wells's invasion from outer space.

MARS AT THE MOVIES

Given its hold on the midcentury imagination, it is hardly surprising that Mars and invading Martians dominated movie serials and B movies from the 1930s to the 1960s. Audiences watched updated Martians devastate Los Angeles in George Pal's 1952 version of *The War of the Worlds,* which won an Oscar for its special effects. Earthlings retaliated by sending legions of B movie actors and actresses to invade the red planet. They found a Mars that, almost inevitably, was as hostile as the deserts of Southern California. In a very real sense, the backlot monsters from outer space in movies such as *Angry Red Planet* (1959) are papier-maché realizations of the creatures that populate the novels of Burroughs and Brackett. H. G. Wells's suggestion that evolution on Mars would produce creatures "different beyond the most bizarre imaginings of nightmare" (1975, 177) was transmuted in the 1950s into a license to interpret Martians as embodiments of absolute sociopolitical, racial, and evolutionary otherness. "Martian" became a synonym for all that was strange or inexplicable. A full account of Mars in the movies would be—let's face it—a repetitive study in itself, but these films in crude but often disturbing ways project onto Mars and its inhabitants the anxieties that had characterized the red planet since Lowell: planetary devastation filtered through the lenses of nuclear war, communist infiltration, and crises of masculine identity when confronted by threats—technological and sexual—to its power.

In the 1950s, as Telotte suggests, science-fiction films presented refracted images of a seemingly unthinkable future and therefore could violate, or tiptoe across, the limits of sociopolitical commentary that were policed more rigorously in mainstream media (Telotte 2001).To conclude this chapter, I want to focus on two films: the *Rocketship X-M* (1950), which was produced, directed, and written by Kurt Neumann as a quickie competitor to *Destination Moon*; and *Devil Girl from Mars* (1954), a British knockoff of American invasion films that ratchets up camp to new levels. Filmed in the California desert, *Rocketship X-M* brings to the screen visual analogues for Lowell's vision of a planet far gone toward ecological disaster. Starring Lloyd Bridges, Osa Massen, John Emery, Noah Beery Jr.

and Hugh O'Brian as the crew of humankind's first expedition into space, Neumann's film has pretensions to transcend the generic limitations of bug-eyed monsters from outer space and the evildoers of Flash Gordon serials. Full of scientific absurdities and inconsistencies, Neumann nonetheless produces a "realistic" scenario to explain the technological failure of an experimental vehicle and offers viewers a reasonably coherent vision of 1950s Mars.[12]

Launched from a secret base at White Sands, New Mexico, the rocketship is the brainchild of Carl Extrem (Emery), who captains the first expedition to the Moon. Although the physics are wacky (the rocketship gradually accelerates from 3,000 mph to 25,000 mph after it has attained its 300-mile-high orbit) the attention to detail effectively dramatizes the obsessions of science-fiction fandom since the 1930s—imagining plausible scenarios to depict spaceflight. In a long opening scene devoted to a press conference, Extrem diagrams on a blackboard the trajectory of the rocketship as it uses the Earth's gravitational rotation to boost it toward the Moon and explains cutaway diagrams of the rocketship before and after its first-stage "engine" is jettisoned. For viewers half a century later, the technology is antiquated camp. The rocketship's controls are lifted from 1950s aircraft: an altimeter, a bank and climb gyro, and a massive throttle are shown repeatedly; orbital and fuel consumption calculations are done with pencils and pads by Extrem and Lisa Van Horn (Massen); and navigation in space depends on critical commands such as "Prepare to turn!" and "Turn!" and on everybody-lean-to-the-right special effects. Imagined in the images of postwar technology, spaceflight has the emotional feel of a World War II bombing run; the "realism" of the opening thirty minutes, in some respects, prepares the audience for the unanticipated disasters that strike the mission.

Before the rocketship is sent careening through space toward Mars, the dominant visual images and dialogue center on the Earth as seen from outer space. Ever since Percy Greg sent his hero across the zodiac toward Mars, questions of how the Earth would appear against the blackness of space had fascinated science-fiction novelists and readers. Pages of description and illustrations, for example, fill the first editions of Bogdanov's novels, and at crucial junctures in his defense of the canal thesis, Lowell and his followers had invoked roads, railways, and crop fields as evidence of the linear transformations of the landscape that could be read from space as evidence of terrestrial intelligence. In *Rocketship X-M*, Harry Chamberlain (O'Brian), the navigator and mission astronomer, joins Dr. Van Horn by the porthole as the capsule climbs into the atmosphere and the arc of the Earth's surface becomes visible. This "marvelous sight," he suggests, has to be seen to be believed: "You can study maps and globes and try to visualize, but the actual experience—it's hard to express it." Comments by other crew members underscore the film's effort to imagine what humans will see when they first ascend into space. Major William Corrigan (Beery), an engineer from Texas, drawls his way through the comic relief—"Boy oh boy, ain't she purty?"—before he takes offense at the observation by Chamberlain that Texas is "a mere speck." Floyd Graham (Bridges), who shortly begins flirting with the stereotypical ice-maiden scientist, Van Horn, manages, "How about that? Something no human being has ever seen before." Ibsen this isn't: the sublimity of the imagined visual experience can call forth only the premises of the first spaceflight stories from the pulps: humankind will have to develop new imaginative and aesthetic as well as a scientific vocabularies to appreciate wonders beyond this world.

But at this juncture, Van Horn offers a comment that opens up a richer and more troubling set of allusions that echo throughout the film. "You know," she says, "it's funny. One never thinks of the Earth that way, as a dying planet with nothing to give out but reflected light—a mere moon." There is no antecedent clause for "that way"; no one has mentioned anything about the moonlike quality of Earth. For those of us whose image of the Earth has been filtered through thousands of images delivered by satellite photography, the comparison of the Earth and moon seems counterintuitive: cloud patterns alone suggest an ecological and meteorological complexity that renders the static, cratered satellite alien, dead, and finally of interest only to geologists and astronomers. But with black-and-white images of the Earth and Moon visible in portholes on opposite sides of the capsule, our planet can be imagined only as a back formation: it shines only by the "reflected light" of a prior concept of planetary wholeness. If Van Horn's characterization of the Earth as a "dying planet" seemed merely odd to some viewers in 1950, for the "ten million" who knew their Burroughs, the allusion is clear-cut: visualizing Earth from space can be imagined only from—and through—the fictive vantage point of Mars. The analogy between the two planets structures the rest of the film: after an improper fuel mixture derived from Extrem's incorrect calculations hurls the rocketship past the Moon and sends the crew into a deep sleep, they awaken days later to find themselves approaching what the commander calls "our most congenial planetary neighbor, Mars."

Rocketship X-M takes viewers to a post-Barsoomian Mars; as the planet looms in the porthole, there are splotches that might be canals and oases, but the landscape through which the five adventurers trek is pure California desert, "sand and rock, rock and sand," as Corrigan puts it. Shot in sepia tones, the scenes on Mars depict a nearly lifeless surface, earthlike for explorers in oxygen masks, but stark, inhospitable, and drearier than even Antoniadi had imagined. Lowell's theory of desertification, however, is given a typically 1950s twist, one that is picked up by (among others) Philip K. Dick: the film transforms ecological devastation into the result of a catastrophic nuclear blast. Nature has not been the foe—the Martians have bombed themselves almost out of existence. Yet Neumann's Mars, as shot by Karl Strauss, depicts the red planet as the bleakest of terrestrial deserts; implicitly, the prospect of nuclear devastation is itself a displacement for anxieties about ecological degradation that are an integral part of Martian science fiction. Until the climactic confrontation with the Martians, there is no plant life visible; the sepia tones defamiliarize an environment within a few hours drive of Hollywood. At the same time, the audience can recognize, in an era before computer-generated special effects, that the "devastation" of Mars exists on Earth. The worst imaginings of postnuclear "blast-effect" are evident in the "nameless horror" of the desert. As the camera pans over rock formations and sand, the California desert figures an uninhabitable landscape of the imagination. A complex ecosystem is reduced to a visual analogue for a dead world. The southwestern deserts that helped to structure Lowell's vision of planetary evolution, even in 1950, represent the values and assumptions that have outlived the canals and assumed an existence of their own.

Before the expedition encounters the remnants of Martian civilization, the terms of planetary destruction are laid out explicitly. The sequence of shots of the party traipsing across the desert lacks much of a dramatic payoff; and scientific study is depicted as a series of authoritative pronouncements, usually by Extrem, with little or no evidence to support them. Extrem looks at the bleak, black-and-white wilderness through binoculars and sees "abundant coloration: green, orange, yellow, black," which he promptly identifies from afar as "manganese, copper, nickel, pitchblane. Immense deposits of pitchblane." Corrigan chimes in with, "You figure on stakin' out a claim here?"—a comic version of the speculation about the mineral wealth of the red planet that has persisted since the nineteenth century. Neumann's Mars possesses abundant mineral wealth, the elements necessary to fuel an advanced civilization. Yet the film stakes its claims to a sociopolitical didacticism by foregoing cheap sets and Martians with "pale faces, and pin heads, and fishy eyes," as Corrigan describes them. Distinguishing itself from a tradition of extravagant fantasy, the film extrapolates from the desert conditions on Mars to offer a cautionary parable for moviegoers.

In the desert, the Earthlings see a vast ruin from a distance, dead tree trunks are visible in the sand, and they uncover an ebony head strikingly reminiscent of African art—dignified, alien, austere. On this skimpy data, Extrem again makes definitive pronouncements about both the vanished culture and the conditions on the planet as a whole: "The mind that conceived this" he intones, "must have been of a high order of intelligence, at least the equal of Earth['s], perhaps considerably above ours." Rejecting the possibility that the Martians were wiped out by a meteorite, he concludes that the wasteland stretching out before them "wasn't caused by a meteor. This is definitely blast-effect, coupled with intense heat." The scene then drives home its didactic lesson for Earth.

"Ironic isn't it?" Extrem asks rhetorically. "The mind of man, wherever you encounter it, Earth or Mars, the highest attainments of human intellect always diverted to self-destruction. Perhaps the entire surface of the planet is one vast ruin like this." The tradition of analogy between Earth and "our most congenial planetary neighbor" allows Neumann to take shortcuts. As writer and director, he can presume that a one-to-one correspondence between the two planets will render intelligible the logic—and deep-seated fears—that link scarcity, violence, and ecological devastation. Although night falls and the explorers find a cave in which to sleep, the commander's moralizing continues: "What a lesson here for our world. One blast—thousands of years of civilization wiped out." If this kind of political commentary is standard fare in the pulps, it is rare in post-code Hollywood. This comparison is fraught with a universalizing tendency: Martians become "human," their desires and aggressiveness mirror our own. But in the tradition that stretches from Bogdanov to Bradbury, viewers can never be certain that human beings are not simply being assimilated to and judged by Martian standards. The shock value of a nearly lifeless Mars depends on theatergoers' intuitive recognition that social and planetary evolution conspire to make the red planet a nightmarish anticipation of Earth's future.

The Martians are primitive mutants, bombed back into a stone-age of aggressiveness and fear. Before they appear, Extrem declares that he "should hate to think" that anyone survived the nuclear holocaust. If they had, he claims, "research" on Earth has shown that radiation would produce "mutations, malformations, disfigurements, blindness." The Martians confirm this "research." With one exception the audience never sees their faces, but their heads are bald, their backs scarred and hunched, and they exhibit the basic semiotics of 1950s cavemen: animal skins for clothes, spears for weaponry. After the landing party divides in two to follow them, a female Martian falls down a gulley in front of Corrigan and Extrem. Getting to her feet, she stretches both arms out as if trying to feel her way in darkness. When the Earthmen approach her, the pupils of her eyes glow as if reflecting the horror of the nuclear blast. She screams, and the males in her party begin raining rocks from above on the Earthmen. Corrigan is killed, and Extrem gets a stone-age ax in the back just before he reaches his friends. As their commander lies dying in their arms, Graham offers the characteristic technomodernist response to back-lot primitives: "Murdering savages." But Extrem responds, "No, Floyd. Poor fear-crazed, despairing wretches. Pity them; pity them. You must get back to Earth and tell them what we've found. Maybe this will . . ." And he dies. Rocketship X-M does not make it back to Earth. Graham and Van Horn get to declare their love for each other on the doomed journey back, the wounded Chamberlain offers only muttering about humankind's overreaching its place in the cosmos, and the rocket—relying on incomplete data and the limitations of pen-and-paper calculation—burns on re-entry into the Earth's atmosphere. The cautionary tale is never told, and the fear-crazed wretches of Earth in 1950 are left to confront the unthinkable future of nuclear devastation, a future already realized on Mars.

Martian invasion movies are a staple of 1950s science fiction. To the extent that "Martian" had become a kind of shorthand for "extraterrestrial," even non-Martian invaders fall into one of two generic categories that hark back to Lasswitz and Wells: cautionary messengers (*The Day the Earth Stood Still*) or lobotomizing aliens who lust for the Earth's resources, including human bodies (*Invasion of the Body Snatchers*). Pal's 1953 version of *The War of the Worlds* is the most well known of the Martian invasion films; but the genre also could accommodate less restrained variations on this topos. In *Devil Girl from Mars,* a superintelligent Martian (Patricia Laffan), clad in a miniskirt, thigh-high boots, cape, and helmet, all of black leather, lands in the Scottish Highlands to repair her spacecraft. She spends much of the seventy-seven-minute film symbolically castrating Earthmen: the threat she poses to Earth's resources is explicitly an assault on masculine identity. Directed by David MacDonald and scripted by James Eastwood, *Devil Girl* filters Hollywoodesque special effects through what sometimes seems a touring company production of *Petrified Forest:* stranded locals, an escaped criminal, a threat from beyond.

Interspersed with the melodramatic entrances of Nyah, the Martian, into a Highlands hotel, two couples find or rekindle romances—an American reporter falls for a bad-girl, adulterous fashion model, and an escaped convict, who accidentally killed his wife, reunites with the barmaid he truly loved all along. These subplots provide romantic ballast for the horrors of an interplanetary dominatrix, an avatar of womankind "freed" from masculinist tyranny. "Many of your Earth years ago," Nyah tells her captives, "our women [on Mars] were similar to yours today. Our emancipation took several hundred years and ended in a bitter, devastating war between the sexes—the last war we ever had. . . . After the war of the sexes, women became the rulers of Mars but now the male has fallen into decline. The birth rate is dropping tremendously. Despite our advanced science, we have still found no way of creating life." By this point in the scene, Nyah stands face to face with Miss Prestwick (Hazel Court), the fashion model who has fled to this remote inn to hide from her married lover and is in the process of being reclaimed by the hard-bitten American newspaperman, Michael Carter (Hugh McDermott). This fallen and glamorous woman becomes the spokesperson for a normative sexual politics when she articulates the threat that the Martian poses: "So you've come here for new blood?" she demands. Nyah declares that she "will select some of the strongest men to return with me to Mars" in a spacecraft built entirely out of an "organic," self-regenerating metal. The implication is clear: technology can reproduce itself; the Martian gynocracy cannot. The loss of political power by Martian males has rendered them sexually impotent, and the helplessness of the Earthmen to counter her paralyzing and death rays gives Nyah repeated opportunities to mock them by demonstrating that she "can control power beyond [their] wildest dreams." Her organic technology, however, is itself feminized: it is "unstable," unreliable, and in need of constant maintenance in order for her to get her spaceship to fly again.

The dominant threats in *Devil Girl from Mars,* then, are explicitly sexual. Although Nyah hypnotizes several people who try to interfere with her, the loss of will, individuality, and "freedom" that equate interplanetary invasion with godless communism in American films of the 1950s is replaced by the threat of castration: at one point, Nyah grabs a gun from the reporter's hand, after he has fired several shots at her from point-blank range, and scoffs, "You poor demented humans—to imagine that you can destroy me with your old-fashioned toy." The impotence of the males, however, is itself both an inversion and displacement of traditional identifications in European and American literature of the fertility of the land with the body of the indigenous woman who willingly surrenders herself—and the resources of her people—to white men. This symbolic interlacing of female sexuality and natural resources waiting to be exploited offers a powerful justification for ongoing intensification. At the threatened end of the world, devil women convert masculine sexuality itself into a resource that—at least on Mars—has become increasingly scarce. In this respect, *Devil Girl* recasts the image of Mars as a dying planet to imply that, on a world of self-regenerating, organic metals and predatory females, the male has become both politically extraneous and vitally necessary as breeding stock. Until the escaped murderer nobly sacrifices himself by volunteering to get on the spaceship with Nyah and then somehow destroy its mode of power—"nuclear fission on a negative condensity"—the Martian gynocracy threatens the symbolic identifications that underwrite the endless exploitation of natural resources.

The film, in this regard, is emblematic of its genre—scientific doubletalk, cheap special effects, and howlingly bad dialogue (True love: "You killed her." Escaped convict: "It was an accident." True love: "Was it an accident that you married her instead of me?"). Nevertheless, *Devil Girl from Mars* offers a glimpse into an alternative, nightmarish world where the anesthetizing comforts of ideology have ceased to function. Wells's Martians wanted human blood; Nyah wants human sperm. Even as the dominatrix from Mars fails in her mission and Earth is saved from the horrors of women's "emancipation," the very need to reassert masculine dominance discloses the possibilities that the core values of Western civilization are anxiety-laden modes of self-preservation. Decades after John Carter, the red planet still haunts matinee visions of the fate of the Earth.

SIX

The Missions to Mars: Mariner, Viking, and the Reinvention of a World

RED AND DEAD

In summer 1964, as Mariners 3 and 4 were readied for launch, participants at a NASA-sponsored institute developed worst-case scenarios of the Martian environment, then tried to imagine what sort of hypothetical life-forms could endure such bleak conditions.[1] The conference proceedings, published two years later, testify to the seductive force that Mars still had for would-be exobiologists. The contributors—including Carl Sagan, Joshua Lederberg, and Wolf Vishniac—agreed that life could exist on the planet, and a few participants, notably the ever-optimistic Sagan, speculated that polar bear–sized creatures could roam the cold, arid landscape. Because most of the scientists were skeptical about the possibility of Martian megafauna, they spent much of their time debating what kinds of life-detection experiments to design for future missions. Although they considered the possibility that Martian organisms might be radically different from anything on Earth, they could envision no plausible alternatives to the fundamental chemistry of terrestrial life: carbon, hydrogen, oxygen, and nitrogen. The search for life, it was agreed, would have to assume that Martian biology was carbon-based, required water and sunlight to survive, and would need to metabolize various nutrients as sources of energy. The seasonal changes in the dark areas of "vegetation" seemed to indicate that organisms on Mars would respond to nutrients and waterlike organisms on Earth (Lederberg 1966, 124–37). When Mariner 4 flew by Mars in 1965, however, such assumptions appeared moot: the photographs of a cratered, barren surface suggested a planet more inhospitable to life than even the most pessimistic participants at the NASA institute imagined. Still, given the results from the Mariner 2 mission to Venus, the red planet remained "the best opportunity, in the immediate future, for the study of extraterrestrial life" (Shklovskii and Sagan 1966, 259).[2] If the assumptions about exobiology in the 1950s and early 1960s were based on analogies to terrestrial life, the Mariner and Viking missions provoked wide-ranging reassessments of what scientists knew about Mars and what they could surmise about the origins and evolution of life on Earth. Between 1965 and 1977, analogies between the two planets were scrapped, revived, and redefined, as planetary astronomers tried to make sense of photographs and data that grew more complex and puzzling with each mission.

Even before John Kennedy in 1961 made landing on the moon a national priority, the Space Science Board of the National Academy of Sciences and, after 1959, the new National Aeronautics and Space Administration were discussing an ambitious program of planetary exploration. Despite public skepticism about the feasibility of spaceflight, the aeronautics industry and many scientists were determined that the New Frontier would extend into space (McCurdy 1997, 29, 58). Although exploring the Moon, Venus, and Mars were the primary goals of the proposed programs, Congress, the president, and the media were told repeatedly that a key justification for this national effort was "the similarity of Mars to earth" and the "likelihood of finding life" on its surface (Tatarewicz 1990, 75; Walsh 1964, 1025–27). Yet only 2 percent of astronomers in the United States in the 1950s identified themselves as planetary astronomers, and there was little scientific or institutional preparation to support the ambitious scenarios that were drawn up between 1958 and 1964 to explore Venus and Mars (Ezell and Ezell 1984, 12). Plans developed in 1958 to land on Mars within six years quickly fell prey to technological limitations and budget constraints (Ezell and Ezell 1984, 25–39; Tatarewicz 1990, 46–49).[3] NASA scientists were intent on sending sophisticated scientific payloads throughout the solar system, but their ambitions depended on rockets developed by the military. Military planners, for their part, were enamored of brute force launch vehicles and large payload capacities that eventually could put troops and weapons into orbit. In practice, interservice rivalries among the Army, Navy, and Air Force over boost payloads and launch vehicles complicated the design specifications for proposed civilian spacecraft. Much of NASA's earlier planning had a back-of-the-envelope quality: budgets for planetary exploration were based on guesswork, the technical problems of spaceflight were underestimated, and grand schemes that made for good reports often took precedence over solving basic engineering problems. The naturalized godfather of the

American space program, Werner von Braun, after all, had boilerplated his futuristic vision of planetary exploration in *The Mars Project* in the early 1950s.

These drawing-board military and civilian space programs of the late fifties and early sixties were, in part, the progeny of the science-fiction novels, stories, comic books, and films of the interwar and postwar periods. If Kurd Lasswitz inspired von Braun and Carl Sagan had a map of Barsoom on the wall outside his office door, the conceptual groundwork for NASA's plans had been laid by science-fiction writers such as Arthur C. Clarke. Howard McCurdy notes that in 1961 NASA adopted the seven-point plan first put forth by Clarke in 1951: unmanned satellites to the Moon and the planets, human missions in Earth orbit to develop and test new technologies, human missions to the Moon, and finally human landings on Mars and Venus (McCurdy 1997, 34–35). NASA scientists, no less than readers of Ray Bradbury, were part of a culture than half anticipated a future of humans excavating impossibly ancient ruins on Mars. An animated spaceflight to the red planet was broadcast on ABC television's *Walt Disney Show* in 1957 and featured cameos by scientists, including Wily Ley and von Braun, whose work on getting the first U.S. satellite into orbit was depicted as a patriotic response to the Soviet Union's launch of Sputnik (McCurdy 1997, 48–49). Although the race against the Soviet Union to put humans on the Moon dominated American justifications for its space program in the 1960s, planetary exploration was also defended on patriotic grounds. The Mars Mariner program and its successor, the Voyager project (which lasted, as a series of planning meetings and developmental projects, from 1960 to 1967), were conceived as first steps in a strategy of planetary exploration that would usher in a future worthy of Buck Rogers. In retrospect, though, the drawing-board interplanetary space programs of the United States and Soviet Union in the 1950s and early 1960s read like a wayward genre of science fiction—brainstorming an imagined future that neither nation could engineer or afford. The Russians envisioned massive landers on Mars by the 1960s, and some American plans (shades of von Braun's *Mars Project*) imagined huge space vehicles on their way to colonize Mars by the 1980s.

NASA's early plans for landers descending to the surface of Mars in 1964 quickly ran into difficulties. In 1962, Homer Newell, director of the Office of Space Sciences, wrote that " 'the successful entry of the capsule into the Mars atmosphere hinges upon the restriction to very light, simple instrumentation and direct transmission to earth,' " but the first plans for Martian landers had to be redrawn because they envisioned a craft "too ambitious for its time, representing too large a technological jump" (Tatarewicz 1990, 43, 49). Two years later, new estimates of the surface pressure on Mars forced major changes in spacecraft design and mission planning; because the atmospheric pressure on Mars was only 10 millibars (1/1000 of the atmospheric pressure at sea level on Earth) rather than the 80 mb previously estimated, landers needed to be redesigned for an atmosphere that would offer comparatively little resistance (Tatarewicz 1990, 61; Ezell and Ezell 1984, 96–101; Kaplan, Munch, and Spinrad 1964, 1–15). Such redesigns ran into technical problems and cost overruns that hastened the end of the planning for Voyager. The Mariner project was scaled back to a series of planetary flybys and an orbiting mission to select landing sites; mission launch dates were delayed and then delayed again. No U.S. spacecraft would touch down on the planet's surface until 1976.

The Soviet Union also pursued a grandiose program of interplanetary exploration during the 1960s; unlike NASA, however, Russians scientists launched one Mars mission after another. Their numerous failures indicated just how difficult interplanetary spaceflight would prove to be. At the oppositions of 1960 and 1962, the Soviet Union launched six different spacecraft to Mars; all of them failed, and only one, Mars 1, made it out of Earth orbit before its communication system went dead, sixty million miles into its journey. Although the exact nature of many of these Soviet missions is based on educated guesswork, one of the four craft launched during the Cuban Missile Crisis in the autumn of 1962 apparently carried a lander. Like NASA, the Soviet Union's planetary program in the early 1960s had vast ambitions; unlike their American counterparts, Soviet scientists had the opportunities to demonstrate that their imaginative reach exceeded their technological grasp (Tatarewicz 1990).

The United States, too, opted for two missions during each launch window, the periods that occurred every twenty-six months. Although each NASA spacecraft had different mission profiles, the instrumentation on board each of the twinned Mariner probes (Mariners 3 and 4 in 1964, 6 and 7 in 1969, and 8 and 9 in 1971) was redundant; unlike the Soviet spacecraft, the Mariner modules also could accept some changes to their programming radioed from mission control to the spacecraft to take into account changing circumstances. The success of Mariner 2, which flew by Venus in 1962, demonstrated that interplanetary exploration was feasible. In November 1964, the two countries launched a total of four Martian probes. The Russian spacecraft lost communications on their way to Mars, and NASA's Mariner 3 never made it out of Earth orbit. Its companion craft, Mariner 4, continued on to Mars, and passed within seventy-four hundred miles of the planet on July 14, 1965. As the spacecraft flew by the planet, Bruce Murray, the Caltech geologist who later would become director of the Jet Propulsion Laboratory (1976–1982), recalls that "the expectation of an earthlike Mars was still very high" (1997, 38). At the age of thirteen, watching as the first photographs began to appear one pixilated line at a time on the nightly news, I confess that I half expected to see the outlines of ruined citadels by brackish canals. The nineteen photographs transmitted in sixteen-tone grayscale, however, needed to be decoded by computers and then deciphered by NASA scientists.[4] For nonscientists, the first photographs were a kind of Rorschach test—gray blobs and outlines that fit few preconceived notions of canal-laced deserts or ancient ruins.

The photographs covered only a small fraction of the planet's surface, but what scientists saw led to significant revisions of the consensus views of Mars. NASA's interdisciplinary TV team charged with interpreting the data from Mariner 4 drew five related inferences from these photographs: (1) Mars seemed far more lunar than earthlike, and (2) given the lack of any visible evidence of erosion or weathering, geologists surmised (3) that the planet's cratered surface must be two to three billion years old. Moreover, (4) the state of preservation of craters and other features suggested that conditions had changed little in the eons since the surface was formed. Since no volcanism had reshaped the surface, it was thought that (5) Mars had to be geologically inactive (Murray 1989, 37–45; Raeburn 1998, 57–59). Yet even as the planet was compared to the Moon rather than to Earth, planetary astronomers still described Mars in a vocabulary that harked back to Lowell. Some astronomers studying the photographs, among them Sagan, thought that they could discern "linear topographical features known to geologists as lineaments." Even though these features seemed "much narrower than the canals" and "could not have been detected from Earth by the best telescopes," they conceivably could be related to the infamous waterways (Glasstone 1968, 128; Shklovskii and Sagan 1966, 289). In his NASA-sponsored book on the Mariner 4 mission, Samuel Glasstone lamented that the angle of the sun, the camera's two-hundred-mile-wide frames, and the disappearance of the canals during the Martian winter had worked against determining their "nature" (127). In effect, two vocabularies interpenetrated in such characterizations of the Mariner 4 photographs: an older rhetoric based on terrestrial observations ("canals" and "lineaments") and a vocabulary borrowed from lunar science, one that invoked metaphors of death and stasis. Although many scientists suggested that the photographs "neither demonstrated nor precluded" (Shklovskii and Sagan 1966, 289) the existence of at least primitive life on the surface, Mars looked far less interesting than anyone had imagined.

Paradoxically, the success of the Mariner 4 mission "dealt a sharp blow to the planetary [exploration] program" (Tatarewicz 1990, 74). Even as NASA went ahead with its plans for future missions, the rationale for exploring Mars was much less compelling in late 1965 than it had been a year earlier: the close analogy to Earth and the romantic hope of finding life seemingly were gone. So too were visions of huge budgets and high political priority for planetary exploration. The lander program was pushed back, eventually by a decade, and the escalation of the war in Vietnam forced science teams to scale back the size and cost of their proposals for instrument packages. The initial designs in 1963 for the Voyager missions were grandiose: two landers weighing a total of twenty-five tons, which would cost $4 billion (Cooper 1980, 97).[5] Although the Johnson administration supported a limited program of planetary exploration, missions to Mars lacked the support of vested interests outside of the aerospace industry. In 1967, with the Mariner '69 mission already being scaled down, Congress cut the Mars Voyager and Mariner Mars '71 missions, which had been scheduled to deliver the first lander to the planet's surface. Rechristened Viking, the lander program ran into more budgetary difficulties two years later. NASA's budget was cut by fifteen percent to $3.3 billion, and the Viking mission postponed from 1973 to 1975 (Tatarewicz 1990, 106). With the success of the Apollo program in putting men on the moon ahead of the Russians, NASA had achieved an unambiguous goal that swelled national pride; in contrast, the exploration of Mars lacked the urgency that had motivated scientists and canal seekers before 1965. Amid concerns that manned missions to return to the Moon were being given priority over robotic exploration of the planets, NASA was left with two sets of Mars missions for the launch windows of 1969 and 1971: another flyby to photograph about 10 percent of the surface and study its atmosphere, and an orbiter to map the planet, take temperature readings, and measure ultraviolet radiation (Tatarewicz 1990, 97–103).

The Mariner Mars '69 project succeeded in sending two spacecraft to Mars: Mariner 6, launched February 24, 1969, and Mariner 7 a month later. These spacecraft carried more-sophisticated instrument packages than those aboard Mariner 4, including television cameras that returned two hundred photographs. Using computers with a data rate two thousand times faster than those on Mariner 4, the two craft sent back detailed information about the chemical composition of the planet's atmosphere and surface. Both craft came within two thousand miles of the planet and returned fifty-nine close encounter photographs (those taken within six thousand miles) that allowed NASA to map about one-fifth of the planet's surface (NMR 1: 60–61; Ezell and Ezell 1984, 156–59; Raeburn 1998, 61–62).[6] Their flight paths took Mariners 6 and 7 over the southern highlands: they missed the massive shield volcanoes of the northern hemisphere and the three-thousand-mile-long Valles Marineris. Spectrographic analyses of the surface and the atmosphere confirmed the presence of ice, but temperature readings indicated, somewhat misleadingly, as it turned out, that the polar caps were largely frozen carbon dioxide. Because the atmosphere was composed of carbon dioxide rather than an inert gas such as nitrogen, scientists concluded that Mars lacked the water necessary to sustain an earthlike ecology (NMR 1:56–57). These scientific results did little to modify the view of Mars that had emerged five years earlier. Yet if the canal theory was dead, a Lowellian rhetoric persisted as a spectral presence in official descriptions of the planet. The NASA press release on September 11, 1969, began: "The Mariner 6 and Mariner 7 last month revealed Mars to be heavily cratered, bleak, cold, dry, nearly airless and generally hostile to any Earth-style life forms" (NMR 1:52). The language of this release mixes objective observations about the topography, implicit comparisons to terrocentric norms of temperature and moisture, and emotive responses to preliminary scientific data—"bleak" and "hostile," after all, imply the existence of human observers operating on an everyday scale of experience. In this sense, the NASA author integrates the vocabulary of 1950s science fiction into an interpretation of photographs shot from 2,130 miles above the planet's surface.

Although the Mariner 6 and 7 photographs revealed mostly heavily cratered and chaotic terrain, they prompted some scientists to compare some Martian features to terrestrial landforms (Markley et al. 2001, "missions to mars" 3). Earlier in the century, the analogies between the two planets had focused on global phenomena—the wave of darkening, for example—that could be detected from Earth. In the age of spaceflight, the evocative ecology of Lowell, Antoniadi, and Slipher gave way to discussions of planetary geology—or an emergent *are*ology—that changed each time a spacecraft returned data from Mars and modified, or shredded, previous assumptions about the planet. Analogic reasoning could be applied to specific surface features and an areological history—or bits and pieces of it—extrapolated from what was known about terrestrial geology. After studying the Mariner 6 and 7 close encounter photographs,

Donald Belcher, Joseph Veverka, and Carl Sagan argued that "in the absence of a deductive physical theory of [Martian] topographic characteristics . . . the geological and meteorological interpretation of surface features . . . depends heavily on analogy with known terrestrial features. Not only is such an analogical approach useful," they maintained, "it is difficult to see what alternatives exist" (1971, 241). Writing before the photographic bombshells provided by Mariner 9, these scientists suggested that dark streaks around craters indicated that Martian landforms had been reshaped by wind erosion; the chaotic terrain implied possible volcanic activity in the planet's past. By emphasizing analogies to Earth landforms, Belcher, Veverka, and Sagan countered the impression that Mars was a lunarlike body of comparatively little interest. This analogic strategy was important: even as Apollo astronauts visited and revisited the Moon, television audiences and many scientists became bored: Sagan aptly described the moon as "a static, airless, waterless, black-sky, dead world" that offered few spectacular payoffs (Sagan 1994, 257; see also Shapiro 1999, 21–23). Mars, in contrast, still posed questions that scientists could not answer, and, from two thousand miles away, the Mariner photographs had not disproved the possibility that life might exist on its surface.

By 1970, interplanetary spaceflight had led scientists to extend and reframe questions about exobiology. Since the seventeenth century, solar system astronomy had centered on the likelihood of finding life, intelligent or otherwise, on the planets. With the widespread acceptance of the nebular hypothesis in the nineteenth century, planetary astronomy had developed an analytical vocabulary that shared important assumptions and values with the rhetoric of evolutionary biology. Even as the canals began to vanish from maps of Mars, the fourth planet became a speculative laboratory for questions, philosophical as well as scientific, about the origins, distribution, and evolutionary course of life. Even those in the space science community skeptical about exobiology shared a range of values and assumptions that were shaping studies of the origins of life on Earth in the 1960s. Several scientists, including Nobel laureate Harold Urey and Sagan, were conducting experiments in laboratories to create artificially complex organic molecules, then using energy sparks to generate amino acids, essential building blocks for RNA and DNA (Jakosky 1998, 54–71; Davidson 1999, 150–54; Hartmann 2003, 417–21). Precise questions about the chemical origins of self-replicating molecules were intertwined with speculations about exobiology and with larger, philosophical questions about humankind's place in the cosmos.

In his popular writing, Sagan was explicit about the implications of planetary astronomy: "In every culture," he writes, "the sky and the religious impulse are intertwined." Shifting to the rhetoric of the scientific sublime, he connects his experience to the wonder he imagines ancient stargazers must have felt: "I lie back in an open field and the sky surrounds me. I'm overpowered by its scale. It's so vast and so far away that my own insignificance becomes palpable" (1994, 120). In passages such as this one, Sagan's value as a kind of barometer of scientific opinion becomes apparent: the question of finding life on Mars or elsewhere in the universe is one aspect of the quest for an extraterrestrial standard against which to calibrate the (in)significance of humankind. In this context, the photographs from the first three Mariner missions serve as a limited but powerful data set that, like Lowell's photographs of Mars at the beginning of the twentieth century, provoke dialectic alternatives, both with philosophical and even theological implications: humankind is alone in the universe or humankind is one of many intelligent species, even if none of them inhabits the red planet. Each alternative, in turn, produces another dialectic: on one hand, if humankind is unique, our cosmic isolation can be taken as evidence of either a special creation by a divine being (Alfred Russel Wallace) or of a random process that seems closer to farce (White 1990, 91–112). On the other, a universe teeming with intelligent life could turn human civilization into a sideshow (Lowell, H. G. Wells) or promise a distant future of a federated cosmic identity (Bogdanov, *Star Trek*). Given these alternatives, the Mars missions threatened to occupy an excluded middle, an antiromantic quest for microbes that could not answer the questions that contemplating the night sky traditionally had provoked. On Earth, nature and life are one and the same; on the Moon and on Mars, nature and life seemed by 1970 to be distinct terms. In part, then, justifications for continuing the hunt for life elsewhere in the solar system became a hedge against the disappointment provoked by a nonbiological nature on Mars.

A DYNAMIC HISTORY

Throughout the late 1960s, the possibility of discovering life continued to motivate NASA's plans to land on Mars. The Mariner '71 project, which again included two spacecraft, was designed to "search for 'exo-biological activity or . . . an environment that could support exo-biological activity' " (Ezell and Ezell 1984, 161). The spacecraft were intended to orbit the planet for approximately four months, providing more sustained and detailed data on the atmosphere and surface than had been obtained by earlier flyby missions. More than double the size of Mariners 6 and 7, each of the '71 craft included upgraded scientific instruments to measure surface temperatures and to study the surface and atmosphere, and an improved all-digital television camera (NMR 1: 81–87; Ezell and Ezell 1984, 162–63). Chief among the mission goals was an ambitious effort to photograph almost all of the surface as a prelude to site selection for the Viking landers.

Mariner 9, launched on May 30, 1971, followed three weeks after Mariner 8 had failed shortly after launch. Although some experiments in celestial mechanics that required two spacecraft had to be scrapped, NASA scientists quickly reprogrammed Mariner 9 to assume many of the tasks of its lost twin, including photographing most of the planet's surface. Mariner 9 entered its orbit around Mars in November during a planet-wide dust storm that obscured the planet's surface. Only the white patch that Schiaparelli had dubbed Nix Olympia (snows of Olympus) was visible above the dust. About the same time, two Soviet craft, Mars 2 and Mars 3, began to orbit the planet and both sent landers to the surface: the Mars 2 lander failed; its companion transmitted twenty seconds of data from the surface before it went silent (Burgess 1990, 39).[7] When the dust storm finally began to subside early in 1972, Mariner 9 began returning photographs that stunned scientists and forced them to jettison their view of Mars as a geologically dead world (Raeburn 1998, 64–65).

Over the course of 146 days in orbit, Mariner 9 returned over seven thousand photographs, many at a resolution that allowed identification of features to one kilometer. Astronomers, geologists, and even climatologists now had to explain a complex history of volcanism, flooding, and weathering etched on the planet's surface. Massive shield volcanoes, a three-thousand-mile-long canyon system, and layered polar caps forced the mission science teams to reassess the surface morphology and geological history of Mars. Surprised by "the photographic evidence of [the] huge volcanoes," Murray characterized himself as "a victim of . . . the prejudices that had grown up in my own mind about the planet [so] as to have great difficulty in accepting and understanding the significance of the new data when it arrived" (Bradbury et al. 1973, 15). The dead Mars of Mariners 4, 6, and 7 gave way to a new and often puzzling mosaic of scientific data, theories, and partial explanations. The volcanoes, splosh craters, outflow channels from massive floods, and teardrop-shaped islands in dry channel beds offered strong presumptive evidence that the "deductive" analogies between Earth and Mars offered by Belcher, Veverka, and Sagan

(1971, 241–52) could provide a geohistorical framework for understanding water flow, volcanism, and wind erosion on Mars.[8] Intriguingly, such evidence suggested that the atmosphere had been much thicker and the surface warmer in the past. NASA's final mission report (December 4, 1972) contained no eye-catching phrases about a "bleak" or "inhospitable" world, but simply termed the primary and extended missions "a success" and referred readers to scientific results to be published in specialized journals and JPL technical reports (NMR 1: 108).

The revelation of a more earthlike Mars revived the excitement that had fueled speculation before Mariner 4. In one respect, the differences between past and current conditions on the planet led scientists to ponder questions that could be traced back to Lowell's appropriation of the nebular hypothesis: How had the planet died? What mechanisms could explain the loss of its atmosphere and a greenhouse effect that presumably would have warmed its surface enough for water to flow across it? The data from Mariner 9 produced an unprecedented outpouring of scientific studies, including a NASA publication, *Mars as Viewed by Mariner 9*, that showcased some of the more spectacular of the 7239 photographs.[9] In a chapter in that book entitled "Similarities: Mars, Earth, and Moon," S. E. Dwornik declared that it is "impossible to look through the thousands of images of Mars returned by Mariner 9 without discovering features reminiscent of those on our native planet" (Masursky 1974, 185). He then compared aerial photographs of Martian and terrestrial features, yet refrained from drawing any conclusions. Like other scientists in the early 1970s, Dwornik found himself confronting the limitations of Occam's razor and Newton's rules of reasoning (same effects, same causes) as explanatory strategies: if the geological "similarities" between Mars and Earth have the same "causes," then the history of Mars would need to be rewritten to account for conditions that would have allowed liquid water to remain stable on its surface. Significantly, then, he cautions that it "can be a profound mistake to assume that similar-looking features actually originated and evolved in a like manner" (185). This methodological caution paradoxically cut against the grain of NASA's rationale for continued planetary exploration and hinted at the problems that Mars posed within the historiography of science. Mariner 9 prompted a return to analogical reasoning.

Since the 1950s, NASA had justified its Mars projects by emphasizing the similarity of the two planets, and the first press release for the Viking missions to put landers on the surface (November 1971) maintained that studying Mars would help answer questions about the origin and evolution of terrestrial life and the "changing processes which shape man's Earth environment" (NMR 1: 110). These goals presuppose that similar geological, biological, and climatological effects on Earth and Mars have the same or similar causes. Even as he cautions against the "profound mistake" of assuming that similar features on the two planets are the result of the same processes, then, Dwornik paradoxically justifies the continuing exploration of Mars because further study "will permit us to unfold the lost part of the Earth's history, now largely obliterated by erosion, mountain building, and other processes" (185). In such statements, he voices a consensus view of Mars that emerged in the 1970s: the red planet preserves a geological history of Earth's past, as though it were the control in a vast experiment; therefore scientists can study its features to understand "universal" processes of planetary change. Dwornik thus revives and updates many of the assumptions that had guided the Air Force funded studies of planetary atmospheres in the 1940s and 1950s. Studying Mars will allow scientists to predict the future course of atmospheric evolution on Earth: "A full understanding of the [Martian] past," he maintains, "is a reliable way to accurate prediction of the [Earth's] future" (185). In reiterating the values and assumptions of predictive science, Dwornik articulates a reborn rationale for future missions to Mars—one based on a dynamic conception of climatology and the quest to uncover the lost origins of Earth's life-sustaining atmosphere.

The photographs from Mariner 9 invited narratives that worked backward from contemporary conditions on Mars to a history, almost four billion years earlier, when its surface features were formed and reshaped. By reconceiving Martian geological history as a dynamic record of meteorite bombardments, volcanic eruptions, and massive floods, scientists tried to reconstruct the sequence of events that shaped the surface. To some planetary geologists, the Martian surface preserved a record of ancient catastrophism, a view that had implications for exobiology: few, if any, life-forms would have survived such a tumultuous early history. Biological evolution presumably would require long periods of relative climactic and geological stability. In a speculative article published before the Mariner 9 mission, Sagan suggested such as a possibility: he theorized that the precession (the wobbling of Mars on its axis) might produce "a microenvironment, not in space, but in time" (1971, 511): for relatively brief periods of a few thousand years, the polar caps would melt, thickening a carbon dioxide-rich atmosphere and allowing liquid water to exist in some niches on the surface. After the Mariner 9 photographs, Sagan and his collaborators, Owen Toon and Peter Gierasch, argued for a connection between the Martian landforms—notably the channels that indicated the presence of water—and changes in the planet's obliquity, the angle of rotational deviation from a vertical axis. In speculating about the planet's geological history, they suggested that a long-term "reversible climactic instability" could be driven by precessional variations in the planet's obliquity (in fifty-thousand-year cycles, the angle would change from 0 to 25 degrees).[10] If the polar caps sublimed into the atmosphere and thickened it sufficiently, the warmer conditions could account for "the channels and other features suggestive of extensive liquid water" (Sagan, Toon, and Gierasch 1973, 1048). Such speculation had been a trademark of Sagan's work since he had suggested the possibility of terraforming Venus in 1961. In the years leading up to the Viking launches, Sagan became the most recognizable advocate, at least in the mass media, of a rigorous exploration program to determine if life still existed on Mars.

While most of his peers shied away from speculating in print about exobiology, Sagan employed the rhetoric of scientific skepticism to hold open the possibility that life might exist on Mars: "*a priori* estimates of the probability of life," he maintained, "are not possible in our present state of ignorance, and such probabilities are merely skepticism indices, calibrating the frame of mind of the speaker" (1971, 511). Many NASA geologists, however, were dubious about the possibility of finding any Martian microbes and critical of NASA's investment in biology experiments when scientists still lacked a theoretical context in which to make sense of the data returned by Mariner 9. Yet NASA's publicity for the Viking mission focused on the search for extraterrestrial life, and Sagan became identified as the point man for this interplanetary adventure. His appearances on television talk shows and his strategy of offering provocative gambits in order to stimulate public interest in scientific inquiry bred envy as well as reluctant admiration for his efforts at self-promotion (Cooper 1980, 64–74; Davidson 1999, 276–82). Well before he became a media celebrity on the PBS series *Cosmos* in 1980, Sagan was, in the words of his colleague and friendly intellectual combatant, Bruce Murray, "the single most important person for the search for life on Mars" (1997, 36). In the buildup to the Viking launch, Sagan raised questions that other scientists had ignored or consigned to the margins of inquiry. Thomas Mutch, the leader of the Viking imaging team, called him "a great provocateur" on the science team, pushing for scanning cameras to detect motion, a flashlight to catch any nocturnal biological activity, and even bait for any wandering Martian macrobes (quoted in Cooper 1980, 76).

Sagan defended his insistence on such "'ideas at the boundaries of the plausible'" as a way to force his peers to design experiments that could rule out these (bare) possibilities (quoted in Cooper 1980, 80). In this respect, he voiced what Cooper calls the "collective unconscious" (12) of those scientists who worked on the Mariner and Viking missions, promoting the role of speculation in planetary science and suggesting strategies to uncover or rule out life on Mars.

THE SEARCH FOR LIFE

The problems of designing life-detection experiments that would work millions of miles from mission control took on new urgency after the photographs returned by Mariner 9. Although efforts had been underway since the 1960s to engineer such experiments, evidence of a past epoch during which water had flowed on the Martian surface rekindled debates about the best means to detect or rule out the presence of life on the red planet. No consensus even existed about the value of searching for life on Mars. Although a December 1968 memo by Gerald A. Soffen, head of the Viking science team, stated that the "'*absolute prerequisite*' for membership on the science team was '*complementarity to other members of the team*'" (cited in Ezell and Ezell 1984, 206), scientific and philosophical differences created tensions among geologists, mission specialists, and biologists. Biologists and biochemists hoped to resolve what NASA billed as fundamental questions about the origin and evolution of life, but Murray spoke for many geologists when he suggested, years later, that the "hugely expensive search for Martian microbes" was "scientifically unjustified" (1989, 68). Different conceptions of the science mission were played out in the protracted process for selecting the sites for the Viking landers. Flight engineers sought the safest possible locations for touchdowns; geologists wanted landing sites that showed evidence of water erosion or volcanic depositions; and biologists lobbied for the most likely sites for the life-detection experiments, ancient floodplains or even the polar regions. Such differences continued to be negotiated during the first weeks of Viking 1's orbits in June 1976 as the landing team tried to balance safety concerns and scientific considerations (Ezell and Ezell 1984, 278–86).

The selection process for the life-detection experiments culled dozens of proposals to three that could be rendered simple and economical enough to be included on the Viking landers. In practice, the life-detection and organic chemistry experiments were more difficult to design and implement than the inorganic chemistry experiment intended to detect characteristic chemical signatures when a soil sample was bombarded with X rays (Cooper 1980, 82; NMR 1: 136–37). Sagan suggested that a telltale indication of life would be a thermodynamic disequilibrium: any major departure from the most likely chemical or physical situation could be attributed to biology (Cooper 1980, 73). But discerning such a disequilibrium would be difficult. The life detection equipment had to fit into one cubic foot, the same space allotted to other instrument packages. Chronic funding problems dictated that the Viking life experiments would sacrifice redundancy in order to include three separate experiments, each with its own assumptions about Martian life and the best strategies for discovering it. Moreover, throughout the decade-long process of designing and testing life-detection devices, scientists had to guard, in part, against their own optimism and expectations (Wilford 1990, 95–102). As it turned out, the ambiguous data produced by the three experiments forced members of the Viking biology team to reassess their assumptions about life, the Martian environment, and experimental design.

From the start, these experiments faced practical and theoretical problems. The pyrolytic release (PR), the labeled release (LR), and the gas exchange (GEX) experiments, as well as the gas chromatograph mass spectrometer (GCMS) designed to study organic chemistry, were at the time the most sophisticated instruments ever sent into space; yet their capabilities were limited, and Viking scientists split along disciplinary lines in assessing their prospects for returning definitive answers about life on Mars. Although the biology package was only one of thirteen scientific experiments on the Viking landers, these three experiments were publicized as the centerpiece of the mission; geologists such as Murray were left to complain that the mission "was not instrumented to reveal even elementary mineralogical facts about" the surface (1989, 68). Without a basic knowledge of the chemical and physical properties of the Martian surface, the biology team had to define the basic chemistry of life and then decide what experimental protocols were most likely to detect or rule out its existence. The principal investigators, Vince Oyama (GEX), Norman Horowitz (PR), and Gilbert Levin (LR), acknowledged that they were operating from terrestrial expectations about how alien microbes might respond to water, nutrients, heat, and light. The designs of their experiments encoded different views about how such hypothesized life might metabolize nutrients and respond to environmental stimuli.

To be detected by the Viking landers, Martian microbes would have to be carbon-based organisms that existed close to the surface, though conceivably they might be protected by dust, rocks, or exotic evolutionary adaptations against ultraviolet radiation. Although scientists speculated about exotic life-forms based on silicon or other elements, the ability of carbon to combine easily with other atoms to form complex molecules, and the abundance of carbon on Mars, and in the solar system more generally, made it likely that any conceivable exobiology would be carbon based. Even on a desiccated planet, Oyama and Levin assumed that Martian microbes would require water: water would be incorporated within cell structures, serve as a medium for transportation and the recruitment of other genetic populations, and play an essential role in reproduction. Because all Earth-based amino acids and nucleotides contain nitrogen, most scientists believed as well that Martian microbes would be able somehow to fix or extract nitrogen from their environment, even though this element existed only in trace amounts in the planet's atmosphere. Taken together, then, the chemical requirements for terrestrial life made Mars seem an unlikely site for a robust biosphere. Any Martian organisms would have had to evolve unusual mechanisms to survive on its forbidding surface. For the teams assembling the life-detection experiments in the early 1970s, both the Martian environment and the hypothesized microbes were hypothetical composites, defined by unknown or uncertain variables. Prior to the launch, Viking biologists were divided or uncertain about what they might find. The leader of the biology team, Harold P. Klein, troubled by the lack of water and the problem of ultraviolet radiation that would break down organic compounds, put the odds of finding life on Mars at one in fifty. Oyama, the most optimistic of the team, put the odds at fifty-fifty, but ironically later became the first convert to a nonbiological explanation of the results of his experiment (Cooper 1980, 93). Although the word "ecology" seldom appeared in the press briefings for the Viking missions or in subsequent scientific discussions, the debates that emerged before, during, and after the life-detection experiments concerned competing models and interpretations of the Martian environment, including water content, temperature, energy sources, and soil chemistry. In turn, different expectations about the response of alien life (if it existed) to earthly nutrients colored the ways in which the results of the experiments were debated and explained.

All three life-detection experiments involved heating soil samples and measuring chemical reactions in order to look for characteristic signatures of metabolic processes. To sample material from a few inches below the surface, the Viking landers had scoops attached to robotic arms that could dig trenches and deposit sufficient amounts of soil in the test chambers to run several experiments over a period of time. After receiving the soil samples, each of the test chambers in the three experiments were sealed and swept by an inert gas (helium) to reduce the risks of contamination and false positives. The experiments could (and would) be run several times, and all had different modes to test the samples under different conditions of heat, light, and moisture. They were, in short, classic Baconian efforts to induce Mars to render up its secrets: an alien "nature" would be scooped, moistened, heated, and retested to produce signs of possible life.

As technologies of inquiry, the Viking life-detection experiments were painstakingly designed to detect life, but their designs encoded specific assumptions and values about the data that they were built to generate and the conditions under which they were intended to operate. Despite their ultimate disagreements about the interpretation of the various experimental trials, all members of the Viking biology team shared basic assumptions about the relationship between technologies and experimental design: theories can be tested objectively because correctly designed instruments and experiments will produce reliable, unambiguous data. But no experiments, least of all ones conducted by radio signals from millions of miles away, employ passive devices that test and record scientific facts; all technologies, samples, and operators, as Bruno Latour (1987) argues, are "actants" that participate in the construction of such facts—intervening in nature, shaping what is and what can be understood of the Martian environment. "Construction," I want to emphasize again, does not mean the subjective or ideological imposition of predetermined interpretations on agreed-upon data but the ongoing reinterpretation of that data within complex social, cultural, and scientific matrixes (Lewontin 1991, 140–53). As the history of the life-detection experiments demonstrates, debates about the interpretation of such data depend not only on different conceptions of life and chemistry but on different institutional deployments of a scientific method that itself is subject to different constructions and interpretations.

The gas exchange experiment was designed by Oyama (a chemist at NASA-Ames) to measure either the production or uptake of carbon dioxide, nitrogen, methane, hydrogen, and oxygen. The experiment could be conducted in two modes: assuming that nutrients were available in the Martian soil, only water would be added to stimulate metabolic activity of whatever organisms existed; if the water produced no response, a mixture of nineteen amino acids, vitamins, other organic compounds, and inorganic salts would be added. After an initial incubation period of seven days, an additional nutrient solution could be added to compare the reactions over time. Gas analyses were conducted on samples of the atmosphere above the soil at the beginning of each incubation and after one, two, four, eight, and twelve days. The sample gas was placed in a stream of helium flowing through a coiled chromatograph column and into a thermal conductivity detector. The system was designed to be extremely sensitive, measuring changes in the concentration of various gases to one billionth of a molecule (NMR 1: 133).

The labeled release experiment was designed by Levin, president of Biospherics, a Maryland-based environmental testing firm. Levin was a nonacademic, a pollution-control engineer in California who had designed an automated system to detect water-borne microbes by means of their metabolic release of carbon (Cooper 1980, 96). His Viking experiment sought to detect metabolic activity in a soil sample moistened with a water-based solution of simple organic compounds (a mixture of formate, glycine, lactate, alanine, and glycolic acid). Martian organisms, Levin believed, would break down these compounds into carbon dioxide, and this breakdown could be detected by the release of gases from a mixture of radioactive compounds (carbon 14) supplied during incubation. The atmosphere above the soil sample was monitored throughout the incubation period (eleven days initially) by a radioactivity detector. The release of radioactive carbon dioxide (or other radioactive gases) presumably would indicate that the nutrients had been metabolized by Martian organisms (NMR 1: 132–33; Cooper 1980, 128–29).

The pyrolytic release experiment was designed by Horowitz, a geneticist and biologist at Caltech, to measure the assimilation of atmospheric carbon. Horowitz's experiment differed fundamentally from the other two. It added no nutrients or water to the soil sample, and tried to preserve what Horowitz understood to be the arid conditions on the planet's surface. Unlike Oyama's and Levin's experiments, PR looked for the kinds of metabolic processes associated with plant life on Earth. The experiment incubated soil in a Martian atmosphere with radioactive carbon dioxide added. Then, by heating the soil at high temperatures to "crack" organic compounds (pyrolysis), it sought to determine whether radioactive carbon had been fixed into these compounds. Water vapor could be (and ultimately was) introduced by direct command from the ground. Then the labeled mixture of carbon dioxide and carbon monoxide (CO_2/CO) was added from a gas reservoir and an xenon arc lamp was turned on for the incubation period, initially set at five days, to encourage any plantlike microorganisms to metabolize the radioactive carbon. After a double process of heating, the purged gases passed through an organic vapor trap that separated the compounds by size, retaining organic compounds and fragments but not carbon dioxide or carbon monoxide. The test cell was then moved from the pyrolysis station, and the trapped organic compounds were released from the vapor trap by heating it to 700 degrees centigrade (1290 Fahrenheit), which simultaneously oxidized them to carbon dioxide. These were flushed into the detector. A radioactive peak would be considered an indicator of biological activity in the original soil sample (Cooper 1980, 146–48; NMR 1: 131–32).

Horowitz was pessimistic about the possibility of finding evidence of life on Mars. For the canceled Voyager mission, he and Levin had collaborated on a life detection device called Gulliver, which would roll out long strings coated with an adhesive to pick up microbes that would then be "fed" nutrients, and the labeled, radioactive carbon dioxide they gave off measured (Wilford 1990, 95). But Horowitz was affected profoundly by the results of the Mariner 4 mission which indicated that the atmospheric pressure on Mars was too low to sustain liquid water on the surface (Cooper 1980, 99–100). In the 1960s and 1970s, Horowitz had studied the most extreme, Mars-like environment that he could find—the dry valleys of Antarctica, the driest, coldest environments on Earth.[11] In these deserts, Horowitz and his collaborators conducted a version of his carbon assimilation experiment and found no measurable evidence of metabolic activity; he concluded that the valleys were lifeless and the soil sterile (Horowitz, Cameron, and Hubbard 1972, 242–45). Convinced that his interpretation of the results from Antarctica were correct, he insisted on experimental protocols that would preserve, inside the PR chamber, conditions that obtained on the surface of Mars in order to "maximize the chance of survival of indigenous life and make reasonably sure that whatever events we may detect are not artifactual" (Horowitz, Hobby, and Hubbard 1977, 4659). His experiment raised the bar for detecting any Martian microbes because its "reasonable approximation of local conditions" effectively removed water from the simulated environment. Horowitz would be searching for an exotic biology that seemed almost a contradiction in terms: life without measurable water.

Horowitz's differences with Oyama and Levin, then, existed long before disputes about the interpretation of the Viking data. The MIT scientist conceded that "in its combination of generality and sensitivity, [Levin's experiment] came close to being an ideal life-sensing device for an aqueous planet" (1986, 134), but he doubted the usefulness of such an experiment given the conditions on Mars. In his account of the Viking mission, Horowitz was brutal about what he perceived as the shortcomings of both Levin's and Oyama's experiments: they were "designed more for Lowellian or pre-Mariner 4 Mars than for the real Mars, and serious questions were raised as to the wisdom of including them in the Viking payload" (131). Horowitz seems to have been the one who raised these questions because he felt that the other principal investigators were fixated on terrestrial biochemistry. In this regard, he held fast to a familiar, Earth-based, conception of ecology, one based on his genetic work with terrestrial organisms. Without liquid water, Mars lacked a context for life—and, not surprisingly, he and his collaborators maintained throughout the months of testing that "the absence of liquid water on the Martian surface excludes any possibility of terrestrial types of organisms" (Horowitz, Hobby, and Hubbard 1977, 4659). In contrast, Levin suggested that because its low density allowed the atmosphere to hold small amounts of water vapor, the relative humidity on Mars often approached 100 percent a few inches above the surface, and he reasoned that Martian organisms might have evolved mechanisms to cope with the inhospitable environment. Given their radically different conceptions of planetary ecology, it would have been surprising if Horowitz and Levin did not offer competing interpretations of the Viking data.

In addition to the biology experiments, molecular analyses of the Martian soil were conducted by the GCMS experiment, designed to look for organic compounds. Many of the Viking scientists felt beforehand that this experiment might provide definitive answers to questions about life on Mars. But the GCMS proved the most difficult and expensive of the lander experiments to engineer. As deadlines loomed for launch, the price of testing and redesigning the GCMS instrument package swelled from $14 million to $59 million (Ezell and Ezell 1984, 222–42). The initial experimental design defined two "extremes" that might result: organic compounds produced by nonbiological processes (thermal, radiation-induced, or photochemical) and those "produced by living systems," organic compounds whose chemistry, distribution, and structure might indicate the presence of living organisms (NMR 1: 134). The GCMS tested samples of the Martian atmosphere before and after the removal of carbon dioxide. Surface samples were heated to 200 degrees centigrade (392 Fahrenheit), while carbon dioxide (labeled with carbon 13, a nonradioactive carbon isotope) was used to sweep the vaporized material into a gas chromatograph column. A mass spectrometer provided a complete mass spectrum every ten seconds for the eighty-four minute duration of the experiment. Materials volatile at 200 degrees centigrade were measured; the same sample was then reheated to 500 degrees centigrade (932 degrees Fahrenheit) to study less-volatile materials and to crack those substances not volatile enough to evaporate. The molecular analysis team, led by Klaus Biemann of MIT, was particularly interested in the long chains of carbon atoms characteristic of complex organic compounds that would be needed to form the nucleotides, and the long chains of nucleotides that would provide presumptive evidence of biology (Biemann et al. 1976, 72–76; Biemann et al. 1977, 4641–62). Although polymer chains of nucleic acid, which can duplicate themselves and transmit genetic information, likely would break down under the ultraviolet radiation that bombards the Martian surface, Biemann guessed that there was a 50 percent chance that his team would find complex organic compounds (Cooper 1980, 83). The difficulty was that the GCMS could not detect any long polymer chains directly because the intense heating of the sample would break them down; organic compounds therefore would have to be deduced from the spectra of their components. Although the GCMS could detect the compounds it was designed to find down to a few parts per billion, it could not identify organic molecules within organisms, which would constitute an "infinitesimal" fraction of the "total bulk" of the soil sample (Cooper 1980, 168). The experiment, then, depended on finding organic compounds that would result from the death of millions of microorganisms, and active microorganisms would need to number at least a million per gram of soil to be detected. In short, GCMS was designed to detect a robust biosphere, not a strange, sparse population of bizarre microorganisms that, as Levin was to suggest, might cannibalize their dead. Such hypothesized extremophiles were among the possibilities suggested during the late summer and autumn of 1976 as the Viking landers began returning data.

THE VIKING MISSIONS

The two Viking missions were launched in the late summer of 1975 and reached the planet nine months later. Both spacecraft carried high-resolution color cameras, and the photographs the first orbiter returned revealed much rougher terrain at potential landing sites than scientists had anticipated. As Viking 1 orbited the planet, the site selection team, charged with getting the landers safely on the planet's surface, scrambled to find a suitable landing area. Many of its members were surprised by the resolution of the photographs. NASA scientists had not realized that during the Mariner 9 mission the dust had never settled after the storms of late 1971. "Instead of a blurred surface, they now saw a fantastic array of geological detail" (Ezell and Ezell 1984, 331), much of it frightening for the dangers it posed for the landers. The original landing site had to be abandoned, and alternatives considered, debated, and rejected (338–49). The eventual landing sites, Chryse and Utopia Planitia, posed few hazards but were far from the most interesting geological features: the polar regions, water-carved channels, craters, or volcanoes. Both craft touched down safely—then, and in retrospect, significant engineering achievements. The first photographs from the surface of Mars were sent to Earth by the Viking 1 lander on July 20, 1976. They revealed no obvious signs of life: no plants, no lichen-like growth on the rocks, no macrobes sniffing the lander. The rock-strewn plains photographed by the two landers, though, seemed hauntingly earthlike. The cameras could not register the thin atmosphere, the cold, the ultraviolet radiation, or the scant traces of water vapor and oxygen recorded by other scientific instruments—only a landscape reminiscent of terrestrial deserts. These evocative photographs became the bases for new representations of Mars in both astronomical literature and in science fiction: the alien, lunarlike landscape imagined by scientists a decade earlier was transformed into a complex topography, one that would be traversed by astronauts and colonists countless times in the future histories of novelists such as William K. Hartmann, Ben Bova, Kevin Anderson, Kim Stanley Robinson, and Geoffrey Landis.

From the start, the life-detection experiments produced unexpected results. The first data from the labeled release experiment were received on July 30, 1976, and indicated a strong positive response; the radioactive count surged to nine times the background level of radiation, a stronger reaction than those usually obtained from biologically robust soils on Earth. Levin originally had been less optimistic than Oyama about finding life on Mars, but now suggested that life was precisely what his experiment had detected. Other scientists were more cautious, even when Oyama's experiment returned surprising results from its first reading: less than three hours after the first incubation, oxygen levels were twenty times the level in the Martian atmosphere (Cooper 1980, 127). The intensity of this reaction led Oyama and two members of the molecular analysis team, Biemann and John Oro, to suggest that the GEX was registering not biological but chemical reactions involving either peroxides or superoxides; both compounds are unstable because they contain "excess" oxygen atoms that can be released readily through contact with small amounts of water. The ultraviolet radiation bombarding the surface of

Mars, they reasoned, would strip water (H_2O) of one of its hydrogen atoms, leaving a free radical (OH), a fundamental constituent in forming other molecules. Such free radicals could react with identical OH radicals to form H_2O_2, hydrogen peroxide. The reddish color of the dust shown in the Viking lander photographs strongly indicated the presence of red iron oxides, such as hematite; an unearthly combination of hydrogen peroxide and surface oxidants would explain both Oyama's oxygen peak and yet another surprising result—the GCMS experiment found no organic molecules, presumably because they were broken down by ultraviolet radiation. A strong oxidant, Oyama argued, would be lethal to any known microorganisms. Levin was skeptical of this explanation, but Horowitz argued that Oyama's oxidant also would explain the LR results: a release of oxygen would combine with the labeled carbon in the organic nutrient and create the carbon dioxide that Levin's device detected. In response, Levin noted that the results of all three experiments were atypical of chemical reactions: a strong oxidant would have continued to react with the carbon and registered much higher counts, and not leveled off as the LR data had done.

Because Horowitz's experiment took readings only after the incubation period was over, his was the last of the four experiments to receive results. The initial data from the pyrolytic release experiment registered a peak well above the baseline for biological activity. Second and third trials also registered readings that seemed consistent with the standards for biology that had been defined before the mission. After receiving these positive results, the PR researchers decided to use auxiliary heaters for three hours to sterilize the soil sample before the experiment was repeated. Horowitz and his team believed that only a significant difference in gas composition between the experiment and the sterilized control sample would be an indication of life. But the results of this trial were again open to differing interpretations: "It appeared that we had found a heat-sensitive synthesis of organic matter on Mars," Horowitz later wrote, "but the fact that 12 percent of the reaction survived the high temperature argued against a biological interpretation" (1986, 138). Two camps, in short, had offered competing interpretations of the data and, in effect, competing versions of "correct" scientific reasoning. Oyama, Horowitz, and Biemann invoked Occam's razor to argue that chemical explanations were simpler and more elegant and therefore had to be preferred to unknown biological processes; Levin, with lukewarm support from Sagan, argued that nonbiological explanations of the data relied on unproved suppositions and necessitated changing the agreed-on experimental protocols to rule out the commonsense interpretation: positive readings meant biological activity. While most scientists eventually came to agree with Horowitz, the Viking press briefings and reports in the media during the late summer and autumn of 1976 offered the public a day-by-day, week-by-week view of the controversy as more trials were run, new protocols developed and implemented for each of the three experiments, and new and differing interpretations of the evidence debated.[12]

The original samples were retested, and the experimental protocols varied: the samples were incubated at different temperatures, given different nutrients, heated, and exposed to light. A second injection of nutrients in Oyama's experiment failed to produce a positive reaction, which he interpreted as an indication that the chemical oxidant had been exhausted. Sagan and Levin, however, suggested that the water vapor in the first injection may have killed off the organisms, and that directly wetting the soil sample was unlikely to have produced a positive response. Ironically, this critique harked back to Horowitz's original dim view of GEX: Oyama was testing for hypothetical organisms in a humid, earthlike environment. A month after the initial results, Levin and Horowitz reran their experiments at Chryse, but with samples heated to 160 degrees centigrade to kill off any putative biological organisms. Both came back negative, a response consistent with a biological interpretation of the data. But members of the biology team who resisted a biological interpretation pointed out that a peroxide or superoxide might be chemically altered by heating, producing another false positive that would mimic biological signals. In articulating the dominant view that emerged among the Viking mission specialists, Oyama and Bonnie Berdahl argued that the release of oxygen could be "ascribed to superoxides in the Martian surface material" (1977, 4675; Raeburn 1998, 75, 90–92). Levin, however, was unconvinced. A third trial produced another presumptive biosignature; meanwhile, at the Viking 2 landing site at Utopia, a second set of tests gave readings that Levin argued confirmed a biological explanation of the data: Utopia was richer in water vapor than Chryse, and the peak was 25 percent higher (Cooper 1980, 188).

At Utopia in the beginning of October, Levin heated another sample to 50 degrees centigrade (122 degrees Fahrenheit) on the assumption that such a gentle heating would not affect chemical reactions involving peroxides, if indeed they were responsible for the release of the labeled, radioactive carbon, but it might well destroy vulnerable Martian organisms (Cooper 1980, 196–97). This procedure greatly reduced the release of carbon dioxide, although unexpected waves seemed to fit neither the hypothesized chemical nor biological theories. Levin and his colleague Patricia Ann Straat maintained that such a gentle heating severely constrained the peroxide theory because no known oxidant would be destroyed at this temperature. A soil sample taken from under a rock, nudged aside by the lander's arm, matched LR's initial reading, again a response that Levin interpreted as consistent with a biological explanation. " 'No one,' " he said, " 'is championing the idea that there is life on Mars, including me. . . . But [the scientists] saying that there is no life on Mars are doing poor science—they are not looking at the data. . . . If you look at our data you cannot say there is no life on Mars' " (quoted in Cooper 1980, 189). This comment is indicative of the testiness that had crept into the biology team's discussions. Questioning the competence of one's colleagues made matters almost as contentious as the canal debate had been earlier in the century.

The lack of a strong positive response to the second injection of nutrients led most scientists to assume that the soils simply reabsorbed the gas when moistened (Goldsmith 1997, 188; Bergreen 2000, 193–97). Levin, however, suggested that the slight reabsorption of gas might indicate that the Martian microorganisms had died; for comparison, he tested the LR with Earth lichen. Lichen responded vigorously to the first injection of nutrients, but then the reaction died away within twenty four hours. This response suggested the possibility that too much water could have been a factor in the Viking and terrestrial control experiments. As other scientists voiced their support for Horowitz's interpretation, Levin continued to maintain that the positive response of the first injection of the nutrient fell within the range of responses from terrestrial samples in extreme environments such as Antarctica.

Yet Levin's incubations never showed an exponential rise in the reaction—a result that most members of the biology team expected if Martian microbes were ingesting the nutrients and multiplying. Meanwhile, throughout the first three cycles of the PR experiment, "Horowitz's results seemed to be taking him farther and farther in a direction he felt was wrong" (Cooper 1980, 178)—toward a biological explanation. Horowitz tried to explain these results by arguing that small amounts of water had released oxygen, which either impeded the synthesis of organic compounds or destroyed them after they were formed (Cooper 1980, 210). Another test run in December, however, put that option to rest: having first moistened and then heated a sample, Horowitz got another positive result that seemingly ruled out both water and oxidants as the active agents in his chemical reactions (Cooper 1980, 218). Using maghemite (a magnetic form of hematite), Oyama reproduced in his laboratory a response that mimicked Levin's labeled release data. A final cycle of experiments with the sample taken four months earlier at Chryse produced a degraded reaction in Levin's experiment; he and Straat argued that this confirmed a biological explanation, while Klein suggested peroxides might also degrade at the 15 degree centigrade temperature of the chamber in which the soil had been kept (Cooper 1980, 220–27). Oyama then improvised a test for oxidizing agents, first heating a soil sample to 145 degrees centigrade (to break down peroxides), then humidifying the soil to destroy superoxides. An oxygen peak, in his mind, confirmed the biology team's assessment: an exotic mix of oxidizing agents on the Martian surface had produced the various peaks (Oyama and Berhdahl 1977, 4669–76; Cooper 1980, 228–29). Even though Horowitz's experiment had produced consistently positive results of varying strengths, the fact that they had been unaffected by light, heat, or water was seen as an argument in favor of a chemical explanation (Klein 1977, 4677–80; Klein 1979, 1655–62). Additional laboratory experiments suggested to Horowitz that a combination of hematite, maghemite, and magnetite could produce Viking-like results.

All three principal investigators for the life-detection experiments presented themselves as operating rigorously within the standards of predictive science: experimental designs, protocols, and operations were checked and rechecked. Horowitz, Oyama, and Levin ran repeated tests to rule out the possibilities that technical malfunctions or contamination had caused the peaks; second and then third trials of the experiments were conducted at both landing sites. The debates became increasingly heated, with Levin maintaining that the oxidants proposed by Oyama and Horowitz were a stopgap measure: no one, he argued, had demonstrated that such agents existed on the surface of Mars. Simply pouring Levin's nutrients on oxidants in a laboratory (as some scientists had done) in order to observe a predictable release of oxygen offered only a suspect analogue, not an explanation of the labeled release results. As with the canal controversy, both sides in the debate invoked Occam's razor as their trump card: Levin maintained that biology was the simplest and most elegant explanation of the data; Oyama and Horowitz, now on the same side of the controversy, argued that the best explanation was an exotic oxidant. Life, they argued, was an explanation of last resort, and therefore nonbiological explanations must be preferred. Given the very different assumptions about the Martian environment and microbiology with which Horowitz and Levin began, the two scientists seemed destined to reiterate their now-entrenched positions and consequently to talk past each other.

By 1977, most Viking scientists were swayed by the negative results of the GCMS experiment, and concluded that the "false positives" of all three tests did not constitute evidence of Martian microbes. Like Horowitz, Murray extolled the "definitive results from the gas chromatograph/mass spectrometer" (1989, 72). Although Martian chemistry must be different from Earth's, chemical reactions seemed a simpler, more plausible explanation for the Viking results than exotic life-forms capable of withstanding the planet's thin atmosphere, bitter cold, and ultraviolet radiation. Mars, in short, had no identifiable ecology. Its surface lacked the organic compounds indicative of the life, death, and recycling of by-products of any recognizable microorganisms. The oxidizing, self-sterilizing soil of Mars became a staple description of the chemistry of the planet's surface.

The Viking life experiments resulted in a barrage of articles in scientific journals. For the first time since the waning of the canal controversy in the 1930s, scientific debates about life on Mars were conducted in professional venues as well as in the popular press. The debate, however, was one-sided. Biemann and his colleagues reported negative results in their analyses of the Viking samples (Biemann et al. 1976, 72–76; 1977, 4641–42). In what became a standard refutation of Levin's interpretation, Horowtiz's team argued that the carbon assimilation (PR) experiment had to be considered definitive because it was "designed to perform a biological test on Mars under Martian conditions of temperature, pressure, water activity, and atmospheric composition" (Horowitz, Hobby, and Hubbard 1977, 4659; Horowitz 1986, 131). In his overview of the biological experiments, Klein conceded that the labeled release experiment "has consistently yielded presumptive positive biological results," but these "ambiguous" findings were countered by the pyrolytic release results which had to be considered "non-biological in origin" (1977, 4679). Although he admitted that "it is not possible . . . to exclude the presence, in the Viking samples, of exotic forms of life not based on carbon chemistry" (1979, 1661), Klein suggested that the biology team's findings signaled a dead end for the search for terrestrial-like life on Mars. Most scientists concurred with Horowitz, who regarded the results from the GCMS as definitive: "The absence of organic matter in the Martian surface at a parts-per-billion level of detectability," he wrote, "was the most important single biological finding of the Viking mission" (1986, 129; see also Ezell and Ezell 1984, 398–412).

In seconding this view two decades later, Bruce Jakosky, a geologist and expert on the climate of early Mars, maintains that "the complete ensemble of [Viking] experiments provided a much clearer picture than did any single experiment," and this picture might be titled "Landscape without Life" (1998, 138). Like Klein before him, Jakosky presupposes that, in theory, the four Viking experiments might produce equally valid results, but, in practice, scientists must adopt one experiment, as Horowitz did, as definitive: the absence of organic compounds means that the oxygen peaks registered by all three experiments must be explained by nonbiological agents. The GCMS experiment functions as a control: it allows for, even demands, nonbiological explanations for results that otherwise might be considered evidence that Martian microbes were feasting on the nutrient solutions. While Jakosky is correct to suggest that the "ensemble of experiments" on the Viking landers can provide a coherent, if speculative, picture of the chemistry of the Martian surface, that coherence is formed paradoxically from incommensurate interpretations of data from life-detection technologies that operate in different ways and make different assumptions about what Martian microbes might be like and how their presence might be inferred. Because the three biology experiments began with different values and assumptions, their individual results do not provide a "clearer picture" of reality but offer the means to generate heuristic views of reality, views that then must be adjudicated by scientists in a variety of disciplines and with a variety of professional backgrounds, experiences, attitudes, and commitments. As Jakosky's comments suggest, this process of adjudication that resulted in NASA's more-or-less official interpretation of the Viking data had (and has) significant institutional implications: the peer-review process in choosing the three life-detection experiments worked as it should; the overall weight of the evidence—interpreted through a rigorous experimental control—has proved self-correcting for anomalies detected by individual experiments.

THE VIKING RESULTS IN THE AGE OF EXTREMOPHILES

Given the publicity generated by the life-detection experiments, it is hardly surprising that the negative findings dampened NASA's enthusiasm for more missions to hunt for microbes. Horowitz seized on the Viking results to promote his dim view of exobiology. Generalizing from his interpretation of the data, he maintained that the geochemical similarities between the two Viking landing sites, thousands of miles apart, were "a consequence of processes that are planetwide in their operation," and he concluded that "the findings made at Chryse and Utopia are probably typical of the entire planet" (1986, 141). In response to scientists such as Sagan, who noted that the landing sites were hardly the most promising areas on the planet to search for life, Horowitz dismissed the "daydreams" that microbial life may exist in eco-niches on Mars or that there are microorganisms in the soil that his experiment failed to detect (145). Throughout *To Utopia and Back,* Horowitz uses his experience on the Viking mission to argue against what he sees as the ill-conceived search for extraterrestrial life. As a generic hybrid—part discussion for lay readers of the Viking mission, part philosophical exploration of the significance of life—his book is a latter-day version of Wallace's *Man's Place in the Universe,* although one committed to a kind of cosmic existentialism rather than a Christian exceptionalism: "Since Mars offered by far the most promising habitat for extraterrestrial life in the solar system," Horowitz argues, "it is now virtually certain that the earth is the only life-bearing planet in our region of the galaxy. We have awakened from a dream. We are alone, we and the other species, actually our relatives, with whom we share the earth" (146). Anticipating the objections of skeptics in the debates over the putative microorganisms in Martian meteorite ALH84001 (a topic I treat in chapter 8), Horowitz holds fast to a version of the argument that he made in the early 1970s: a definitive indication of extraterrestrial life would have to meet rigorous criteria, including evidence of RNA and DNA. Scientists would have to catch the microorganisms in the act of cellular mitosis. A quarter century after Viking, however, few scientists are willing to go as far as declaring that earthlings are "alone in the universe." With missions either underway or on the drawing board to search for life on Saturn's moon Titan and on Jupiter's moon Europa and intense interest in subterranean and polar environments on Mars, the eco-niche argument is alive and well, at least as a possibility that needs to be explored (Jakosky 1998, 138–41; Boston, Ivanov, and McKay 1992, 300–8; Priscu et al. 1999, 2141–44; Hartmann 2003, 195–96).

Even in the 1970s, though, Horowitz's view of the Viking experiments was not unanimous. Ben Bova, the science-fiction novelist and popular science writer, declared that the "confused and puzzling results" of the biology experiments indicated that "Martian life is different from ordinary chemistry—and also different from Earth-type biology"; he also predicted confidently that "manned (and womanned) expeditions" would be sent to Mars before the end of the twentieth century (Rovin 1978, 3).

Mainstream scientists, particularly some biologists and microbiologists, were less convinced than Horowitz that the Viking data could be explained by peroxides. Even though he agreed that "life is a hypothesis of last resort," Sagan admitted that he believed the results of the experiments, particularly Levin's, were ambiguous; and James Lovelock suggested that the "experiments had little chance of finding life on Mars, even if the planet were swarming with it" (Sagan 1994, 150, 241–42; Lovelock quoted in Hansson 1997, 8). Robert Shapiro, a professor of chemistry at NYU, and the author of a book on extraterrestrial life, characterizes the Viking results as "ambiguous and confusing" and "surprising and provocative" (1999, 187, 197). Skeptical about the design of Horowitz's experiment, he remains an unabashed supporter of Levin's interpretation of the data: as he points out, the labeled release experiment was tested exhaustively in different Earth environments and in every case detected life. When terrestrial samples were heated to 160 degrees centigrade (320 degrees Fahrenheit), hot enough to kill any microorganisms, no positive response was detected: the experiment worked as advertised. "The Labeled Release experiment," Shapiro concludes, "clearly stands as evidence for present life on Mars, just as the ALH84001 data serve as evidence of life in the past" (209–10). Shapiro's invoking of the Martian meteorite controversy suggests something of the half-life of the Viking experiments. As developments in extremophile microbiology have changed fundamental conceptions of life and molecular evolution, "ecology" has become a more inclusive descriptor that now encompasses a range of environments in which organisms—unknown and unsuspected in 1976—can thrive: thermal vents on the ocean floor, basalt formations a mile below the Earth's surface, and freshwater lakes buried miles beneath Antarctica (Taylor 1999; Karl et al. 1999, 2144–46). As Lawrence Bergreen argues, the Viking experiments had the bad timing to be designed and conducted just before microbiology underwent a revolutionary paradigm shift (2000, 180–89).

In retrospect, both the "chemistry" and "biology" debated in 1976 and 1977 by Viking scientists might be viewed as terrocentric metaphors for processes on Mars that remain opaque. The experiments themselves depended on these embedded, historically contingent analogies in their designers' efforts to register local anomalies that would confirm the operation of "universal" biological processes. Because they had to interpret the Viking data within the context of 1977 conceptions of microbiology, Horowitz, Oyama, Klein, Biemann, and even Levin constrained "life" to narrow ranges of environmental, morphological, and evolutionary possibility.[13] By the 1990s, such evidence was being reassessed within new conceptual and scientific frameworks: an emerging paradigm of extremophile microbiology and more sophisticated, provocative, and puzzling views of Martian climatology, hydrology, and geological history than those that had been gleaned from the Viking missions. By 2003, the *New York Times* was running articles in its science section describing the experiments as "costly failures" that produced data which were "frustratingly ambiguous" (Broad 2003, D4,1). Recently, a team of researchers reran versions of the Viking experiments at different sites in the Atacama Desert of Chile; the results suggest a "dry limit" for microbial life, even as they indicate that the GCMS experiments would have missed some organic compounds on Mars (Navarro-Gonzalez et al. 2003, 1018–21). If one resists the certainty born of such twenty-twenty hindsight, it seems that the achievement of the life-detection experiments was to provide a range of data that remains open to ongoing reassessments.

For the better part of two decades, Levin and Straat were pretty much lone voices arguing for a biological interpretation of the labeled release data. To some extent, they were victims of what we might call the asymmetry of explanation that is applied to exobiology. Most scientists involved in the multidisciplinary search for evidence of extraterrestrial life have seconded Sagan's dictum that extraordinary claims require extraordinary proof. There is, however, no corresponding claim made for nonbiological explanations—extraordinary peroxides apparently do not require extraordinary proof, only probabilistic scenarios. This asymmetry of explanation assumes that the question of life is invested with such scientific, metaphysical, and even theological significance that any claims about the existence of life on Mars, Europa, Titan, or elsewhere in the galaxy must be supported by ironclad evidence. Exobiologists can frame (or feign) no hypotheses. In the scientific literature on both the Viking experiment and the controversy over possible nanofossils in the Martian meteorite ALH84001, laboratory results that mimic in some, though often not all, aspects the data presented in defense of biological interpretations are frequently cited as though they were definitive.[14] Biological explanations can be refuted, even discredited, by demonstrating that a given phenomena might have been produced inorganically. The requirement for extraordinary proof in exobiology—the asymmetry of explanation—is intended, in part, to banish the speculative methodology and extraordinary claims that some proponents of life on Mars have made for a century: "Proof," Lowell declared in 1906, is "a supposition advanced" that "explains all the facts and is not opposed to any of them" (Lowell

1906b, 160). Such "supposition[s]" no longer count as evidence in exobiology, even if, paradoxically, they remain a legitimate counterstrategy to the arguments for present and past Martian microbes. For most scientists, life on Mars remains (at best) a plausible scenario rather than a probabilistic inference from existing data. Increasingly, however, it seems that Levin has had some success in making agnostics out of formerly true believers in a lifeless Mars.[15]

Following Horowitz, Oyama, and their collaborators, a number of scientists in the 1980s and 1990s offered nonbiological interpretations of Levin's data; these included both alternate hypotheses to account for the chemical reactions his instrument measured and reasons for a broader skepticism about the existence of life on Mars. Critics argued that the response during the initial peaks of the LR experiment was too much too soon for a biological response because the second injection of the nutrients into positive samples failed to produce a similar peak, as one would expect with biologically active soils on Earth. The weaker peak then could be explained inorganically: most of the oxygen had been exhausted by the initial chemical reaction. Ultraviolet light from the sun could produce the anomalous peaks under certain circumstances, and it would also destroy any organic compounds on the surface. Hydrogen peroxide formed in the upper atmosphere might descend to the surface and account for the oxidation in Levin's experiments; certain clays might produce the positive response as well. The clincher, of course, was the argument on which Horowitz and Klein had relied: no organic compounds were detected by the GCMS. More generally, skeptics maintained that Mars was too dry and the environment too extreme to support life; and ultimately a nonbiological explanation seemed simpler than invoking exotic Martian microbes. Even scientists who speculated that early Mars might have produced microbial life suggested that the possibilities of its continued survival were minimal. Without liquid water, Mars seemed to lack the means of transport and recruitment of new genetic material that would sustain bacterial colonies indefinitely (McKay and Stoker 1989, 189; McKay et al. 1992, 1234–45; Boston et al. 1992, 300–8; McKay 1997, 263–89). In almost all of these articles that counter Levin's interpretation of the LR results, Occam's razor is brandished to justify the view that inorganic processes occurring on the planet's surface produced the Viking results.

In the mid-1990s four factors contributed to rekindle a debate on what most planetologists had considered a settled issue. First, post-Viking studies indicated that Mars potentially had enough subsurface water to support some form of microbial life, even if that life died out millions or billions of years ago (Burgess 1990, 114–51; Carr 1995; Doran et al. 1998, 28–36; Malin and Edgett 2000a, 2330–35; 2000b, 1927–37; Lobitz et al. 2001, 2132–37; Baker 2001, 228–36; Hartmann 2003). In fact, for its temperatures and pressure, the Martian atmosphere often holds as much water as it can, a relative humidity of 100 percent (Burgess 1990, 44). Such suppositions hardly constituted proof, but they eliminated one of the key objections made in 1976 and 1977 to the possibility of living microorganisms on or near the surface. Second, the controversy over possible fossilized nanobacteria in meteorite ALH84001 led to heated debates about the Martian environment. These controversies alerted scientists in a variety of disciplines to the implications of extremophile microbiology, a field that had not existed in the 1970s. The discovery of colonies of microbes in terrestrial environments that previously had been considered sterile encouraged biologists to speculate about the possibility that fossilized nanobacteria—or living microorganisms—might exist on Mars. Third, the Pathfinder mission in 1997 generated renewed interest in the planet, and NASA once again explained its mission objectives as an important step to the ultimate goal of hunting for life or its relics on Mars. Finally, the advent of the World Wide Web allowed Levin to make his case to a broad Internet readership, appealing to a new generation of scientists, enthusiasts, and browsers who had not been part of the Viking consensus in the 1970s.[16]

Levin's frustration with the standard interpretation of the LR data, vented in a series of articles between 1981 and 2004, stemmed largely from his perception that other scientists did not respond to his counterarguments or new data but simply reiterated the consensus view of 1977. Levin countered the objections of his critics, and his efforts to keep alive a debate that other scientists considered settled could be seen as a case study in the retrospective revisiting of scientific data. With his background in environmental engineering, he recognized that developments in extremophile microbiology could be used to buttress his views. He also seized on the chemical and meteorological results from Pathfinder to describe the dynamics of a Martian ecology that could support his putative microbes. By 2000, he had developed three interlocking strategies to counter his critics.

Much of Levin's criticism was directed at the GCMS experiment, which required ten million bacterial cells to obtain a response in its one hundred milligram soil sample; the LR experiment could and did detect as few as fifty bacterial cells in the same amount of soil (Levin 1997, 151;

Levin, Kuznetz, and Lafleur 2000, 48–62).[17] The hydrogen peroxide theory, developed to explain the oxidizing response of Martian soils, had been questioned by scientists who failed to detect any signatures for this compound in the Martian atmosphere (Krasnopolsky et al. 1997, 6525–34). Levin drew on their arguments and noted that hydrogen peroxide and "any of its proposed derivatives do not approximate the thermal sensitivity of the martian agent that caused the labeled release responses" (Levin 1997, 153). Other candidates for the oxidizing agent, such as the "UV formation of peroxonitrites," he maintained, were problematic because they required more nitrogen than is available in either the soils or atmosphere of Mars; they too fail to explain the thermal response detected by LR (155). Moreover, the existence of hydrogen peroxide in ancient permafrost cores indicates that it did not destroy microbial life on Earth; the extremophile organisms that were discovered in the 1990s could survive the Martian environment that Horowitz had declared uninhabitable. Although various clays were proposed as possible agents for the "false" positives of the LR experiment, Levin objected that the laboratory results obtained by A. I. Tsapin and his collaborators did not replicate precisely the Viking data; the laboratory reactions were not thermally sensitive and no tests were done at the lower temperatures that had given Levin the initial positive responses in 1976. Finally, even the most sophisticated laboratory tests on candidates for the exotic oxidizing agent were not calibrated to the Viking instrument (Tsapin et al. 2000, 68–78; Yen et al. 2000, 1909–12; Levin 2002, 266–67; Tsapin, Goldfeld, Nealson 2002, 268).

The Pathfinder findings that water was present in small amounts in the soils strengthened Levin's rhetoric and weakened objections that the planet was too dry to support microbial life. This data, Levin and his son claimed, "eliminat[es] all explanations but a biological interpretation of the LR Mars results" (Levin and Levin 1998, 30). Pathfinder also returned evidence that there are significant differences between temperatures at the surface and those less than a meter above it. Because the Martian atmosphere is so thin, the ground does not cool as efficiently as it does on Earth so that warm air, in effect, bubbles off the surface (Schofield et al. 1997, 1752–58). This "diurnal concentration and conservation of heat in the soil" provides Levin with a mechanism for arguing that the thin columns of water vapor in the atmosphere are swept downward to the warmer surface, accounting for the hoar frosts photographed by the Viking landers (Levin and Levin 1998, 34; Levoy 2001, 245–49). At night, the cooling atmosphere reaches 100 percent humidity, and vapor condenses and freezes on the surface. During the day, the evaporating frosts quickly saturate the atmosphere with water vapor, inhibiting the sublimation of ice. Under an insulating layer of translucent ice, Levin maintains, the dark surface absorbs heat and conducts it to the subsurface ice. This mechanism allows Levin to imagine a thin layer of warm air above the ground (the colder air above this layer cannot accept the water vapor) and transient melting and refreezing of water in the surface soils. This insulating effect creates a daily hydrological cycle that requires relatively small amounts of water and solar and chemical energy. Evidence of such thermo-chemical changes, in fact, was registered by the GCMS as well as the life detection experiments. In the wake of recent results from the Mars Odyssey mission indicating abundant water ice just below the surface in large areas of the planet, Levin's account of the Martian environment conforms to current scientific knowledge of surface chemistry, hydrology, and thermal disequilibria (Boynton et al. 2002, 81–85). In a paper delivered in August 2004, Levin uses photographs from the Spirit and Opportunity rovers to bolster his claim that there is sufficient water in the Martian soil to sustain biological processes. He suggests that white patches observed in the soil disturbed by the landers' airbags and wheel treads are thin films of ice that have formed as water, literally squeezed from the ground, has frozen (David 2004a). While the cycle Levin describes is not "proof" of microbial life, it is a now-credible "supposition advanced." As one planetary geologist (who asked to remain anonymous) said to me in late 2003, "the hydrological cycle's an interesting idea; maybe it's even provocative. Even though I haven't changed my mind and can't disprove the possibility [of microbial life on the surface of Mars]. There's sure enough water there."

As conceptions of life broadened to include extremophile organisms in the 1990s, some microbiologists began to take a second look at the Viking data. In addition to Levin's crusade to get others to admit that the LR data did meet the criteria for a biological response, other scientists raised the possibility that the design limitations of the GCMS experiment had skewed interpretations of the Viking experiments. In 2001 researchers at the Scripps Oceanographic Institute reran the GCMS experiment, but began with different assumptions from Horowitz about Mars as an abode of extremophile life: microorganisms in the regolith could secure the energy needed to survive from chemical reactions in the atmosphere; there was sufficient water in the soils for many known terrestrial extremophiles to exist; and the thermal differences between surface temperatures and those just a few centimeters above it could drive robust processes that might support up to fifteen million bacterial cells per cubic gram of regolith—about the same amount as terrestrial soil in arid climates. Re-creating the GCMS tests on powdered rock that had been heated to remove organic contaminants, these scientists concluded that as many as thirty million cells per gram may have escaped detection on Mars in 1976 (Glavin et al. 2001, 1–5). Such revisionist interpretations of the Viking apparatus and the results they produced have raised the possibility that future experiments focused on the chemical reactivity of the soil might reach similar, suspect conclusions by discounting biological activity (Zendt and McKay 1994, 108–46).

To avoid such a scenario, Levin lobbied unsuccessfully for years to include more sophisticated life-detection experiments on future NASA landers. His proposal to send a modified version of the thermal and evolved gas analyzer (TEGA) on the failed Mars polar lander went nowhere. In brief, Levin suggested a low-cost modification of his Viking experiment to test for chiral activity, or chemical handedness. Chiral activity is the preference for one handedness of a molecule in the presence of both. For reasons that are not well understood, life-forms on Earth produce and use only left-handed (L-) amino acids and right-handed (D-) carbohydrates; therefore, a preference for L-amino acids would seem a surefire way of distinguishing biological activity from chemical reactions (Levin, Miller, Straat, and Hoover 2002, 78–86; Jakosky 1998, 104–5; Levin 1997, 146–47). While NASA showed no interest in such an experiment, Levin received a more favorable hearing in Europe. In 1996, the Russian spacecraft *Mars 8* was scheduled to rerun a version of the labeled release experiment; the Mars oxidant experiment (MOX) consisted of thin films of chemically reactive biomaterials, including amino acids, which apparently were "intended to look for molecular-handedness; enzymes (indisputable evidence of life) . . . show a preference for left-handed amino acids and right-handed sugars" (Hansson 1997, 7). The Russian mission failed, however, and the MOX experiment was never run. As straightforward as such a test might seem, it is not on NASA's flight manifest for future missions. The European Space Agency, however, included two biology experiments on its lost Beagle 2 lander, one to detect methane in the atmosphere or soil, the other to measure the ratios of carbon isotopes.[18] Both experiments take issue with the party-line view of planetary scientists that the negative results from the GCMS still satisfy the key methodological criteria: the simplest explanation for the labeled release data is chemical, not biological. The loss of Beagle 2 means no experiments designed specifically to look for biological activity are currently planned, although NASA plans to include a new GCMS and a laser spectrometer on its Mars Science Laboratory rover, scheduled for launch in 2009.

Although Levin continues to publish in journals such as *Science* and *Icarus,* in response to new, nonbiological interpretations of the LR results his favored outlet for his extended reconsiderations of the Viking data has been the annual proceedings of the Society for Optical Engineers, in a section devoted to instruments, methods, and missions for astrobiology (Levin 1997; 2001).[19] The Viking labeled release data, which Levin has argued for years would vindicate his views, recently have been made available on a NASA-sponsored Web site (http://wufs.wustl.edu/missions/vlander/lr.html) in response to arguments that the responses evidence circadian rhythms that correspond to daily temperature variations (Miller, Straat, and Levin 2001). The Web makes it difficult to ignore debates within scientific communities: Web sites sponsored by organizations such as the Planetary Society, the Mars Society, and the Astrobiology Institute of NASA, not to mention related listservs and chat groups, can disseminate quickly abstracts, images, and full-scale articles from refereed scientific journals. The availability of such information can fuel controversies that otherwise might be relegated to footnotes in specialized journals and give space to unorthodox opinions that might be marginalized by traditional means of scientific communication. Judging from the scientists rerunning versions of the Viking experiments and analyzing the Viking data sets, Levin has succeeded in producing a generation of agnostics, scientists who believe we lack sufficient data to resolve questions about Martian biology, natural history, and potential habitability. Yet a substantial majority of planetary scientists, when asked, still invoke the results of the GCMS to argue that Viking detected no life on Mars. The Viking results, says Chris McKay, are "inconsistent, ecologically, with what we know about Mars' surface environment" (NMR 2: 223). Paradoxically, the conclusions drawn by Horowitz, Oyama, Biemann, and Klein still stand even though their assumptions about Mars and the limit conditions for life are no longer tenable (Rothschild and Mancinelli 2001, 1092–1101; Nisbet and Sleep 2001, 1083–91).

MARS ON THE BOOKSHELVES

The Viking missions contributed to the ongoing transformation of the genre of popular science writing. Mars always has been an intriguing object for science writers delving into questions about the possibility of life on other worlds. The works of Percival Lowell, Agnes Clerke, William

Pickering, E. M. Antoniadi, Mary Proctor, H. Spencer-Jones, Hubertus Strughold, and Gerard de Vaucouleurs described the limited data they had available, discussed general principles, and offered accessible accounts of what was then known or surmised about orbital mechanics, atmospheric composition, hydrological cycles, calculations of surface pressure, and, in the case of the canals, the physiology and psychology of visual inference. All responded, implicitly and explicitly, to the environmental concerns of their time, and all had to cobble together narratives from a few experiments, spectrographic measurements, and analogies to phenomena on Earth. The planetary astronomers who wrote books or collected various papers under a single cover, as Pickering did, argued about data; those who wrote general histories commenting on their work, such as Clerke (1902) and Spencer-Jones (1940), had to extrapolate from this limited data in order to fill pages speculating about exobiology or, in the 1950s, the future of humankind in space (Ley and von Braun 1956).

With the success of the first three Mariner missions, Mars became paradoxically a less interesting and more complicated object for scientific writers. The first popular expositions of post-Mariner Mars by Shklovskii and Sagan (1966), Wily Ley (1966), and Glasstone (1968) included sections devoted to interpreting the photographs from Mariner 4, but the cratered surface, in one respect, seemed almost to speak for itself: no canals, no water, nothing but grayscale images of an lunarlike landscape. Yet these writers had new data to explain about the atmosphere, composition of the polar caps, surface insolation, and radioactivity, which required at least rudimentary explanations of interplanetary science. After Mariner 9, planetary astronomers and popular science writers alike had vast amounts of data to digest, explain, and try to fit into a coherent historical narrative: a cold, nearly airless, and cratered surface had been reshaped by volcanism and water. Rather than the relatively coherent—if incommensurate—pictures of Mars offered sixty years earlier by Lowell and Wallace, scientists such as Sagan and Murray offered a mosaic of partial or incomplete portraits that would require decades of additional exploration to modify, reject, add to, and rewrite. In response to the scientific literature of the early 1970s, popular accounts of Mars favored generic hybrids, such as Martin Caidan's *Destination Mars* (1972) and Jeff Rovin's *Mars!* (1978), part scientific exposition, part pop-culture catalogue of science fiction, part overview of the history of astronomy, and part anticipation of planned and imagined missions to the red planet in the last decades of the twentieth century.

The Viking missions, however, changed generic expectations about what a "science" book could and should do. The ambiguous results of the life-detection experiments led to an expanded market for books that explained, in accessible language, why experiments that had given initial positive results for biology were ultimately rejected by most NASA scientists. Books by Patrick Moore (1977), Richard Lewis (1978), Eric Burgess (1978), and David Chandler (1979) approached this task in a variety of ways, although all sought to affirm the Horowitz-Klein consensus and to forecast the future course of NASA exploration. Vikings 1 and 2 also produced studies by prize-winning science writers, notably Henry S. F. Cooper Jr. (1980) and John Noble Wilford (1990), that offered a kind of scientific anthropology, one that relied on in-depth interviews with the NASA researchers who had worked on the life-detection experiments and that tried to give readers a sense of "science in action," planetary science as heuristic, dialogical, or emergent rather than absolute. Even as some of the key figures in the Viking missions, Norman Horowitz (1986) and Bruce Murray (1989) as well as Sagan, wrote popular accounts of their scientific work, their writing itself had been transformed by the search for life on Mars. In general, popular writing about Mars had begun to separate itself from progressivist narratives of science and consequently become more attuned to the tensions, pressures, and uncertainties of planetary exploration. In the 1990s, these pressures and uncertainties were dramatized in numerous science-fiction novels that sent humans off to a red planet far different from the imagined worlds of Edgar Rice Burroughs and Ray Bradbury.

SEVEN

Transforming Mars, Transforming "Man": Science Fiction in the Space Age

DEAD MARS AND THE NOSTALGIA FOR BARSOOM

The cratered, lunarlike surface of Mars revealed in the Mariner 4 photographs in 1965 had profound and paradoxical effects on the science fiction of the late twentieth century. Although in some novels, such as Lin Carter's *The Man Who Loved Mars* (1973), the warring tribes and ancient civilizations of Burroughs and Brackett lingered on, Martian superraces usually were relegated to the planet's remote past, banished to the outer reaches of the galaxy, or, as in Tim Burton's 1996 film *Mars Attacks!*, sent packing to a universe of camp and parody.[1] Yet post-Mariner science fiction exhibits an almost obsessive nostalgia for the myths of Lowellian Mars, and this nostalgia, in turn, suggests why the planet continues to fascinate those who dream about colonizing or terraforming an alien world (Fortier 1995, 36–43). In Philip K. Dick's novel *Do Androids Dream of Electric Sheep?* (1968), the Mars of pulp fiction haunts the miserable inhabitants of a colony on the red planet. For colonists stranded on a planet that "wasn't conceived for habitation, at least not within the last billion years," the lure of a nonexistent Martian past is almost irresistible: "There's a fortune to be made in smuggling precolonial fiction, the old [science-fiction] magazines and books and films, to Mars. . . . You can imagine what it might have been like. What Mars *ought* to be like. Canals" (Dick 1996, 150, 151). The science fiction smuggled to Mars reveals the desperate need of the colonists for an alternative to the devastated and largely depopulated Earth of Dick's novel. More generally, Dick's invoking of Lowellian Mars represents a nostalgic desire among authors, readers, and moviegoers to reimagine the dying world as a new frontier, a stage to act out fantasies of humankind's destiny as a spacefaring civilization, or as a bolt-hole to escape ecological—or nuclear—devastation on Earth. The dying world never quite dies in the science fiction—and the scientific speculation—of the late twentieth century. Although post-Mariner Mars no longer harbors utopian Bolsheviks or feudal warlords, it reasserts its significance in hundreds of science-fiction novels, films, and stories as a testing ground for Gerald O'Neill's high frontier—a site for the next generation to test its mettle as, in Carol Stoker's words, a "*Star Trek* civilization" (quoted in Markley et al. 2001, "interviews" 1; O'Neill 2000). As an emblem for humankind's interplanetary future, Mars assumes a paradoxical significance in science fiction between 1960 and 2000: it is both a dead world that resists human efforts to understand, colonize, or transform it and the site of humankind's next giant leap in its technoscientific, and even spiritual, evolution. As Arthur C. Clarke puts it in his introduction to Jack Williamson's *Beachhead* (1992), Mars remains "*the* hope for science fiction" because it is still the most likely site where humankind will have to redefine its terrocentric conceptions of human experience (10).

Nostalgia for an imagined Martian past and speculation about an imagined future, then, are dialectical responses to the ambiguities that Mars represents after 1972. For science-fiction writers as well as planetary astronomers, the liquid-carved channels, floodplains, and volcanoes first revealed by the Mariner 9 and Viking missions raise as many questions as they answer. The persistence of the romance tradition of a Lowellian Mars of lost civilizations suggests that such dreams mark a liminal space of theory or speculation—a space that marks the absence of scientific certainty. In the authors' biographies appended to the 2002 collection of original short fiction, *Mars Probes,* the award-winning novelist Michael Moorcock declares that "bit by bit and very slowly—probably even a bit reluctantly—the scientists are beginning to find increasing evidence that [Burroughs, Brackett, and Bradbury] described pretty much the planet as it is" (Crowther 2002, 310). Other contributors to this volume, Gene Wolfe, Paul McCauley, Ian McDonald, Allen Steele, Brian Aldiss, and Stephen Baxter among them, all pay homage to the imaginative continuity between pre-spaceflight and post-Viking fiction. As Ray Bradbury declared to an NBC interviewer on the morning of the Viking landing, "there is life on Mars—and it's us!" (quoted in Crowther 2002, 307). Such comments indicate that the ghostly image of the dying planet still serves as a placeholder in complex systems of scientific and cultural knowledge making: it is a metaphor for the uncanny that continues to define perceptions of the red planet.

As a liminal space between scientific knowledge and exotic fantasies, Mars becomes a testing ground for new metaphors—and new and repackaged narratives—to describe humankind's relationship to the cosmos. If metaphors of entropic decay are still prevalent in the novels of Dick and his contemporaries in the 1960s, they tend after the 1970s to be replaced by or set against visionary images of planetary transformation and the adaptation of colonists to a brutally harsh environment. Rather than being locked into a Lowellian future of inevitable ecological collapse, Mars becomes a still-dynamic world that harbors the raw materials needed to engineer it (back) into a viable biosphere. During the post-Viking era, the ecological and narrative constraints imposed on the dying planet give way to scenarios that reanimate exobiological speculation: exotic life-forms thrive beneath the Martian surface (Benford 1999; McAuley 2001); alien builders of the so-called Face on Mars leave behind mysterious and potentially world-changing technologies (Steele 1992; Stabenow 1995; Douglas 1998); and indigenous civilizations reclaim their planet during the relatively clement periods predicted by Carl Sagan's "long winter model" of Martian climatology (Watson 1977; Williams 1986; Sagan 1971, 511–14). The complex relationships among still-powerful narratives of a Lowellian dying planet and those that envision various forms of ecological, political, physiological, and even spiritual transformation knit together three strains of Martian science fiction—realistic descriptions of future missions, the discovery of alien artifacts or races, and the possibilities of terraforming the planet—into generic hybrids. In the last case, science fiction shades into and is informed by mainstream scientific speculation: the possibility—suggested by numerous scientists, including Martyn Fogg (1995), James Oberg (1982), Paul Birch (1992), and Chris McKay, Owen B. Toon, and James F. Kasting (1991, 489–95)—that "everything we know about Mars seems to suggest that [terraforming is] possible" (McKay, in Markley et al. 2001, "interviews," 3; see also Shirley 2004).

The two hundred or so novels about Mars published between 1950 and 2000 register not only changing scientific views of the planet but also competing conceptions of science fiction and its social significance. The novels that I consider in this chapter reinforce the notion that no literary genre, least of all science fiction, simply dispenses with one set of assumptions and values in favor of another; older forms are constantly reincorporated and redefined so that the literary past becomes a dynamic horizon of expectation, a tradition that is always being reshaped by current efforts. As Samuel Delany argues, science fiction historically has been characterized by its self-referentiality and allusiveness, by its efforts to signify not a supposedly hard-and-fast "real" world but the simulations and "what ifs" that redefine and contest commonsense notions of "reality" (Delany 1994; see also Freedman 2000, Malmgren 1991, Pierce 1987, Rose 1981, Suvin 1979). In the case of post-Mariner science fiction, the nostalgia for an older, romanticized Mars self-consciously leads writers to superimpose the planet's past and present, and to revisit older traditions of science fiction both parodically and respectfully. Three short stories published in 2002, for example, revisit the traditions they have inherited: Mike Resnick and M. Shayne Bell have a stand-in for Burroughs's John Carter encounter peace-loving hippies on Mars ("Flower Children of Mars" in Crowther 2002, 87–100); Paul Di Fillippo offers a comic sequel to Stanley Weinbaum's "Martian Odyssey" (Crowther 2002, 42–61); and Eric Brown reimagines Wells's Martians as the peaceful remnants of a dying race that reached nineteenth-century England and provided the factual basis for *The War of the Worlds* ("Ulla, Ulla" in Ashley 2002, 3–26). In Brown's story, astronauts discover a fleet of cylinders and tripods in underground caverns on Mars; on a peaceful country estate, a twenty-first-century eccentric guards his family's secret: the Martians did make it to Earth, uttered their lament for their now-dead planet—"Ulla, ulla"—and died with their tripods unbuilt and their desire for a peaceful coexistence with humankind twisted by Wells into a tale of merciless invasion.

The intertextual relationship in such parodic tributes to Burroughs, Weinbaum, and Wells extends to ongoing reassessments of the relationships between a (seemingly) dead planet and a still-living Earth. In his 1999 *Rainbow Mars,* Larry Niven puts together a compendium of Martians and different visions of their planet: Schiaparelli, Lowell, Wells, Burroughs, Weinbaum, Lewis, and Bradbury all are invoked. What unites these parodic strains is Niven's envisioning a future Earth in the image of science-fiction Mars. Global warming and intensification leave the Earth a dying and desiccated world—"the oceans were small blue patches on a world mostly gone red"—dominated by canals: "Blue threads wriggled over the Earth . . . rectilinear networks branched out from tiny cubistic pumping stations" (1999, 189–90). The dying planet of Lowell retains its vigor as an image of looming environmental crises on Earth, and the warring menageries of Martians in the novel, from Wells's cannibals to Weinbaum's barrel creatures, measure our distance from an older tradition of science fiction even as they embody the consensus of a century of thought experiments: dying planets mean battles over dwindling resources.

The literary parodies that are endemic in contemporary Martian fiction return repeatedly to the kinds of images that Brown and Niven employ, and these images have a complex double function. A remote Martian past of "tall trees and . . . rich Earth" (Brackett 1953, 141) haunts our present knowledge of a planet characterized by its poisonous atmosphere, searing cold, and sterility; but the idealized Mars of terraforming literature and future colonization—one that depends on resurrecting "the relics of [a] biosphere" (McKay in Markley et al. 2001 "interviews," 2)—is reciprocally haunted by our vision of a stillborn planet inimical to terrestrial life. My choice of the main verbs is deliberate: the relationships among past, present, and future Mars and between any and all of these visions of the planet and Earth can be characterized by the double hauntings that figured prominently in earlier science fiction. Like the "ghosts of the past" and the "phantasms in a dead city" (Wells 1993, 163, 164) who stalk the ending of *The War of the Worlds* or the Lotharians in Burroughs's *Thuvia, Maid of Mars,* Mars itself is temporally dislocated, haunted by suppositions of what it may have been and what it may become.[2] In this regard, rather than an ontology, a structure that can be analyzed to classify and predict the behavior of, say, the climactic instability of Mars, the planet is defined in the cultural imagination by what Jacques Derrida (1994) terms a *hauntology*. This neologism registers the uncertainty and disorientation that comes with recognizing that seemingly bedrock assumptions and values are based on suppositions, fictions, and illusions about individual identity, social existence, and material reality. As Niven implies, Mars experiences various hauntings in the age of spaceflight, and it is the ongoing return of the dying worlds of Mars and Earth that characterizes much of the science fiction from 1965 on.

HEINLEIN AND DICK

Even before the Mariner missions, two major science-fiction writers of the postwar period, Robert A. Heinlein and Philip K. Dick, exploited the ambiguities of a planet that existed simultaneously in different conceptual and temporal time frames. Their novels were extremely influential, and, while neither of their careers is identified as closely with Mars as, say, Bradbury's, their works shape much of the subsequent science fiction that probes the boundaries between real and imagined worlds. Paradoxically, even as Lowellian Mars was marinered out of existence, Heinlein's and Dick's novels, notably *Stranger in a Strange Land* (1961) and *Martian Time-Slip* (1964), became benchmarks for future histories of the red planet.

In Heinlein's *Double Star* (1956), human colonists and indigenous Martians coexist by cultivating diplomatic rituals and (comparatively) benign indifference. The Martians are " 'a very old race and they have worked out a system of debts and obligations to cover every possible situation' " (Heinlein 1986, 63–64). The humans, at least the noble ones, embody the tenets of liberal imperialism: "freedom and equal rights must run with the Imperial banner" so that "the human race must never again make the mistakes that the white subrace had made in Africa and Asia" (158). Coexistence with the ahuman Martians and their elaborate culture of ritual, propriety, and rational thought leads enlightened characters to reject the mistakes of the past, including imperialist clichés such as the Humanity Party's belief that humans "have a God-given mandate to spread enlightenment through the stars, dispensing our own brand of Civilization to the savages" (1986, 160). At this stage in his prolific career, Heinlein appropriates the rhetoric of frontier individualism and (trans)human rights to depict contact with the ancient race of Martians as a form of intercultural exchange—a fictional revisiting and setting to rights of "the mistakes [of] the white subrace." The economics of colonization—and the war for land and mineral rights that had characterized European and American imperialism—are shunted to the background in order to focus on the transformation of the hero from out-of-work actor to statesman (Franklin 1980, 103–5). While this novel looks back nostalgically toward the common-man heroism of Burroughs, it also refashions that individualism into an agent for the forces of a new-look imperialism. By the time he wrote *Double Star,* Heinlein already had produced several scripts for films, including *Destination Moon* (1950), that pitted resourceful heroes against the agents of authoritarian bureaucracies (Carter 1988, 113–15). In his subsequent novels, the myth of the moral individual resisting governmental corruption became an obsession.

Heinlein's cult classic, *Stranger in a Strange Land,* parades his libertarian and antifeminist political views as a form of social protest against gray-flannel conformity. In his commentaries on science fiction from this period, Heinlein delighted in taking potshots at contemporary literary taste. Excoriating the "sick literature" and "ash-can realism" of modernists such as James Joyce and Alberto Moravia, he declared that he was "heartily sick of stories about frustrates, jerks, homosexuals, and commuters who are unhappy with their wives" and celebrated "the mature speculative novel [as] the only form of fiction which stands even a chance of interpreting the spirit of the times" (Heinlein 1959, 42–43). The Martians who remain offstage in this novel are the mystical antithesis to the "frustrates" and "jerks" who, in Heinlein's eyes, have corrupted the United States. The popularity of *Stranger in a Strange Land* in the 1960s and 1970s owed a good deal to its attacks on conformity; the novel celebrates an unfettered individuality defined by fantasies of defiance, masculine sexual freedom, and a self-congratulatory rejection by the novelist's alter ego, Jubal Harshaw, of cardboard totalitarianism (Franklin 1980, 126–40). If group sex, four decades ago, could present itself as a rebellion against puritanical restraint, it now seems more an adolescent *Playboy* fantasy in which women are invariably beautiful, willing, and unendingly available. The authoritarianism and ideological implications of hero worship go unremarked. Bruce Franklin notes that Charles Manson modeled his "family" on Heinlein's Martian cult, even including water-sharing ceremonies that celebrate the hero's spiritual authority (1980, 127).

Heinlein's hero, Valentine Michael Smith, is the only survivor of the first expedition to Mars. After his parents are killed, he is raised from infancy by Martians, now transformed into mystical sages with the god-like ability to manipulate matter at will. Smith returns to Earth with Martian psychic and physical powers; both of and beyond the human race, he becomes the leader of a cult that offers a masculinist wet-dream of sexual "freedom." The split between mind and body that plagues humans is unknown to Smith, who, like the Martians, apprehends—"groks"—through a somatic understanding of and union with others: grokking promises knowledge of, in, and through an unalienated body. To grok is to experience, believe, and know simultaneously, and in the 1960s was employed rather earnestly by some in the counterculture as a talisman against conformity, consumerism, and sexual repression (McNelly in Hipolito and McNelly 1971, 327–32; Rovin 1978, 100–2). Forty years later, grokking seems less a rebellion against or a transcendence of conformist ideology than a cracker-barrel philosophy intended to skewer "ash-can realism" and other aspects of a postwar ethos of corporatism and consumerism.

At the heart of *Stranger in a Strange Land* lie the ceremonies of water sharing that underlie the sociosexual idealism of the ancient race of Martians. Heinlein both etherealizes and mystifies the planet on which *Double Star* was set. All of the novel's action is set on Earth, while Mars functions as the site of an unalienated and, to humans, inaccessible communion: a realm of spiritual and sexual rebirth. In this respect, Heinlein's mythology of individual power appropriates the communal ethos, symbolized by the water-sharing ceremony, and makes it the centerpiece of his hero's rejection of bourgeois ideology and sexual morality. The problems of Martian colonies that Kornbluth and Merril, Bradbury, and del Ray had used to critique militarism and political corruption are transmuted by Heinlein into an indictment of bourgeois liberalism. Once again Mars becomes the Valhalla of strange gods. Terrified of the multifaceted liberation that Smith embodies, "frustrates, jerks, homosexuals, and commuters," in the throes of their political and sexual repression, ultimately martyr the hero on the altar of small-minded conformity. The "mature speculative novel," it appears, can capture the "spirit of the times" only by casting the "stranger" as a being who embodies a radical, individualistic ideology, yet paradoxically rejects the conventions of midcentury individualism. As significantly, such speculation no longer focuses on a "realistic," scientifically plausible Mars; the red planet has become, in effect, an emblem for an idealized state of heightened consciousness.

In Dick's novels, this tendency to turn Mars into a marker of psychological difference or alienation becomes a means to depict the wasteland of midcentury existence. Dick wrote frequently about the planet, often as a site of failed human endeavors at colonization to illustrate the gulf between capitalist propaganda and the brutal realities of oppression. Before the hallucinogenic intensity of *Martian Time-Slip* (1964) and *The Three Stigmata of Palmer Eldritch* (1965), Dick's short stories make pulp-fiction Mars the setting for a nightmarish surrealism. In this regard, Dick is concerned less with changing scientific perceptions of the planet than with its potential for reflecting the psychic dislocation and confusion—the "gubble"—of life in postwar America.[3]

In his short stories of the 1950s, such as "The Impossible Planet" (published in *Imagination,* October 1953), Dick returns repeatedly to the nightmare images of ecological devastation wrought by war and resource exhaustion. These tales employ Mars both as a reflection of contemporary problems and as an intimation of Earth's future. In "Martians Come in Clouds" (*Fantastic Universe,* June-July 1954), peaceful refugees from a planet that is "a vast desert without limit or end" descend on Earth and communicate telepathically that they wish only to live on great disks on Earth's oceans; the last of these "buggies"—"an ancient gray shape"—is killed (Dick 1987, 2: 123, 122) by ignorant and fearful small-town Americans. In "Survey Team" (*Fantastic Universe,* May 1954), the survivors of an endless war on Earth that has left the planet uninhabitable blast off for

Mars. When they arrive, the crew find a planet stripped of its resources, the devastation worse than that on war-torn Earth. There are no Martians, just the wreckage of an industrial civilization that has cannibalized an entire world. " 'They've *used* Mars up,' " declares Mason, a scientist. " 'Used up everything. Nothing left. Nothing left at all. It's one vast scrap heap' " (2: 371). Even for these veterans of a twenty-first-century war, the fate of the Martians inspires horror: "Beyond the ruined city stretched out what had once been an industrial area. Fields of twisted installations, towers and pipes and machinery. Sand-covered and partly rusted. The surface of the land was pocked with great gaping sores. Yawning pits where scoops had once dredged. Entrances of underground mines. Mars was honeycombed. Termite-ridden. A whole race had burrowed in and dug in trying to stay alive. The Martians had sucked Mars dry and then fled" (2: 372). This nightmare collage of images brutally undermines midcentury visions of industrial power and prosperity. Mars is imagined in explicitly terrestrial images of decay: the oxidation of rust, the imaginative yoking of its inhabitants with termites.

Dick's ascientific planet is thus doubly haunted: by its own past as an analogue of enterprising twentieth-century Earth and by what it represents for the survey team: an image of their future and a reflection of their own bankrupt values. Dick's characteristic narrative turn comes when the survey team realizes that the planet to which the Martians had fled six thousand years earlier is Earth: "We're back where we started. Back to reap the crop our ancestors sowed" (2: 376). In effect, civilization is defined by its blind faith in the failed values and cannibalizing technologies of the mid-twentieth century. The story ends with the survey team, despite Mason's protests, seeking a new frontier beyond the solar system, "a virgin world. A world that's unspoiled" (2: 377), so that they can begin the self-destructive cycle of ecological aggrandizement and violence again. The destruction that in Bradbury's *Martian Chronicles* was displaced onto the horrors of nuclear war becomes, for Dick, an irrevocable consequence of human greed and irresponsibility—insatiable desires for more and more resources. If the dead planet of "Survey Team" extends the critique of unchecked capitalism evident in novels such as *Outpost Mars,* Dick's "vast scrap heap" of an exhausted world recasts the specter that has haunted Mars since the novels of Wells and Lasswitz: "used up" Mars represents the dark underside of a modernist faith in mines, factories, and progress.

While "Survey Team" offers a sardonic take on postwar science fiction, many of Dick's critics have argued that the significance of his work lies in its disrupting the narrative logic and the ideological underpinnings of realistic fiction.[4] In one sense, the juggled realities of novels such as *Martian Time-Slip* reflect a process of defamiliarizing the colonialist and corporatist narratives of the pre-Mariner era. The planetary environment is that of the 1950s—a quasi-breathable atmosphere and an aggressive program of human colonization—while the hallucinogenic quality of overlapping and competing realities belong to the 1960s and, many critics contend, to the advent of postmodernism. It is worth considering, though, the inverse of this view: Dick does not "invent" postmodern strategies and transport them to various science-fiction settings so much as he transposes familiar science-fiction images of Mars as a dying planet into the "interior" narratives of psychological representation. "Survey Team," in this regard, depicts an early landscape of postindustrial ruin: the devastation associated with the atomic bomb is not a consequence of political evil but the "reality" that underlies the logic and rhetoric of capitalist exploitation.

In *Martian Time-Slip* (1964), Mars is both colony and frontier, American hucksterism superimposed on "habitable Mars, this almost-fertile spiderweb of lines, radiating and crisscrossing but always barely adequate to support life, no more" (Dick 1995, 10). Humans have begun to settle a planet that still harbors the remnants of an "old civilization," the Bleekmen, "poverty-stricken, nomadic natives" on whom "time ha[s] run out" (11, 27). In a novel that models schizophrenia as both an effect of and an escape from the corruption and coercion of modern society, the Bleekmen serve as intermediaries between objective "realities" and subjective experience (Freedman 1984, 15–24). Heliogabalus, a Bleekman servant, tells the union boss Arnie Kott that schizophrenics "take a brave journey. They turn away from mere things, which one may handle and turn to practical use; they turn inward to *meaning*" (93). This "meaning" blurs reality and illusion, the human and the nonhuman. If the novel insists that schizophrenia, in the words of the protagonist, Jack Bohlen, "is one of the most pressing problems that human civilization has ever faced" (85), it is because such psychic and temporal dislocation transcends medical and psychiatric regimes of diagnosis, classification, and treatment. The "gubble" of schizophrenia mirrors the moral confusion and political incoherence that underlies humankind's exploitation of the Martian "frontier"; at the same time, it marks a resistance to the dehumanizing effects of a corporatist ideology. Schizophrenia is an "artifact . . . deliberately constructed by the ailing individual or by a society in crisis" (75) to ward off the Kafkaesque nightmare of life on Earth. In effect, Dick's characters internalize the tensions and contradictions of existence on a dying world. *Martian Time-Slip* thus extends the thematic of "Survey Team" by implying that the experience of Mars marks a return of the repressed, a reinscription on the psyche of the terrestrial conditions—"too many people, too much overcrowding"—that have made "mental illness" a "sign of the times" (124). The schizophrenic's "utter alienation" from "the outside world" might stand metaphorically for the breakdown of the analogical relationship between Mars and Earth and their life-forms: the frontier ideology of the United States is revealed, on Mars, to be an amalgam of self-destructive tendencies: corporate bad faith, political repression, and consumerist self-interest.

In *The Three Stigmata of Palmer Eldritch* (1965), as Peter Fitting (1983, 223-28), Scott Durham (1988, 173-86), and David Golumbia (1996, 90-98) have argued, Dick calls into question the very bases of reality itself. Earth in the twenty-first century has become a hothouse (180 degrees in New York City), and colonists have been drafted and "required to begin new, alien lives on Mars or Venus or Ganymede" (10). To cope with the otherwise unbearable gloom and boredom of trying to raise mutated crops on a planet where native insects "had been waiting ten thousand years, biding their time, for someone to appear and make an attempt to raise crops" (39), the colonists on Mars take the drug Can-D to induce their "translation" into the Barbie-like sets of Perky Pat layouts. During these drug-induced efforts at "transubstantiation," the colonists try to forget "that gloomy quasi-life of involuntary expatriation in an unnatural environment" (68, 44). Mars provides Dick with a crucial image for the dead landscapes that pervade his fiction: the inhospitable planet is both a projection of the individual's schizophrenia onto a seemingly "external" reality and an introjection of the meaninglessness of the world. Mars is the dead world that haunts the Earth. If the socionatural environment both reflects and is reflected by "this gloomy quasi-life," this psychological "tomb world," according to Katherine Hayles, is at once "a delusion and necessary purgation" (1999, 178). When Palmer Eldritch returns from a ten-year odyssey to another star system with a more powerful hallucinogen, Chew-Z, and plans to market it on Earth, Mars morphs into a site marked by the involutions of reality and illusion; the setting itself destabilizes any sense of a coherent objective reality, any firm boundaries between "inside" and "outside." This tension in Dick's novel between visionary hallucinations and "realistic" depictions of a hostile, pre-Mariner Martian environment is offered by multiple narrators with disparate voices and perceptions. Like *Martian Time-Slip, Three Stigmata* thus offers what Durham calls "a radically contestatory politics of experience" (1988, 174).

Yet even in the indeterminate, competing realities of *Three Stigmata,* Mars retains its totemic quality as an ecological and sociopolitical dead end—a nightmare vision of the failed dreams for the consumerist good life on Earth. As the self is radically divided from its sense of bodily integrity and, as Hayles argues, "the deadness inside [is] projected onto an exterior landscape" (1999, 176), Mars is reinscribed as the site of that alienation: a planet haunted by its past of dead seas and vanished life-forms and by the fantasies of colonial aggrandizement. This meaningless existence on Mars produces a desire for transcendence, for faith in *something,* of which Can-D and Chew-Z are symptoms. Such faith, however, marks not the birth of the frontier but its negation, the emptying out of meaning by the resistance of a (Martian) reality to humankind's efforts to conceptualize it. Dick's Mars finally suggests a kind of collective, semiconscious recognition that all the hope, and all the hype, of scientific adventurism has come up against an ecological and psychological impasse that it can neither avoid nor overcome. The contestatory politics of experience in these novels undoes visions of Mars as the setting for utopian dreams, morality plays, or heroic adventures. Having internalized the vistas of a dying world, the individual can experience the cost and consequences of Martian colonization only as "gubble."

MARS AFTER MARINER

Dick's novels of the mid-1960s are typical of the disenchantment that the disappearance of the canals provoked. The desolation of Mars becomes a crucial metaphor in the 1950s and 1960s for sociopolitical and psychological repression, though often employed more crudely than in *Martian Time-Slip.* In *Not in Solitude* (1959), for example, Kenneth Gantz, a U.S. Air Force colonel in charge of "top-secret Air Force duties" (dust jacket) and editor of *The United States Air Force Report on the Ballistic Missile,* describes a nightmare ecology on Mars: a single life-form—an intelligent and omnivorous plant—has spread across the planet and destroyed all other life. By making the harshness of the Martian environment the product of a malevolent intelligence, Gantz demonizes the hostility of the planet to terran biota and American colonization. Cold war fears are projected into the form of a totalitarian ecology.

Those novelists without top-secret duties explored the implications of a dead world more compellingly. In his underrated classic *Farewell*

Earth's Bliss (1971) the British novelist D. G. Compton depicts the socio-political and psychological effects of the isolation, loneliness, and suffering that the Martian environment represents. Prisoners from Earth exiled to the arid, nearly lifeless planet find themselves confined in an autocratic penal colony. Jacob, a newcomer, lands on Mars and barely survives a dust storm that rages for thirty-seven days and kills many of his fellow prisoners. When the dust clears, he views a landscape that mirrors his psychological state: "The dust was dry and dead, seamed with yellow and black on the cut cliff faces of the ridges. And between the red horizon and the aching sun there was nothing but cold. Cold he could feel now through the soles of his shoes, cold that scorched the desert naked and dry, cold that screamed in the ancient stillness, cold that was red, red cold deep in a mile of stiffened dust" (48). Compton's description paradoxically pays homage to and rejects romanticized visions of Mars. The harshness of the environment becomes an apt image of both external forms of repression and internalized mechanisms of self-policing. As with oxygen regulators and heat coils, so it is with human will, imagination, and freedom on this Kafkaesque Mars.

The dark underside of colonization is not the interplanetary capitalism imagined by Kornbluth, Merril, and del Ray, but the "stark horror" of a forgotten colony ruled by a governor who uses the harshness of the environment to justify draconian controls on prices, wages, speech, sexual relationships, and religious beliefs (132). In this regard, the dying planet of Burroughs and Brackett exists as the trace of a vanished past that the prisoners can sense only as a memory frozen into the landscape itself: "They were moving across the bottom of an ancient sea, a vast egg-shaped hollow where high above their heads waves had once scrambled up the sky and fallen and scrambled again. And shrunk and dried and died, with nothing now to break the old, indestructible, icy silence" (85). The silence and cold of the landscape mirror the authoritarianism and religious fundamentalism that keeps the colonists in check. Jacob learns by the end of the novel that dust storms are raised intentionally by the governor each time a ship lands in order to kill the weak and pacify the survivors. Having survived a month of terror and deprivation trapped in their spacecraft, new prisoners can be manipulated into confessing their crimes and accepting the harsh regime of endless self-denial. The "red cold" forces Jacob to internalize the self-denial necessary to survive in this alien environment: he relinquishes his qualms about the colony and, at the end of the novel, helps to raise another dust storm when the next ship of exiles arrives.

As *Martian Time-Slip* and *Farewell Earth's Bliss* suggest, even prior to the Mariner missions Mars had become a site of blasted romanticism, an image of despair tinged with a sense of betrayal born from the failure of Mars to be what "it *ought* to be." Yet in much of the juvenile fiction of the 1950s and 1960s, such as Donald Wollheim's *Secret of the Martian Moons* (1955) and Lester del Ray's *Marooned on Mars* (1962), interplanetary adventure survives. On the eve of the first Mariner mission to Mars, James Blish begins *Welcome to Mars* (1967) with a foreword (written in 1965) that rejects the "transplanted Arabia called 'Mars' " in favor of a "semi-real Mars" of craters and deserts, a setting that offers "a new theatre for a human hero" (10).[5] Yet the plot of Blish's novel suggests Tom Swift meets Buck Rogers: the teenage hero and heroine travel to Mars in homemade antigravity capsules and eventually discover a vanished civilization in a city beneath the planet's surface. The last of the dying Martians bequeaths to them—in a semiregal rhetoric of property, legal succession, and inheritance—the secrets of "high and ancient Mars" (146) as well as the care of the dune cats, a semisavage but loyal race of servants who roam the planet's surface and offer the humans the prospect of colonizing a new world by profiting from the knowledge of the old (Ketterer 1984, 284–90). As this summary suggests, Blish's novel is a hybrid: it combines a "semi-real Mars" with all sorts of scientific shortcuts and improbabilities. Without them, there would be no novel. His hybridization self-consciously presents its fictional Mars as a faithful rendering of up-to-date knowledge of areography and anticipates a narrative strategy that prevails after the Viking missions—finding a way to reinvest Mars with some form of meaning, to justify writing a future history in which the exploration of the planet is high on humankind's agenda. Blish's lost race of Martians, in this respect, marks an always-incomplete transition from romance to realism. If the hybridization of Martian science fiction, in part, is a response to a new scientific paradigm, the goal for novelists after Mariner is to recast the genre of adventure fiction to accommodate a planet that had become a rocky wasteland.

For the seven years between Mariner 4 and the startling photographs returned by Mariner 9 in 1972, Mars was a world as dead and possibly as uninteresting as the Moon. Even Compton's dreary landscape seemed almost verdant in comparison to the areas of the surface that Mariners 6 and 7 photographed in 1969. For science-fiction writers, the first three Mariner missions created a Mars that Harry Harrison in "One Step from Earth" (1969) described as "a planetary stillbirth of boulders, coarse sand, jagged rock" (Hipolito and McNelly 1976, 253). The most important novel of this period registers a profound disillusionment with Lowellian Mars and with American adventurism by playing on the tragic differences between still-enticing visions of a living world and the realities, as they were imagined, of a sand-covered hell.

Ludek Pesek's *The Earth Is Near* (1975) won the German Children's Book Prize in 1971, but its stark, often terrifying description of the first human mission to Mars and the psychic and physical toll that it exacts on the twenty-man crew owes more to Compton's sensibility and the inhospitable landscape revealed by the early Mariner photographs than to adolescent fiction about dune cats or hollowed-out moons. Pesek's is a classic of post-Mariner fiction, the most chilling and evocative of the failed or troubled mission novels that abound in science fiction after the 1960s. In contrast to American first-mission novels, such as Gordon Dickson's *The Far Call* (1978), *The Earth Is Near* demystifies the heroic, masculinist narratives of the conquest of space by subjecting the nationalism and idealism of space flight to a grim, futuristic quasi-realism. All of Pesek's characters have generic American names (O'Brien, Norton, Glennon, McKinley, Compton, and so on) and the novel critiques the tendency to translate Mars into a twenty-first-century American—or Americocentric—frontier. The narrator, one of the mission's two doctors, remarks that despite the knowledge gleaned from unmanned probes, "each of us cherished some of that secret, irrational romanticism" that the astronauts will discover extraterrestrial life or the remains of an ancient civilization. "After all," he continues, "dreams are essential for the progress of science—the most serious of scientific hypotheses is really nothing but a dream" (113). This "irrational" romanticism surfaces as the persistent desire to explore the planet even as the astronauts fall victim to technological failures, madness, and death.

Having landed hundreds of miles off target, the mission bogs down in a series of futile and fatal efforts to traverse impassable deserts during howling sand storms. The rovers break down one by one, and the last of the gliders that provide mobility for the mission is lost over what its pilot seems to have believed was a site of indigenous vegetation. The crew deteriorate psychologically in searing but unspectacular ways during their eighteen-month stay on the planet. In short, Pesek's grim account of this failed effort to find life on Mars rejects the science-fiction traditions he has inherited: "We didn't find green monsters crawling about, or ancient canals reflecting the roofs of Martian cities in their dark waters, not even a Martian Atlantis covered with sand. Our time on Mars was a long record of endless and monotonous toil" (186). As one exploration scheme after another fails, the astronauts become less stubborn survivors than automatons. Pesek's novel deheroicizes space exploration and subverts the easy identification between the final frontier, national pride, and masculine identity so crucial to the first years of Soviet and American manned missions (Carter 1988; Penley 1997; McCurdy 1997; Bryld and Lykke 1999, 92–117). This demystifying of space exploration extends to the planet as well. Mars has no canals, no cities, no ruins, no ghosts. *The Earth Is Near* becomes a kind of tragic farce: the hollowness at the center of the novel depicts a space-age mentality without a goal or rationale besides self-animating "dreams" of "progress."

BOLDLY GOING WHERE NASA HAS YET TO GO: FICTIONAL MISSIONS

The photographs from Mariner 9 and the Viking orbiters and landers, as I discussed in chapter 6, profoundly altered scientific conceptions of the planet and, with them, the possibilities for science-fiction writers and their audiences. The photographs of the surface of Mars from the Viking landers, as Carl Sagan remarked, rendered the rock-strewn terrain of the planet a place rather than a light in the sky or a fuzzy image in a telescope. The Viking landers and, twenty years later, the Pathfinder mission provided photographs that allowed millions to project themselves imaginatively into frames that looked eerily reminiscent of terrestrial deserts (Morton 2002). With their extended mission continuing, the 2004 and now 2005 rover missions are adding exponentially to the photographic inventory of the Martian surface. By the end of 2004, Spirit and Opportunity had returned over 50,000 photographs. As a mimetic art, landscape photography both freezes time and offers the prospect that one can enter into a frame that is coextensive with a material, embodied reality. In contrast to the orbital photographs returned by Mariner 9, the Viking orbiters, and even Mars Global Surveyor, the images from the five vehicles to land on Mars create the illusion of an embodied experience of the Martian terrain: time is frozen, the world remains unchanged as though awaiting human explorers, surveying the land and posing for photographs. The surface photographs of Mars, in this regard, are so arresting because they seemingly reveal an essence of the place that photography always promises to reveal. The three-dimensional images of the Pathfinder site (complete with disposable 3-D glasses) included in the National Geographic book on Mars (Raeburn 1998) take us through the looking glass that had served as a barrier for writers between 1964 and 1976. The surface takes on a third dimension of small undulations, rocks in the foreground and rocks that recede toward an abrupt horizon. This flood plain topography appears through the minimalist technology of early 1960s science fiction, the red and green plastic lenses in the cheap cardboard frames that movie theaters passed out for a few horror films (the original *Thirteen Ghosts* [1960]). The Spirit and Opportunity rovers ratchet upwards the tendency to project oneself imaginatively into and—crucially—over an alien planetscape. The rovers' mobility and microscopic examination of rock samples offer a kind of prosthetic visual semiotics that seem uncannily akin to the perspective of a crawling infant. The crater interiors, "blueberry"-strewn plains, and striated hills that gradually have been revealed by the Spirit and Opportunity photographs have unfolded at an insectile pace—a slow crawl across a landscape that has begun to reveal itself in geological and experiential analogues: rocks show evidence of standing water; the robotic toddler begins to orient itself in a new world. Spirit and Opportunity trigger visual and mythic associations that already exist in the collective imagination: the survivor's view as she crawls across the desert looking for water, for example. Intimations of Geological Time from the (Subconscious) Recollections (Triggered by the Visual Stimuli) of Childhood.

If the Viking, Pathfinder, Spirit, and Opportunity photographs offer windows onto an entire world, then fiction about Mars after 1976 fleshes out imaginatively a logic and aesthetics of areography. In particular, the photographs from and of Mars offered the possibility of adapting a tradition of nature writing to an alien world; the descriptions such as Compton's and Pesek's from the pre-Viking era are replaced by often-extended set pieces that fictionalize an embodied knowledge of the Martian surface. In novels by Frederik Pohl (1976), Kevin Anderson (1994), Geoffrey Landis (2000), and others, readers encounter a space-age version of the sublime—one that engenders awe, respect, love, and fear reminiscent of the nature writing of the nineteenth and twentieth centuries. But such projections of landscape description onto the red planet require acts of imaginative possession—a possession that can take place only on a world that is always and already partially humanized. This possession, then, invokes a dialectical response: a rekindling of awe in a semiexplored and (mostly) unexploited world—a "giant mountainous wilderness," as Robinson calls it (Markley et al. 2001, "interviews," 9)—and a recognition that possession means exploitation and some degree of terraformation, of the inhabitation of Mars by an alien species. The Viking photographs thus mark an important step in the acculturation of the Martian landscape to human conceptions of the environment. In this respect, the desert vistas returned to Earth by the Viking landers offer two different kinds of invitations to the human imagination: on the one hand, Mars becomes ahuman, evoking the sense that to understand areography—ancient flood plains, immense canyon systems, and gargantuan shield volcanoes—one must extend history itself into the geological, or rather areological past, an imaginary that extends three or four billion years back in time. But this ahuman quality provokes as well the desire to impose human desires on alien landscape, to remake Mars in the image of an unspoiled Earth.

The Viking photographs also inspired a generation of science-fiction writers to recast the old-fashioned adventure novel as a hi-tech confrontation with the unearthly nature of vast canyons, ancient riverbeds, and massive craters. After 1976, the future mission novels—Ben Bova's *Mars* (1992), Stephen Baxter's *Voyage* (1997), and Gregory Benford's *The Martian Race* (1999)—depict the exploration of Mars as a generic overlay: part epic journey, part heroic quest, and part "realistic" depiction of the best-laid plans gone awry as astronauts encounter crisis after crisis on an alien world. Such novels, most written by Americans and many by scientists, declare their faith in the future of space travel while offering cautionary tales about the perils that astronauts may face on Mars and the dire consequences for humanity if we fail to open a new frontier on the red planet. Yet if this strain of science fiction advertises self-consciously its antiromanticism and often veers toward polemics for a vigorous program of interplanetary exploration, the simulated "realism" of these near-future histories is always impure, shot through with the generic values and assumptions of the dying planet that had fascinated authors and readers from Burroughs to Bradbury. "Realistic" scenarios almost invariably cannibalize and subsume the romance of the red planet. In Robert Zubrin's *First Landing* (2001), the historian-hero's first dispatch from the surface promises "to make amends" for the Barsoom "destroyed by the Mariner probes" by filling the planet with "new life, love, adventure, and unlimited potential" (24). This projection of a fictional past into an imagined future is suggestive of the problems a critic encounters in trying to construct a taxonomy of Martian science fiction. The future histories of Martian exploration mark an unstable boundary between what seems feasible with present-day technologies and fantasies of colonization and terraformation.

If these fictional hybrids reinvigorate as well as adapt the conventions of the dying planet, they also bring space-age adventure within the strictures of scientific plausibility. A key staple of the future mission novels are long, hardship-filled journeys across a forbidding planetscape. Epic treks across the dead sea bottoms of Mars had been a staple of early Mars novels since Burroughs's *Princess of Mars,* and these odysseys across a wilderness of hostile aliens and natural dangers invariably test the hero's strength, courage, and honor. In the post-spaceflight literature of the 1970s, 1980s, and 1990s, astronauts and early colonists embark on forced marches and seemingly impossible journeys to reach cached provisions, way stations, distant settlements, or, quite often, the last working spacecraft on the planet. The traditional quests of pulp-fiction heroism are transformed into fables about the perils and tenuousness of the very technologies that have enabled humankind to reach Mars. Humans are stripped of or betrayed by their equipment through bad luck, bad planning, or deliberate sabotage; and they must reassert space-age versions of the values that defined their fictional forbearers—Natty Bumpo, Allan Quartermain, and Tom Swift as well as John Carter. Their journeys force them to confront the limits and capacities of the physiological, technological, and psychological resources that define the American (and Russian) myths of the heroic human exploration of space. In a variety of novels after Pesek's *The Earth Is Near*—Sterling Lanier's *Menace under Marswood* (1983), Ben Bova's *Mars* (1992), Ian Douglas's *Semper Mars* (1998), Kim Stanley Robinson's *Red Mars* (1993) and *Green Mars* (1994), and Geoffrey Landis's *Mars Crossing* (2000)—technological disasters force humans to march across the planet's surface, dealing with psychological conflicts, the hostility of the environment, and political and philosophical questions about humankind's future as an interplanetary species. In transforming nature-writing into the areological sublime, these novels reflect complex responses to the prospect of traversing an alien world: awe at the vastness and strangeness of Mars, a compulsion to exploit its resources, and a desire to transform that world into another Earth. The Martian landscape takes the measure of the human will to explore and survive.

Almost as a matter of course, these future-mission novels offer "realistic" scenarios for exploring Mars even as they project the ecological pressures and the conflicts of late-twentieth-century politics into the future. This investment in a near-future technological "realism," however, is only one half of a dialectic: set against descriptions of new and adapted technologies for human survival on the planet are plot devices that mark their authors' reinvestment in the uncanny—often discoveries of the technologies or artifacts of ancient civilizations. After the Viking missions, novels authored by scientists, such as William Hartmann's

Mars Underground (1997), Gregory Benford's *The Martian Race* (1999), and Paul McAuley's *The Secret of Life* (2001), make the discovery of life or its remnants on Mars essential to "realistic" future histories that reflect new scientific paradigms of extremophile biology. Where Gantz imagined a voracious and intelligent plant overgrowing Mars, McAuley makes the discovery of a Martian cousin to Earth microorganisms, one capable of integrating the DNA of terrestrial life-forms, the center of his cautionary tale of the future of biotechnology. Ben Bova's *Return to Mars* (1999) and many other novels hunt larger game, relegating various Martian races to the remote past. The artifacts of these vanished civilizations register another version of the sublime—a lost knowledge that resists human understanding, appropriation, and exploitation. In different ways, all of these novels indulge in the fantasy—or the alternative history—that Mars matters in crucial ways to the future of humanity. In this respect, they imagine a future (or a past) in which the space program does not fall prey to changing national priorities, Reaganomics, and the losses of the Mars Observer, the Mars Climate Orbiter, and the Mars Polar Lander. Baxter's *Voyage,* which describes the personal, political, and budgetary struggles to get to Mars in a 1980s that never happened, is typical of a genre conditioned by fifteen years of economic recession, budget cuts, and a pork-barrel shuttle program. His novel is a fictional rendering of criticisms leveled by scientists such as Bruce Murray, who decried the shortsightedness of an administration that canceled or pared down Mars missions while it committed billions to a "vain search for [an] economic justification" for the shuttle program and "flawed theatrical visions" of the "militarization of space" (1989, 23). Similarly, Jack Williamson's *Beachhead* (1992) describes an underfunded, shoestring effort to explore Mars, but one that nonetheless depicts colonization as a beacon of hope for an Earth in the throes of environmental and demographic crises and the resulting sociopolitical conflicts. Ancient aliens and human terraformers are fantasy figures who offer means to overcome imaginatively the frustrations of a stalled space program and the deferral of the *Star Trek* frontier.

MARS AS THE ABODE OF ANCIENT ALIENS

The alien-artifact novels reanimate the mysteries of Mars by drawing on myths of a lost race of technologically advanced, even quasi-divine, beings, who have left behind only traces of their prowess. Like Lowell's canal builders, these civilizations are unapproachable, though separated by vast differences in time rather than the distances of interplanetary space. Nevertheless, these traces confirm the technological backwardness and evolutionary insignificance of humanity in the cosmos. Terry Bisson's *Voyage to the Red Planet* (1990), Allen Steele's *Labyrinth of Night* (1992), Dana Stabenow's *Red Planet Run* (1995), Hartmann's *Mars Underground,* and Ian Douglas's *Semper Mars* (1998), to name only a few, center on the discovery of alien artifacts that resist human efforts to uncover their meaning and suggest a fundamental ambiguity in humankind's response to its loss of prominence in a biologically robust universe. On the one hand, the prospect of benevolent aliens offers a New Age alternative to both a traditional Judeo-Christian deity and a mechanistic, secular science (Bryld and Lykke 1999, 50–64). Yet on the other, the absence of these aliens suggests that they have succumbed to the entropic fate of a dying planet and are either extinct or have left for greener planets elsewhere in the galaxy. In either case, their artifacts remain inscrutable, even though the humans who discover their machines or decipher their cryptic messages become subject to the very desires that the alien technologies seem to have been designed to satisfy. Frequently, these novels borrow from Paul Verhoeven's film *Total Recall* (1989), itself an adaptation of Dick's short story "We Can Remember It for You Wholesale" (*Fantasy and Science Fiction,* April 1966).

At the end of *Total Recall,* the hero Quaid (Arnold Schwarzenegger) places his hand in the four-fingered handprint of the alien makers of a miraculous technology and triggers a physics-defying process that instantly terraforms Mars. The 500,000 year-old atmosphere generator, buried inside a mountain, obviously has never been used because Mars has no earthlike, oxygen-rich atmosphere. There is no explanation why superbeings would build a vast subsurface technology that can benefit only oxygen-breathing creatures and then never activate it. This technology exists only as an expression of human desires to transcend the environmental constraints and political obstacles that confine the film's colonists to isolated domes and a dismal existence. The *machina ex dei* acts as a means to reassure humanity that the vanished race who built this technology cares about and for oxygen-dependent humans; a half million years earlier they had known that the colonization of Mars could open a new, seemingly Edenic frontier. This reassurance paradoxically is necessary because the existence of this miraculous alien technology suggests that these vanished aliens have anticipated and indeed *determined* human desires long before humanity could conceive of even the rudiments of such godlike, technoscientific power. The alien's four-fingered handprint in *Total Recall* is an apt image for what is ultimately an ascientific belief in transcendence. Yet this belief in 1990s science fiction is always colored by an anxiety that such benevolence is a function of an absolute power and knowledge: if the absent aliens stand in for a traditional deity, they can also represent contrasting, if dialectically linked, fears. These aliens may prove as remorseless as Wells's invaders, as totalitarian as any Nazis, or as unconcerned about the insignificant inhabitants of the third planet as their artifacts sometimes suggest. Many of the alien artifact novels of the 1990s explore this dark underside of the fantasy of transformative technologies. In these works, the lost technologies of Mars seem the ghostly traces of an absolute that defies representation: the machine without instructions, the message that cannot be decoded, the voice that cannot be heard.

Hartmann's *Mars Underground* concludes with the autoactivation of an alien mechanism buried in the southern polar cap. Rather than the magical—and scientifically impossible—atmospheric generator of *Total Recall*, the alien machine is "a castoff" neither "waiting, or remembering, or dreaming," a forgotten relic that has only "the same meaning as an arrowhead in an Illinois cornfield, a stone axe lost in a French cave, a fragment of a painted pot, staring into the sun every day from a cobbly desert in Iraq or Peru or China or Arizona." Its significance cannot be understood outside of its lost technological and cultural context. "Filled with 3.2 billion years of silence" (427), the alien artifact is a reminder of the limits of human knowledge, a totemic stand-in for the questions that science cannot answer. Like the bits and pieces of terrestrial archeology, it is indifferent to humanity's conceptions of time and meaning. This meaninglessness, however, always exists in at least an implicit dialectical relationship to the abundance of meaning represented by the magical technologies of *Total Recall*—the superhuman wisdom of advanced civilizations that transcends and mocks the limitations of twentieth-century knowledge and experience. In this respect, the mysterious alien technologies are the hauntological other of terraforming—the dream of an ultimate material, even spiritual, redemption on Mars for the sociopolitical and environmental shortcomings of Earth.

In many respect, though, it is difficult to take these lost-civilization novels as much more than updatings of *The Devil Girl from Mars*. Even the novelists who employ the narrative device of lost civilizations on the red planet often distance themselves from making truth claims about the planet they represent.[6] And the genre as a whole occasionally verges on self-conscious parody. In his hilarious send-up, *Voyage to the Red Planet*,

Terry Bisson satirizes both the heroic pretensions of space exploration and the quasi-theology of benevolent aliens leading humankind to a promised land of peace and prosperity. After the demise of NASA, a ragtag film crew out to shoot a Martian blockbuster on location commandeers a mothballed vessel, the *Mary Poppins,* and blasts off for Mars. After various misadventures, they discover the remnants of a vanished civilization that had made humanity its "project" as well as a mysterious message from these "Creators." At the end of the novel, the message finally is decoded to reveal a wisdom reminiscent of a *Mad* magazine parody: "Good luck!" (231).

Comedy, though, usually is in short supply in novels that use Mars as a way to escape from the ecological and sociopolitical problems that threaten to overwhelm a near-future Earth. The earnestness of much of this literature reflects the widespread view that science fiction is a form of dead-serious speculation about, if not an intimation of, possible futures. Brian Aldiss's *White Mars* (1999) ends with a plea for his organization APIUM (Association for the Protection and Integrity of an Unspoilt Mars) intent on protecting the planet against the (future) ravages of colonization (323); Zubrin's *Mars Landing* concludes with an afterword that explains his Mars Direct Plan and encourages readers to join the Mars Society. If Burroughs's John Carter novels depict a nostalgia for Bronze Age values of male heroism and an alienation from an increasingly bureaucratized modern world (Stecopoulos 1997, 170–91), the alien artifact novels paradoxically reveal that the dreams of returning humankind to its originary wholeness are a function of the profound alienation of contemporary mind, body, and spirit. The exploration and colonization of Mars in Hartmann's *Mars Underground* are set against a background of environmental degradation and capitalist exploitation: "Earth's last chance at sustainability was being squandered" by corporations, "like the ancient gods," who "hovered permanently but unseen in some economic exosphere around each planet [and] permeated the very air of Earth and Mars, like smog" (45, 217). A prominent planetary scientist and author, Hartmann yokes implicitly the lack of human missions to Mars in the late twentieth century to a critique of multinational corporatism similar to that voiced by Kim Stanley Robinson in his Mars trilogy.

In Douglas's *Semper Mars* and Steele's *Labyrinth of Night,* the United States battles forces of evil (the United Nations!) to control alien technologies found inside the Face on Mars; these technologies promise their possessors almost unlimited power to overcome the problems of overpopulation, energy generation, food shortages, and ecological collapse. In this respect, the very presence of such artifacts from a technologically superior race often signals or promises an apocalyptic moment in the (future) history of humankind (Leib 1998). In *Semper Mars* and its sequels in Douglas's Heritage Trilogy, the alien technologies ultimately link human evolutionary and sociopolitical history to the machinations of malevolent aliens out to destroy competitor races when they near the threshold of interstellar travel. Running wild with Richard Hoagland's claim that the Viking photos of the Cydonia region of the planet's surface reveal a humanoid face a mile long, as well as various "pyramids" and "plazas," Douglas remasculinizes the high frontier (Hoagland 1996). Shootouts on Mars become do-or-die battles to escape the fate of a dying world—Earth.

Significantly, these depictions of an eco-apocalypse in the twenty-first or twenty-second century cut across party lines. In Douglas's space opera paean to the U.S. Marines, *Semper Mars*, the only hope for a polluted and overcrowded Earth is "the wholesale industrialization of space" and the exploitation of alien artifacts (31). In contrast, Stabenow's *Red Planet Run* veers toward New Age mysticism in yet another plot that involves the mysterious lost civilization of Cydonia. Terraforming in her novel becomes the desire of the universe to bring or return life to a dead planet rather than the imposition of human desire and terrestrial biota on an inhospitable landscape. In both cases, however, the diagnosis is similar: in the mid-twenty-first century, a population crunch plagues a planet beset by global warming and its attendant consequences:

> Rising sea levels—and disastrous storms—had killed millions, and driven tens of millions more into refugee status. Precious agricultural land in Bangladesh and coastal China and the US Gulf Coast had been swallowed up by the advancing tides. As nations like the United States and Russia finally ended their dependence on fossil fuels and polluting industries, the poorer nations of the world had finally reached the point where they could become rich . . . but only by embracing the air-, land-, and sea-destroying industries that the US and others were now abandoning. Acid rain generated by China was destroying forests in western Canada, and the Amazon rain forest was all but gone now, replaced by ranches and acid-pissing factories. The last of the oil was vanishing; other raw materials vital to civilization—copper, silver, titanium, cobalt, uranium, a dozen others—were nearly gone as well. (Douglas 1998, 30–31)

Although this passage seems as though it could have been lifted out of a dozen left-leaning dystopian novels of the near future, Douglas uses this threat of the collapse of western civilization to spin a right-wing fantasy of a tyrannical United Nations rationing resources and stifling American initiative, power, and scientific know-how. The threat of environmental collapse becomes the motivation for the exploitation of Martian resources and the *machina ex dei* left by aliens under the Face. The militarization of Mars, in this regard, plays out the fantasy of capitalism and the endless exploitation of resources: the struggle for alien technologies is a politicized version of the desire to transcend the problems of resource extraction, production, and pollution.

Yet this fable of techno-militarism as a means to shore up a (masculinized) American identity is shot through with the same sort of scientific mumbo jumbo that had characterized science fiction a century earlier. In a bizarre updating of the atmosphere factory at the end of Burroughs's *A Princess of Mars,* the so-called D & M Pyramid at Cydonia is identified as "some sort of titanic apparatus for creating a warm and breathable atmosphere over a large portion of the northern hemisphere; its destruction had resulted in the rapid and inevitable bleed-off of the artificial atmosphere into space, the freezing of the oceans, the suffocation, freezing, and mummification of humans [slave workers for the genetically engineering superrace that had built the complex] suddenly faced with Armageddon" (368). This device, in effect, reveals, yet again, that the dominant metaphors of such science fiction are neither those of progress nor political change but of miraculous transformations: the return to life of a dead world. Douglas projects the manichaean conflicts of right-wing politics back in time and out into the galaxy; interstellar space looks like the Barsoom of *The Gods of Mars:* warring races, heroic virtues, and humans or their stands-in fighting to throw off the yoke of slavery. In this respect, *Semper Mars* registers the frustrations of a soon-to-be spacefaring race in dry dock; the novel's alien artifacts are outlandish metaphors for the desire to jump years into the future and transcend the failures, delays, and slow pace of NASA's space efforts. These miraculous technologies mark the gap between the decades-old promises of a Flash Gordon civilization and the political, economic, and technological difficulties of getting humans to Mars.

TERRAFORMING MARS

By the early 1950s, scientific assessments of Mars had made the colonization of an earthlike twin seem unlikely. Although the composition of the atmosphere would not be understood until the Mariner era, best-guess estimates of available water and oxygen placed the inventories of those resources far below what would be necessary to sustain human life. Pulp fiction writers in the 1940s, notably Jack Williamson, had raised the possibility of re-engineering entire planetary environments, and after Heinlein described the terraforming of Ganymede in *Farmer in the Sky* (1950), other writers—including Arthur C. Clarke, Isaac Asimov, and Walter M. Miller—depicted a near-future Mars in the process of being converted to an earthlike home for colonists (McCurdy 1997). Terraforming—the speculative science of planetary engineering—offered readers in the 1950s a Martian future far different from that envisioned in Dick's "Survey Team." Recognizing that Mars likely was a near-dead world, science-fiction writers began to speculate that its chemistry—notably the water and oxygen thought to be locked into the polar caps—would permit artificial interventions that eventually could render the planet habitable for future colonists. Rather than conceding that environmental destruction is an inevitable consequence of human colonization, early terraforming fiction holds open the possibility of a godlike redesign of the red planet. In this respect, these thought experiments redefine the moral valence of the anticolonialist novels of the 1950s. Terraformed Mars offers the hope of redressing sociopolitical and environmental failures on Earth by inverting the effects of humans on their environments. The technologies that threaten to destroy Earth—nuclear weapons in the 1950s and 1960s, greenhouse gases thereafter—are transformed into hi-tech plowshares to sow a new utopian ecology on Mars.

Clarke's *The Sands of Mars* (1952) marks a decisive break with the parablelike quality of Bradbury's *Martian Chronicles* by depicting the "realistic" problems of settling colonists on an Antoniadian planet that lacks sufficient food, water, and air to support human life. In his foreword (dated 1967), Clarke describes his work as "one of the first science-fiction novels about Mars to abandon the romantic fantasies" of Lowell, Burroughs, Lewis, and Bradbury, and it goes on to dramatize the difficulties of manufacturing resources "at the end of a supply line that's never less than fifty million kilometres long" (Clarke 1974, v, 86). In Clarke's novel, Warren Hadfield, the chief executive of the Mars colony, worries not about aliens and evil billionaires but " 'cold, lack of water, lack of air' " (86) and indifferent support from Earth governments demanding profits rather than interplanetary adventure. The problems of labor and self-sufficiency are at the heart of Clarke's novel; they reflect "realistic" speculations about the prospects for settling Mars and provide a conceptual blueprint for plans that remain on the drawing board of enthusiasts in the Mars Society in the twenty-first century. Rather than a straightforward saga of human engineering prowess conquering a hostile world, *Sands of Mars* raises ethical concerns about terraforming the planet: the novel seeks to ensure that humankind will not simply repeat the mistakes of its terrestrial past. The discovery of an indigenous race of animals with well-developed cognitive capabilities leads Clarke's settlers to conclude that terraforming Mars depends on humankind's "duty always to safeguard the interests of its rightful owners" (199). This language of property rights and ownership both remakes Mars in the traditional image of an idealized commonwealth and distinguishes *Sands of Mars* from contemporary science fiction (that of Kornbluth and Merril, for example) that depicts the red planet as a battleground for the social and economic conflicts that characterize Earth. Morally as well as scientifically, terraforming marks a new beginning for the human race, a chance to resolve the environmental problems and political conflicts that beset Earth. For Clarke, enhancing the conditions for life on Mars becomes a measure of humankind's moral fitness, an indication that understanding and justice have progressed in concert with space-age technology.

In the fifties, then, terraforming fiction redefines space-age heroism as something more than interplanetary swordplay and conquest. It suggests both a faith in technology and the promise of a human destiny that transforms or transcends the dead-end cycles of exploitation envisioned by Dick and Bradbury. Isaac Asimov in "The Martian Way" (*Galaxy* 1952) places the politics of water at the center of future conflicts between colonists on Mars and Earth. His story describes the heroic venture to capture a huge ice asteroid from the rings of Saturn and return it to Mars so that the colonists no longer will have to import expensive water from Earth. The following year Walter Miller's "Crucifixus Etiam" appeared in *Astonishing Science Fiction.* Best known as the author of the post-apocalyptic science-fiction classic *A Canticle for Leibowitz* (1959), Miller offers the first—and one of the most thought provoking—explorations of the values that drive the dream of terraforming Mars and the sacrifices that such a project requires.

In 2134, Manue Nanti, a Peruvian laborer on Mars, works long, dreary hours, swinging "the heavy pick into red brown sod." Like his fellow workers, he is a cyborg, physically augmented to survive conditions on an inhospitable planet: with "plastic areator valves stitched into his chest," Nanti depends for his survival on an external "mechanical oxygenator [that] served as a lung, sucking blood through an artificially grafted network of veins and plastic tubing, frothing it with air from a chemical generator, and returning it to his circulatory system" (Miller 1980, 49). The Mars he inhabits is alien enough to require such technologies but thoroughly familiar in its sociopolitical and economic structures. Although his labor contract calls for a five-year stay on Mars, the "oxygenator" is the stigmata of a lifelong commitment. Although Nanti fights desperately—and futilely—to retain the ability to breathe on his own, his fate is sealed by the technological augmentations that make him, in effect, a Martian: "The wasted, atrophied chests of the men" who have been on Mars for a few years signify the loss of the ability to breathe on their own; if they ever return to Earth, they will "still need the auxiliary oxygenator equipment" (49). Despite what seems a conventional set up—an exploited and undereducated worker serving the ends of a project he does not, at first, understand—Miller resists the leftist, antiestablishment tenor of Kornbluth and Merril; *Gunner Cade,* published a year earlier, ends with a revolt of Martian miners against a totalitarian colonial government. In contrast, Miller yokes terraformation to the transformation of "man."

Nanti and his coworkers learn after a year of backbreaking labor that the huge hole they have been digging is not a well or mine but one of three hundred shafts for controlled nuclear explosions intended to blast helium and oxygen into the atmosphere. The terraforming project to give Mars "a breathable atmosphere" is, to say the least, long-term; at a mass meeting a half hour before the explosion, the workers are told: " 'Three hundred wells, working for eight centuries, can get the job done' " (64). Mars, for Miller, is neither a place to get rich nor a dumping ground for an overpopulated Earth but "an eight-century passion of human faith in the destiny of the race of Man" (68). Nanti quells an incipient rebellion among the other workers and comes to embrace his fate as a permanent resident of his new home planet and a spiritual commitment to a new future for humanity. A devout convert to Catholicism, Miller identifies terraforming as the resurrection of a "dead" planet; the passion of Nanti is the faith necessary to build the space-age equivalent of a medieval cathedral. The planet itself becomes a monument to the "destiny of the race": religious faith guarantees the survival of a "passion" a thousand years into the future and underwrites the subsequent eight centuries of terraforming. Nuclear blasts on a dying world are planetary stigmata, signs of humankind's collective commitment—composed of thousands of individual sacrifices—in the name of a transcendent passion.

Asimov's and Miller's short stories anticipate two of the terraforming strategies that find their way into the scientific literature that emerged in the wake of the Mariner 9 mission. If the seemingly lifeless planet photographed by Mariners 4, 6, and 7 dampened enthusiasm for terraforming, the evidence of water erosion and ancient channels suggested that science-fiction scenarios such as nuking Mars to create an atmosphere were not scientifically impossible. Water on Mars in the past indicated that the planet's climate might be unstable, driven by changes in the precession of the planet's axis of rotation over the course of thousands of years (Belcher, Veverka, and Sagan 1971, 241–52; Sagan, Toon, and Gierasch 1973, 1045–49). Carl Sagan's 1960 dissertation had considered the possibility of terraforming Venus in an era before the 800 degree temperatures of its surface had been measured (Davidson 1999); with the data from the Mariner missions in hand, *Icarus* (which Sagan then edited) devoted several articles to speculation that both subsumed some of the generic postulates of earlier science fiction and promoted terraformation as a means to speculate about the relationships among planetary atmospheres, hydrological cycles, and long-term climate changes.

In contrast to the Christian ethos of Miller, the scientific speculation about terraforming voiced (at least implicitly) a different kind of reverence and envisioned a different future for humankind: imprinting human desires and civilization on the cosmos (Bryld and Lykke 1999, 92–117). Joseph Burns and Martin Harwit calculated that it would be possible to induce a change in Mars's precession by putting an asteroid in orbit around the planet to prolong its "spring"; the less the angle of deviation from a vertical rotation, the warmer and thicker the atmosphere would become (Burns and Harwit 1973, 126–30). Sagan suggested two possible methods for terraforming Mars: scattering dark material on the polar caps to decrease their albedo, and thereby increasing their capacity to absorb solar radiation, and crashing an ice-rich asteroid into the planet's atmosphere. As Mars became warmer, the carbon dioxide locked in the polar caps would sublime, thickening the atmosphere. As the atmosphere began to retain more solar heat, more carbon dixoide would be outgassed from rocks and the regolith, thickening the atmosphere even further and triggering a runaway greenhouse effect. Calculating that the polar caps contained enough carbon dioxide and water to create an atmosphere as thick as Earth's, Sagan speculated that the reborn capacity of the planet to retain heat would end its millennia-long winter (1973, 513–14).[7]

Writing in the 1970s, Burns, Harwit, and Sagan offer their plans as extensions rather than transformations of an emerging ecological ethos. The rhetoric of both articles is redolent with the concerns of overpopulation and environmental impact statements. Burns and Harwit note that "there is always something repugnant about man pushing his own interests and fixing nature" (1973, 127), but suggest that their scheme for transforming Mars into a world ready for large-scale colonization is the least intrusive. Sagan presents his plan as a kind of prosthetic aid to control the planet's "natural climactic instability," and urges that planetary engineering take place "only after a thorough and ecologically responsible program of unmanned planetary exploration has been completed" (1973, 513–14). The greenhouse effect that Sagan's plan would induce transforms the unwanted consequences of fossil fuel consumption on Earth into a means for planetary rebirth. His vision of a terraformed Mars differs from Miller's: rather than a testament to a human passion for the "destiny of the race," Sagan gestures toward an ethics of stewardship and ecological responsibility. In applying the concept of ecology to Mars, Sagan invokes both a scientific (and science-fiction) tradition of a "life-bearing planet" (Lederberg 1966, 133), and anticipates a future in which "unmanned planetary exploration" will end and the entwined projects of human exploration, colonization, and terraformation will begin. In this early scientific literature on terraforming, then, Mars retains its traditional, even over-determined role as a site for speculating about humankind's place in the universe, even as it serves as a thought experiment, a second chance to avoid the errors and tragedies of the European colonization of the Americas and Africa. On Mars, twenty-first-century humanity can begin to correct or transcend the errors—overpopulation, pollution, and resource exhaustion—that have characterized our inhabitation of Earth. In this regard, terraforming holds open the possibility of revivifying an idealized version of the American frontier (Bergreen 2000, 4–5).

For novelists in the 1970s and after, terraforming opens narrative possibilities too tempting to resist. Several novels, such as Greg Bear's *Moving Mars* (1992), follow Sagan in positing Martian life-forms that have adapted to the long-winter model and revive periodically. In *Martian Spring* (1986), Michael Lindsay Williams has Mars terraform itself into a living world, although an advanced race of Martians devote their considerable knowledge to augmenting the periodic warming of the planet. Williams's novel cleverly displaces the ecological anxieties that had defined a previous century of science fiction; rather than succumbing to the irrevocable heat-death of their planet, Martians have built their social, technological, emotional, and political existence around the physiological and climatological imperatives of "the Long Sleep [that must] be successfully traversed" (148). The greenhouse effect induced by the Martian spring thus relocates fears of eco-devastation within a cyclical paradigm that naturalizes the effects of otherwise catastrophic global change. In other works, such as Ian McDonald's *Desolation Road* (1988), terraforming approaches magical realism. The transformation of the environment becomes the stage for redefining what it means to be human, and in Ian Watson's *Martian Inca* (1977), terraforming Mars is identified with making the subject whole; the lure of the Martian frontier is the chance to start over by transforming the self. Not all novels, though, accept an easy equivalence of planetary engineering and social progress. In Donald Moffitt's *Crescent in the Sky* (1989), a partially terraformed Mars is the setting for a Martian caliphate, one of the end products of the millennium-long ascendancy of Islam. A brutal and corrupt court—characterized by infighting among palace factions headed by grand viziers and eunuchs—is challenged by Bedouins and disaffected rebels who have transported the nomadic lifestyle of their ancestors, complete with genetically engineered camels, horses, and falcons, into the thin atmosphere of Mars. Moffitt's novel disrupts the association of technoscientific and sociopolitical progress: on Mars, absolute rulers have their heads attached to the bodies of younger clones to prolong "their" lives; thieves have new hands regrown to replace those lost to Islamic law; and genetic engineering, cloning, and microsurgery serve the interests of a medieval court.

Although some terraforming literature verges on interplanetary twelve-step programs, other science-fiction writers concentrate on the physical and emotional costs of surviving on Mars. Two decades after *Outpost Mars* and its tale of feral adaptation to an alien environment, Frederik Pohl in *Man Plus* (1976) and Kevin Anderson in *Climbing Olympus* (1994) center novels on humans surgically altered to survive on the forbidding surface of post-Mariner Mars. Both novels respond to the Viking and Mariner missions by reimagining colonization in terms of surgical transformation and genetic manipulation. In *Man Plus,* the hero, Roger Torraway, undergoes harrowing procedures, including castration, that transform him into a cyborg capable of living on the surface of a hostile, sterile world. This novel and its sequel, *Mars Plus* (Pohl and Thomas 1995), reconfigure the relationship of "man" and nature in order to reimagine both the human and environmental consequences of adaptation. Cyborgs are the future's response to the problems of Martian settlement, a means for Pohl and Thomas to raise the possibility that terraforming Mars "would be a massive boondoggle, lots of effort for very little positive result" (210). Re-engineering humankind to adapt to life on Mars turns Torraway and his kind into mythic embodiments of the consequences of trying to transport terrestrial concepts of the human to a world that enforces new forms of cyborg identity. On Mars as on Earth, human existence remains embedded in complex relationships with its environment.

But not all thought experiments about planetary engineering were as thoughtful as *Man Plus* or *Crescent in the Sky*. By the early 1990s, terraforming had become not only the subject of scientific symposia and articles in prestigious science journals, but also a cover story for *Life* (McCurdy 1997). For its most enthusiastic proponents, terraformation promotes the faith that the same greenhouse effect that is warming Earth can be harnessed to produce a rebirth of Mars: pollution is transformed magically to production. Terraformation becomes a technological gamble that will recapture a spirit of adventurous conquest, naturalized as essential to human—specifically American—identity. In science fiction, planetary engineering signals a rebirth of self and culture as well as the transformation or resurrection of a world's ecology. The irony in much of the popular literature on terraforming is that it is precisely the "risk-taking" entrepreneurial spirit that has led to the debilitation of Earth; to repackage manifest destiny for the solar system is to risk repeating on Mars the mistakes that Dick had dramatized in "Survey Team." In some venues, Mars becomes a crucial site for the projection of these fantasies of limitless transformation—of both our "natures" and an inhospitable nature—and for the repression or displacement of actual material limitations for life on any planet.

The most popular treatment of terraforming hit movie screens in 1990. *Total Recall* "solves" the problem of exploitation that so obsessed science-fiction writers during the 1950s: Quaid liberates the miners and sex workers (many genetically damaged by radiation seeping through the cheaply constructed domes) from having to pay to breathe and thereby "frees" Mars from the domination of an evil monopolist (see Mizejewski 1993; Schmerz 1993; Glass 1990). If, as Robert Miklitsch suggests, the popularity of Verhoeven's film stems from its seductive offer of "a fantastic resolution to [the] real, lived contradictions" of late capitalism (1995, 18), its fantasy of ending what he terms "the *anxiety of ecology*" can be accomplished only by the shotgun marriage of two often antithetical traditions in post-Mariner science fiction: the metaphysical speculation posed by the discovery of the technology of an advanced (usually long-vanished) race, and the "realistic" speculation about colonizing and terraforming Mars.[8] *Total Recall* can be seen as a fantasy in which the film's overt concerns about political freedoms and subjectivity mask the contradictions within the logic of intensification that pits ecology against economics. The terraforming of Mars in the film is presented as an ultimate "solution" to the problems of oppressed labor, exploitation, tyranny, and invasive governmental interference in the lives of colonists. While the scenario for instant terraformation is scientifically absurd, the film shoehorns an alien world into a familiar semiotics. The surreal ending (despite a conventional, throwaway line that raises the possibility that, gee, maybe it was all a dream) transforms Mars into an Edenic amalgam of two planets, a brave old-new world of blue skies, panoramic vistas, and unspoiled terrain. But the significance of the presumed future colonization of Mars, and of humanity's place in the universe, remains anchored in the values and assumptions of humankind's irrevocable alienation from "nature." Verhoeven converts the twists and turns of Dick's story about memory and identity into a filmic mode of double alienation, from one's environment and from one's self: on a planet of mutant workers victimized by shoddy domes and an evil governor's monopoly of breathable air, Quaid, the rebel, is apparently the product of memory implants that override his "original" identity, as the government agent Hauser, so that he can infiltrate the underground without giving himself away to the mind-reading leader of the mutants. The film has attracted so much critical attention because it offers a popcult primer in postmodern notions of identity: morality lies in Quaid's rejecting his "real" identity in favor of a liberatory fiction.

Ultimately, Verhoeven's film offers viewers a double temporal displacement of the relationship between self and environment. On a world seemingly locked into the tyranny of commodifed air, it is an untold narrative of the past—the vanished alien race—that holds the promise of a future. Miklitsch notes astutely that "while the [terraforming] reactor is explicitly a product of an extinct civilization, the alien labor that produced it—and therefore a certain history of Mars—is wholly effaced in the film: the technology is simply *there*" (1995, 23). The one-button solution to the massive engineering problems of planetary transformation not only mystifies the scientific description of how we get to this future, it also simplifies the ethical problems involved in terraforming: the right to terraform an alien environment is presented literally as a life-saving solution for Quaid, about to suffocate on the surface, and for the planet as a whole. In yet another sense, the absent labor behind the film's miraculous terraforming can be seen as a repression of the lung-destroying work depicted in Miller's "Crucifixus Etiam" or the grim mining operations that had a been a staple of leftist science fiction in the fifties—in novels and stories by Kornbluth and Merril, del Rey, Brackett, and Dick. The film's terraformed future, ironically, returns the audience to the values of an idealized past in which workers are less individuals than a chorus of extras for the heroic individual. In this regard, the blue-skied Mars that Schwarzenegger views at the end of *Total Recall* can be both a characteristic Dickean illusion—the ambiguous dream where fantasy and reality interanimate—and a totemic promise that the future governor of California embodies as a paean to Reaganesque America: the heroic individual has freed both atmospheric oxygen and the captive labor represented by the genetically monstrous workers, promising a brave new world ripe for capitalist investment. In the beginning, to paraphrase John Locke, all the (Martian) world was America—a world endlessly open to exploitation. Yet as Dick intuited in "Survey Team," this desire to envision the future as the extension of an idealized past inevitably confronts the limits of such exploitation. *Total Recall* ends before its new frontier can develop into a postmodern version of Ray Bradbury's Mars—"IRON TOWN, STEEL TOWN, ALUMINUM CITY, ELECTRIC VILLAGE, CORN TOWN, GRAIN VILLA, DETROIT II, all the mechanical names and metal names from Earth" (1950, 102–3). Miraculous alien technologies finally express only humankind's boundless desires to benefit from and transcend an industrialized Earth. And Mars remains bound to analogies that were in vogue a century earlier.

EIGHT

Mars at the Turn of a New Century

REINVENTING MARS

For two decades, until the Mars Global Surveyor (MGS) began to orbit Mars in 1997, the results from the Viking missions structured views of the red planet. The photographs and data returned by the Viking landers and orbiters took years to sift through and fit into a more-or-less coherent image of the planet—a dead world, frozen into senescence, that was popularized by science-fiction authors such as Ben Bova and Kim Stanley Robinson.[1] Yet even as this consensus emerged, the scientific disciplines and technical specialties that had contributed to the Viking missions—microbiology, computer science, telecommunications, photography, and planetary ecology among them—were undergoing revolutionary developments that altered the questions that could be asked about past and current conditions on the planet. Consequently, the scientific contexts for the Global Surveyor and Pathfinder missions in 1997 were significantly different from those that obtained twenty years earlier. In part, this chapter examines the fragmenting of the Viking consensus and the often intense debates that have been provoked both by new data and by the changing values and assumptions that shape its interpretation. If "the exploration of Mars [played] out like a detective story" between 1997 and 2004, it had no agreed-on or even foreseeable conclusion (David 1999, 40). In 2001 Kathy Sawyer, writing for *National Geographic,* described planetary scientists as "mystified—and astonished" by the images returned from the Mars Orbital Camera (MOC) on the Global Surveyor (2001, 30). Kenneth Edgett, a mission scientist on MOC, describes his response to these photographs as "bafflement": "'Much of [the data] doesn't add up,'" he acknowledges. "'It's spine-chilling . . . mind-boggling'" (Sawyer 2001, 36). In brief, the results from the Global Surveyor (1997–2004), Mars Odyssey (2002–2004), the European Space Agency Mars Express orbiter (2004), and the Spirit and Opportunity rovers (2004–2005) have proved as revolutionary as the Mariner 9 photographs were in 1972.

As Edgett's comments suggest, writing about current, planned, and future exploration poses historiographic and conceptual problems. "Mind-boggling" photographs, as the canal controversy demonstrates, can prove red herrings, objects of contention, or a set of clues, if interpreted correctly, to unlocking age-old mysteries. With the 2004 missions of NASA and the European Space Agency on or in orbit around Mars, much of what scientists think they know about the planet remains open to the possibility of wholesale revision. This chapter, then, itself must be a work in progress; even if a four-armed Thark emerges from a crater and bounds off with a NASA rover, the 2004 missions to Mars have raised new questions that have altered yet again the conditions of possibility—the paradigm—for interpreting the data returned. On the eve of the failed 1999 missions, Dan McCleese, the chief scientist for Mars Exploration at the Jet Propulsion Laboratory, suggested that "we will find out what we don't know, but more than that, we will find answers to questions we haven't even asked because we don't know enough" (quoted in Bergreen 2000, 237). In the case of the 2004 missions, the answers suggest new questions—notably about life—that Spirit and Opportunity are not instrumented to ask.

MARS AFTER VIKING, 1976–1996

The study of Mars in the 1980s and 1990s took place amid continuing budgetary problems and three major mission failures. The inability of the Viking missions (at least in the eyes of most scientists) to find evidence of life on Mars dampened congressional and public enthusiasm for another billion-dollar mission. Without much hope for microbial Martians, planetary scientists at NASA found themselves losing a war of attrition to fund a new generation of spacecraft. After 1980, billions of dollars were funneled into Ronald Reagan's Star Wars missile defense system and to a space shuttle project that owed more to "pork-barrel politics" and an "ossified" bureaucracy than it did to any scientific rationale (Sagan 1994, 265). In discussing his term as director of the Jet Propulsion Laboratory (1976–1982), Bruce Murray decried both the sacrifice of "robotic planetary exploration" to the Space Shuttle program and "NASA's obsession with the development of massive new space systems" at the expense of scientific research (1989, 21, 49). Mars missions were delayed, scaled back, and delayed again.

Yet for some NASA veterans, the Viking biology results were less a failure than an indication that "the question of life on Mars was by no means settled" (von Braun and Ordway 1979, 198) and therefore an incentive for future exploration. Even as the bulk of NASA's budget went to the Space Shuttle program and then the International Space Station, the search for life remained a crucial selling point for plans to explore Mars. More significantly, the question of "life" itself underwent a complex redefinition during the 1980s and 1990s: extremophile biology greatly expanded the range of environments in which microorganisms could thrive; microbiology was transformed by new understandings of cellular evolution; and controversies about possible nanofossils in the Martian meteorite ALH84001 led teams of scientists to identify previously unstudied biomarkers in mineral formations. While no life-detection experiments were scheduled for any of the NASA missions in the 1990s, all of the spacecraft were bristling with instruments to search for water, inventory the minerals and chemical make-up of the planet, photograph the surface at resolutions of a few meters per pixel, and piece together a coherent history of the planet's climatic history. The ill-fated missions of the 1990s—the Mars Observer that went dead in August 1993 as it approached the planet; the Mars Climate Orbiter that missed its orbital insertion and flew past the planet in 1999; and the Mars Polar Lander that crashed to the surface a few months later—signaled a strategic change in scientific thinking about how to search for life. Rather than force-feeding terrestrial nutrients to putative microorganisms and then roasting them to measure by-products, NASA planned to study intensively the chemistry, hydrology, and meteorology of the Martian environment (past and present) in order to assess the chances that there had been (or still might be) life on the planet.

While mission scientists were (and remain) emphatic that, as Bruce Jakosky puts it, "'the search for [microbial] life is driving the overall science strategy,'" that search is defined in terms that science-fiction writers of the 1950s would approve (quoted in David 1999, 39). The goal of NASA's exploration strategy, according to Charles Elachi, Director of JPL's space and earth-science program, "'is to better determine the biological environment on Mars, its past, present, and the planet's potential future'" (quoted in David 1999, 38). As Elachi's comments suggest, the profile for the 1999 missions linked investigations of a "biological environment"—an ecology—to the possibility of future human exploration and colonization. Even the loss of the Mars Climate Orbiter (MCO), the Mars Polar Lander (MPL), and the two shuttle disasters (*Challenger* in 1986 and *Columbia* in 2003) could not dampen the enthusiasm that some scientists feel for the future of humankind on Mars. Two decades after Viking, as I shall discuss at the end of this chapter, inventories of volatiles in the Martian atmosphere and beneath the surface, particularly carbon dioxide and water, offered a basis in fact for scientific speculation about future colonists living off the land.

For planetary geologists, the results of the Viking missions were more complex and intriguing than they were for biologists. The orbital photographs, in particular, prompted an overhaul of theories about the planet's geological history. Approximately 90 percent of the articles about Mars published in scientific journals between 1976 and 1990 dealt with the polar caps, the evidence of water erosion on the surface, or the planet's climatological history.[2] As a rule, these articles sought to interpret the Viking data as the basis for theories to explain the geophysical history of the planet.

Although many scientists had strong reservations about the biological implications of Sagan's long-winter model of climatic change, they acknowledged that the "periodic and eccentric fluctuations in the shape of Mars's orbit" and the obliquity of its rotational axis that "oscillates wildly between 150 and 350" would likely create "global climatic fluctuations." Murray suggested that this history would be "recorded in the polar-layered deposits" first revealed by the Mariner 9 photographs (1989, 65). Because the complex layered terrains in the polar regions indicate a history of repeated depositions of volcanic and wind-born dust and wind erosion, "oceanless Mars" might preserve climatological data that would be impossible to infer from the polar regions of Earth (Murray, Malin, and Greeley 1981, 306–8). With the variables of "oceans, biosphere, ice sheets, and atmosphere" absent, Murray surmised that Mars's polar regions might serve as a kind of three-billion-year-old laboratory, preserving a record of "extremely sensitive response[s] to slight variations in sunlight" that would allow geologists to understand the processes of glaciation and climate change on Earth (1989, 74, 75). The Viking photographs, in this regard, offered a dramatic indication of what scientists did not know about climatological changes on Earth.

For Murray and many others, climate change became a crucial issue because a global understanding of Martian history seemed the only way possible to begin to account for the evidence that the planet, in the distant past, had been scoured by massive floods. The channels draining into

Chryse, north of the Valles Marineris, indicated that a flood of almost inconceivable proportions had occurred during the Noachian era, some three and a half billion years ago. Based on calculations of their width and slope, most scientists agreed that these outflow channels were carved in a matter of days or weeks "by floods of as much as half a billion cubic meters of water a second—ten thousand times the flow of the Mississippi" River (Morton 2002, 158; see Hartmann 2003, 221–24).[3] Although the source and history of the water necessary to produce such floods were uncertain, most geologists were convinced that Mars once had been warm enough and had an atmosphere thick enough for liquid water to remain stable on its surface, at least long enough to carve vast flood plains and dendritic run-off channels (Murray, Malin, and Greeley 1981, 288–90). Although alternate explanations for these features—pressurized gas and clathrates (slurries of liquid carbon dioxide)—surfaced, the dominant view for the last thirty years has been that Mars long ago was comparatively warm and wet.[4] This consensus, however, raised three sets of questions: What had happened to the water? How had the planet lost much of its atmosphere and consequently the ability to retain solar and thermal heat? Did life exist in the distant past on this warmer and wetter world? The Viking evidence suggested that the answers to these questions were interrelated.

In 1979, James Pollack argued that the polar caps did not contain enough carbon dioxide to account for the lost atmosphere of Noachian Mars, and drew on comparative analyses of the carbon cycles of Earth and Venus to suggest a geochemical explanation for this loss. Carbon dioxide reacts with silicate rocks and water to form carbonates; these eventually are precipitated out of the water as carbonate sediments or limestone. Pollack theorized that as this process leached carbon out of the Martian atmosphere, the greenhouse effect that had warmed the planet began to fail. As Mars cooled, surface water froze and either was trapped underground and covered by dust or sublimed into the atmosphere. He calculated that the condensation of this water vapor, convected to the poles, would maintain the ghostly traces of a hydrological cycle—the waxing and waning of the polar caps that had been observed since the eighteenth century (Pollack 1979, 479–553; Morton 2002, 152–54). As Morton suggests, "Lowell would have loved Pollack's ideas" because they fulfilled the promise of a truly comparative planetology; four billion years ago Earth, Mars, and Venus all were enveloped by thick carbon dioxide atmospheres, then followed different evolutionary routes to their present states. Venus experienced a runaway greenhouse effect as volcanoes spewed carbon dixoide into its atmosphere and the sun's heat split water molecules into hydrogen (which was lost to space) and oxygen (which, through processes of weathering, was locked into minerals in the planet's crust); but Mars had much of its carbon stock trapped in carbonate rocks and the regolith (Grinspoon 1997). Pollack's theory, however, held out hope that if a warmer and wetter Mars had harbored life, fossil evidence of its existence might still be found on this "lithified corpse of a living world" (Morton 2002, 154).

While most scientists acknowledged that much of the planet's atmosphere had been lost to space, many were intrigued by the "persuasive geologic evidence that abundant water (or water-ice) exists near the [Martian] surface" (Carr 1981, 186). Even if this "geological evidence" could be associated with the oldest Martian terrain, the Viking data encouraged scientists, like Michael Carr, to study possible mechanisms for the desiccation of Mars. Carr maintained that significant amounts of carbon dioxide and water existed in underground aquifers covered by subsurface ice and porous basalt rocks (1981, 185–89; 1995, 17–46). If these subsurface aquifers were still heated by a molten core, they could remain in a liquid state, protected from surface temperatures by thickening layers of ice. Estimating how much water remained on Mars, either locked in the polar caps or trapped underground, was difficult; but knowing that the clay minerals in the Viking soil samples contained between .03 and 3 percent water, Carr estimated that there was enough water on the planet to cover its surface to a depth of four hundred meters. His theory offered an elegant explanation for the huge flood plains, run-off channels, and splosh craters. If the aquifers existed below the southern highlands of Mars, the pressure of the water would exert a massive upward force on lowland aquifers and the ice caps that contained them. When asteroids or meteors struck the planet or a significant Marsquake occurred, water exploded across the landscape in massive floods. After a relatively short time, the water would freeze, sublime into the atmosphere, or percolate below the surface to begin the cycle again; Mars would revert to its dry, cold state (Carr 1995; Hartmann 2003, 246–49).

Pollack and Carr were among many scientists who helped to produce a broad and flexible model to explain two conclusions that emerged from the Viking missions—the planet's watery past and its abiological present (Morton 2002, 156–67). This consensus, though, was more a mosaic of rough-edged pieces, best-guess assumptions, and speculative extrapolations than an ironclad framework. It could, and did, accommodate different views of the mechanisms by which climatological change had occurred. One key variable in the scenarios devised by Pollack and Carr was the time frame for such change. If climatic instability were driven by the physics of Sagan's long-winter model, then the water-related landforms on Mars could signify something very different from an irreversible loss of atmosphere and water over eons. As early as 1978, Carr and Dave Scott began to calculate the age of various surface areas on Mars by using cratering statistics derived from lunar geology. The Apollo missions had returned scores of rocks from the Moon that allowed scientists to date its oldest geological formations at almost four billion years. Calibrating crater densities to the age of samples retrieved from different locations on the lunar surface, scientists determined that heavy meteorite bombardment on Mars had ended about three billion years ago (Wilhelms 1993, 208–10, 280–82; Hartmann 2003, 91–101). The lunar surface, then, offered a "crude clock" to calculate the age of the Martian surface, if scientists could assume "that cratering history was uniform at least across the inner parts of the Solar System" (Walter 1999, 9).

To date the surface features on Mars, Carr and Scott had to compare the density of Martian and lunar craters within a given area, assume a "uniform" rate of bombardment, and then correlate that data to the appearance of landforms: the heavily cratered terrain of the southern hemisphere, for example, had to be older than the northern plains, which showed comparatively little cratering. Crater density became a widely accepted means for dating the surface of Mars, and the map that Carr and Scott produced served as a benchmark for subsequent studies of the planet's geological history (Morton 2002, 106–112; Zuber 2001, 220–227). In modifying their approach, William Hartmann suggested that because Mars was closer to the asteroid belt than the Moon, craters formed at about twice the rate that they did on the lunar surface. Calling "the ratio between the cratering rate on Mars and on the moon 'the most important number in the solar system'" (Morton 2002, 132), Hartmann offered three rough measures to date the surface areas of Mars: the Noachian-Hesperian boundary (the intense bombardment of Mars after the massive floods that scoured the surface) occurred between 3.8 and 3.5 billion years ago; the Amazonian period (the flooding and reforming of the northern lowlands, the Vastitas Borealis) began 1.8 billion years ago; and the surface of Olympus Mons, the comparatively uncratered regions of this vast shield volcano, may be as recent as 200 million years (Hartmann and Neukum 2001, 165–94; Hartmann 2003, 29–35). While Hartmann's chronology is widely accepted, it is, in effect, a virtual time line, one based on the accuracy of that "most important number in the solar system," the ratio of meteorite bombardment on the Moon and Mars more than three billion years ago.

THE PATHFINDER MISSION

NASA's missions to Mars in the 1990s were designed for a Martian landscape that had been shaped by the Viking data. No biology packages were included on any of the spacecraft launched by the United States during the decade, and the array of scientific instruments—altimeters, magnetometers, and X-ray spectrometers—represented a broad consensus about the questions that planetary scientists needed to have answered. The Mars Observer, launched in 1992, was instrumented to study atmospheric and surface chemistry, to map the planet, and to provide better-resolution photographs than those obtained by the Viking orbiters. It was an expensive undertaking (almost a billion dollars), but it relied on cumbersome and often antiquated technologies, notably its obsolete on-board computers. It was also, significantly, a single-shot venture. The Mariner and Viking missions each consisted of two spacecraft: when Mariners 3 (1964) and 8 (1971) failed after launch, Mariners 4 and 9 continued successfully on their missions. But budget constraints forced NASA to launch this big-ticket, high-risk mission without a backup vehicle or a contingency plan. When Mission Control lost contact with the Mars Observer as it approached the planet in 1993, it meant, as NASA administrator Dan Goldin put it, the loss of " 'the whole system' " (quoted in Bergreen 2000, 217). With this failure in mind, NASA spent the mid-1990s struggling to redefine its approach to interplanetary exploration.

Under Goldin's direction, NASA adopted corporate strategies in a quest for efficiency and streamlined the way it did business, redirecting its energies to low-cost, redundant missions. "Faster, better, cheaper" became the unofficial slogan of an agency that planned to launch up to ten missions a year. The most spectacular of these, at least in terms of public relations, was Pathfinder in 1997, the first lander to return to Mars in two decades. Pathfinder was designed primarily as a geological mission to demonstrate that a bare-bones approach to planetary exploration would work, yet NASA's public relations team nonetheless justified the mission in the ambitious rhetoric of exobiological discovery. The Pathfinder Press Kit (July 1997) emphasized that the mission's scientific objectives offered the opportunity "to begin answering fundamental questions about the origin of life": chemical analyses of rocks on the surface would shed light on "climatic changes on another planet," and this new-found knowledge of hard rock petrology, in turn, could illuminate "the consequences of natural and human-induced [climatological] changes on Earth" (NMR 1: 211). While the anonymous author reiterates a forty-year-old strategy for justifying the exploration of Mars by linking scientific data collection to "eventual human expeditions," she or he also invokes, implicitly, the problems of global warming and emphasizes NASA's role as an agency devoted to ecological investigations of Earth. The longest section of the Press Kit, however, is entitled, "The Search for Life," and, significantly, its ten paragraphs do not mention the Pathfinder mission. Instead, the author provides thumbnail sketches of extremophile microbiology, the ALH84001 controversy, and the technological developments in electron microscopy and laser mass spectrometry which David McKay's team used to arrive at its tentative conclusion that the meteorite contained fossilized Martian nanobacteria. The section concludes by describing plans, now delayed by six to ten years, for a sample return mission in 2005. Pathfinder, in effect, is depicted as a trial run for NASA's future ventures in exobiology. Searching for life on Mars, two decades after Viking, dominates this brief for media attention, public support, and congressional funding.

Despite NASA's sales pitch, Pathfinder had to be engineered almost from the ground up; few veterans of the Viking missions were still working at NASA twenty years later (Raeburn 1998, 120–21). With a budget under $300 million, the mission developed new strategies and designs for landing scientific instruments on another planet. In place of sophisticated engines designed to maneuver a lander into position for a soft touchdown, the Pathfinder payload was enclosed in a cocoon of air bags. Parachutes were used to slow the lander's descent, allowing the payload to drop to the surface without employing conventional landing gear (Mishkin 2003). The deployment worked perfectly: the cushioned lander bounced more than a dozen times, rolled across the surface, and came to a halt. The air bags that had cushioned the impact deflated, the petals enclosing the Sojourner Truth rover opened, and, after radioed commands from Earth, the six-wheeled, "twelve-inch tall geologist" rolled onto the Martian surface (Bergreen 2000, 67). The technological limitations of this better, faster, and cheaper mission, however, precluded sophisticated biochemical experiments. Pathfinder had only three scientific packages: an imager that returned photographs from the surface, an alpha proton X-ray spectrometer for analysis of rock and mineral compositions, and instruments to record atmospheric and meteorological conditions (NMR 1: 223–25). More importantly, the rover moved, and photographs returned to Earth charted its slow, radio-controlled circuits around a small patch of Martian terrain as it analyzed the composition of various rocks in the Ares Vallis. Pathfinder became the first vehicle to crawl across the surface of another planet.

Over the course of its first month, the rover maneuvered across one hundred square meters and conducted ten separate chemical analyses. During this period, Pathfinder returned over a gigabyte of data, including 9,669 lander and 384 rover images and approximately four million wind, atmospheric pressure, and temperature measurements (Golombek et al. 1997, 1743). Pathfinder was both a geological mission and an Internet event. NASA's web sites posted visually compelling images and reams of data for those who cared to wait for downloads over 28 and 56K modems (Raeburn 1998, 132, 139). By July 7, three days after Pathfinder had landed, the mission web sites had received one hundred million hits; by the time Sojourner stopped transmitting more than a month later, its solar batteries depleted, NASA's sites had received a half a billion hits (Bergreen 2000, 93). The film director James Cameron, a Mars exploration enthusiast, asserted that Sojourner focused humanity's "collective consciousness" and "project[ed] it to the Martian surface" (Space.com, 30 January 2004). If Pathfinder photographs re-created some of the excitement of the Viking pictures sent back to Earth two decades earlier, their availability on the World Wide Web provided a kind of instant forum for mobilizing public support. As Mark Adler at JPL suggested, " 'if we had charged a nickel for everybody who looked at the Pathfinder site, we could fund another [Mars] mission' " (quoted in Gaslin 1998, 18). The geological results were consistent with the theory that Mars had once been warm and wet; the landing site traversed by the Sojourner rover was an ancient flood plain (Raeburn 1998, 166–67). The rocks it studied, given whimsical names by project scientists, were generally similar to those found in the continental crust on Earth; some seemed to be conglomerates, others perhaps sedimentary, suggesting to Matt Golombek, the Project Scientist for the mission, that water had been stable on the planet's surface early in its history (Bergreen 2000, 101). In this respect, Pathfinder's results confirmed the rough consensus that had been emerging since the 1970s: three billion years earlier, Mars had been earthlike in its chemical composition, and the processes that had formed its surface—water and volcanism—were similar.[5]

MARS GLOBAL SURVEYOR AND MARS ODYSSEY

The other three missions launched by NASA during the oppositions in 1996 and 1998 were more ambitious than Pathfinder, part of an overall strategy leading to a sample return mission. Mars Global Surveyor entered orbit in 1997, and although the data it returned was overshadowed, at least in the popular press, by Pathfinder, most scientists recognized that its findings were "infinitely more significant" (Bergreen 2000, 265; 199–220). The Mars Global Surveyor included four scientific instruments: a Thermal Emission Spectrometer (TES) to identify the composition of the surface; the Mars Orbital Camera (MOC); the Mars Orbiter Laser Altimeter (MOLA) to obtain extraordinarily sensitive measurements of variations in topography; and a magnetometer to identify and study remnant magnetic fields on the planet (Hartmann 2003, 74–79, 146–47, 159–62; Morton 2002, 60–64; Bergreen 2000, 113–14, 154–55; Bandfield, Hamilton, and Christensen 2000, 1626–30). The Global Surveyor's key experiments were updated versions of designs that had been included on the failed Mars Observer, and many of the science team members remained the same. Beginning with the first photographs returned in 1997, these instruments revolutionized conceptions of Mars. The revolution, however, was upstaged temporarily by the failures in 1999 of the Mars Climate Orbiter and the Mars Polar Lander.

The MCO included science packages that had been aboard the Mars Observer when it was lost in 1993 as well as a radio relay for the polar lander: experiments included a color imager, a radiometer for mapping the surface, and an array of instruments to study the planet's atmosphere and the distribution of subsurface water (NMR 1: 248–49, 361). The MPL was scheduled for a soft landing near the north polar cap and carried a robust scientific payload including a stereo imager, a microphone, lidar (light radar) to analyze dust hazes in the atmosphere, and a meteorology package designed to detect carbon dioxide and water vapor four to six inches above the surface. Most importantly in the eyes of many in NASA, the robotic arm on the lander could deposit soils in a Thermal Evolved Gas Analyzer (TEGA), which would heat samples to reveal their chemical composition. This search for water in the soil near the pole was not a biology experiment, but, as part of an overall research strategy, it could identify likely sites for future missions seeking to detect extremophile organisms that might exist below the surface (NMR 1: 388–90).

The MCO mission failed when Lockheed Martin, the principal contractor for the spacecraft, provided the navigation team at JPL with measurements in English units (inches, feet, and miles) rather than in the metric measurements that had been specified. Consequently, the team failed to make a final trajectory correction that could have prevented the loss of the spacecraft (Bergreen 2000, 276–78). Reports by the Inspector General's Office and an independent assessment board faulted an understaffed and inadequately trained navigation team, software that had not been fully tested, and a lack of coordination between Lockheed Martin and JPL (Reichardt 1999, 221). An editorial in *Nature* called into question the entire management philosophy that Goldin had embraced. It characterized NASA as "a chronically underfunded programme" legitimized by "an empty slogan"—faster, better, cheaper—"which, after all, simply mirrors corporate culture in the late 1990s: do more with less, and downsize as much as possible" ("Coming to Terms" 1999, 217). This critique, in some respects, recalls those made by Murray and Sagan in the 1980s, and it has been echoed in some European scientific circles. The European Space Agency's orbiter and Beagle 2 lander were designed as a low-cost, high-return option to NASA's missions. Although it had no rover, Beagle 2 carried a robust biology package (described below), while the twin NASA rovers, Spirit and Opportunity, did not.

The Mars Polar Lander was aimed toward an area potentially far more interesting than the landing sites for Viking and Pathfinder. The risks involved in trying to land in the polar regions were well understood. Although photographs from the Mars Global Surveyor showed the target landing area to be relatively smooth, Stephen Saunders of JPL admitted that scientists " 'don't have a lot of experience yet in interpreting [those] images,' " and consequently the polar regions " 'could still harbor lethal hazards too small for Surveyor to see' " (quoted in Lawler and Kerr 1999, 2248). The loss of the Polar Lander compounded the political problems that NASA faced. The Independent Assessment Team suggested that the MPL had crashed into the surface when its lander legs deployed prematurely, probably because a software glitch incorrectly signaled a false touchdown; the lander engines cut off while the spacecraft was still above the surface, and the probe crashed. In cost-cutting measures, software testing for the landing system had been done on the cheap, and the lander was built without the capability to communicate to Mission Control its descent and landing telemetry (Bergreen 2000, 304–10). The lack of telemetry during the descent meant that the design and navigation teams could not analyze crucial data and pinpoint the reasons for the loss of the mission. Like the Mars Climate Orbiter, the Polar Lander was underfunded from the start, and its mission compromised by a command structure that switched project managers and staff after the craft was launched. The next lander (of almost identical design) originally scheduled to touch down on Mars in 2001 had to be postponed to a later launch window, and NASA's investigations of the polar regions will have to wait until the Phoenix mission scheduled for 2007.

Meanwhile, the instruments aboard the Mars Global Surveyor were returning paradigm-shattering results. In April 1999, the scientific team on the magnetometer experiment, headed by Mario Acuna, announced that it had found evidence of magnetic stripes of alternating positive and negative polarizations in the Martian crust. These patterns are analogous to the magnetic striping detected decades ago on the ocean floors of Earth—crucial evidence for the movement of the Earth's crust that led to the understanding of plate tectonics. The remnant magnetic striping implies that Mars once had plate tectonics as well. On Earth, plate tectonics require both a massive energy source—a molten core and volcanism to uplift, subsume, and regenerate crustal plates—and massive amounts of water (Bergreen 2000, 256–57; Stevenson 2001, 214–19; Hartmann 2003, 217–18). The implications for Mars are significant: early in its history, the planet seems to have had enough geothermal activity and sufficient water to reshape its surface: the Tharsis volcanoes (Arsia Mons, Pavonis Mons, Ascraeus Mons) and Valles Marineris may be geological evidence of such activity. Images taken from the ESA Mars Express orbiter during 2004 suggest that Olympus Mons and other Martian volcanoes may be dormant rather than extinct; some lava flows may have occurred within the past two million years (Neukum, et al. 2004, 971–79).

MOLA (Mars Orbiter Laser Altimeter) used short pulses of light emitted by a laser to measure the distance from the spacecraft to the surface within a few meters. Firing the lasers at intervals during its orbits of the planet, MOLA returned data that has allowed scientists to map areas of the planet's surface with an unprecedented precision. Viking orbital photographs, which showed only impressionistic blurs at resolutions smaller than one hundred meters, were replaced by the first interplanetary topographic maps. If the Viking and Pathfinder images had brought the Martian surface within the semiotics of everyday visual experience, MOLA revealed a surface that could be understood by scientists on a human, experiential scale of the sort familiar to readers of science fiction. For field geologists, the study of an environment depends on "hiking around, breaking open rocks, and seeing and touching the ground." Geology, in this sense, is an experiential science concerned with "size, shape, texture, color, pattern, relief, and spatial and temporal relationships" (Malin 1999, 42). These relationships, however, only can be inferred by the detailed topographic maps put together from the MOC and MOLA data. Even detailed maps (of Earth as well as Mars) are, as Morton suggests, "summaries and generalisations" (2002, 221), yet the topographic detail revealed by MOLA has redefined the level of generalization. To more than a few NASA geologists, hiking around Mars has become more a mission goal than a science-fiction fantasy.

As significant as the data returned by the magnetometer and MOLA have proved, much of the scientific and public interest in the Global Surveyor's mission centered on the 140,000 photographs returned by the MOC by late 2003. These images marked a decisive advance over the Viking orbital photographs; with narrow-angle resolution of 1.4 meters per pixel, the MOC revealed both earthlike vistas and strange landforms that led scientists to revisit many of their assumptions about the planet (Malin 1999, 48). Michael Malin, the principal investigator on the MOC and president of Malin Space Science Systems, released these photographs in comparatively small batches, even as he and his research associate, Kenneth Edgett, authored articles that suggested that the surface of Mars provided evidence of water seepage and erosion in geologically recent time. Malin described the "most important findings of the camera during its first year of operation" as "evidence of ground seepage and possibly ponding, evidence of sustained liquid flow, and layering in the upper crust of Mars to an unexpected degree" (1999, 45). By the following year, Malin and Edgett published detailed evidence of recent outflow channels, debris fans, and gullies on crater rims that cut across sand dunes (2000a 2330–35; Hartmann 2003, 354–65). As dramatic as these photographs are, they have other implications as well. In a popular article in *Sky and Telescope,* Malin emphasizes that the value of the MOC images of the cameras he designed for the Mars Global Surveyor lie in their ability to provide geographical and historical contexts for individual sites on Mars (1999, 42). The photographic evidence for groundwater seepage, and, in a striking series of images released on November 13, 2003, of the "distributary fan" of an ancient riverbed, offers small pieces of a larger mosaic, one that prompts new accounts of the history, nature, and scale of the dynamic processes and local environments that have reshaped conceptions of the planet. But these photographs, available on the web (at mars. jpl.nasa.gov/mgs/msss/camera/images) are also cultural artifacts, and therefore they exist within the complex contexts that have been the subject of this study: a history of previous planetary photographs, older paradigms, technological developments in photography, media coverage of the more striking images, and the viewer's own experiences and imagination. The photographs from the MOC and the "hauntingly Earth-like" vistas they reveal recall an archaeology of visual images, such as the balloonist's photographs of London's Hyde Park that Lowell and Morse used as a terrestrial analogue for the Martian canals.[6] For millions of viewers, the MOC photographs brought the surface of Mars within the context of their own experience, looking out from an airplane flying over the desert regions of Earth.

In June 2000, Malin's and Edgett's cover article for *Science* provided photographic evidence of recent groundwater seepage in 120 locales, primarily in the southern highlands: on the interior walls of impact craters, within the walls of two major valley systems, and on the walls of the distinctive pits in the south polar region (2000a, 2330–45). Remarkably, most of these sites lie in some of the coldest regions on the planet's surface. With an average scale of three meters per pixel, these photographs show erosional patterns that mimic terrestrial gullies in desert areas on Earth. Having considered and rejected the possibility that these features could have been caused either by "granular flows" or "avalanches fluidized by atmospheric or soil gases" (2333), Malin and Edgett argue that the most plausible explanation is groundwater seepage from ice or liquid water reservoirs close to the surface. In several photographs, the gullies cut across transient landforms such as sand dunes, indicating that the seepage is quite recent—anywhere from a few thousand years to a few weeks old. Subsequent studies have confirmed that the features are recent, though estimates of their ages describe only a range of geological time (Hartmann 2003). Malin and Edgett conclude that the most logical explanation for this evidence in colder regions on Mars is that ice barriers below the surface prevent water from evaporating directly into the atmosphere, as it may in the warmer equatorial regions; solar heating or other as yet unexplained processes may warm the gullied areas enough to thin or disturb the subsurface ice and permit occasional seepage from these reservoirs.

In a subsequent article, Malin and Edgett examined planetwide evidence of the layered sedimentary terrains that characterize the valley systems and cratered areas of Mars. Noting striking similarities in the composition of these layered depositions across the surface, they reject volcanic activity or meteorite bombardment as explanations for these pervasive features. In brief, they argue that the light-colored outcroppings in eroded terrain date from the Noachian period and the darker material from more recent eras. In contrast to warmer and wetter models, those that imagine a Mars of small oceans and possibly biota and that draw heavily on analogies to Earth, Malin and Edgett suggest that other "uniquely martian explanation[s]" for the layered terrain need to be considered (2000b, 1935). Before the November 2003 release of an MOC photograph provided evidence of a "distributary fan" characteristic of what Malin suggested was an ancient river delta (http://www.msss.com/mars_images/moc/2003/11/13/) the photographs from the MOC had offered no definitive indication for planetwide rivers or water runoff. Therefore Malin and Edgett suggest that the similarity of layered terrains can be accounted for by a model of early Martian history that supposes meteorite bombardment and the existence of shallow seas, perhaps life bearing, occurred at the same time. Suggesting that "astronomical perturbations" could affect atmospheric pressure on the planet, Malin and Edgett posit that as the atmosphere thickened and thinned periodically, material ejected into the atmosphere by impact craters could stay suspended for long periods of time, envelope the planet, and eventually form sedimentary layers. In this view, the varying of thick and thin layers of sedimentary rock record variations in atmospheric density, the formation of shallow "seas," and the retreating "coastlines" as water evaporated—an inference that some geologists draw from the composition of the flat, largely uncratered northern terrain.

Malin and Edgett repeatedly emphasize that the mechanisms for variations in atmospheric pressure as well as the processes by which sedimentary materials may have been transported across the Martian surface are "unclear," and their speculations register the complexities of reconstructing areological history. In contrast to the stately precessional cycles that drive the climatic instabilities of Sagan's Martian spring, Malin and Edgett suggest a kind of punctuated catastrophism: rather than a single planet-changing event, they describe a complex of processes that produce a landscape that is both analogous to Earth's and "unique." This "apparent correlation between the end of heavy impact bombardment and the cessation of the most aggressive and dynamic agents of geomorphic change" suggests that the environment of Mars reflects a history of stochastic bombardment, thickening atmosphere, the transportation of sedimentary materials across the planet, and then a periodic thinning of the atmosphere and a relatively quiet geomorphic history (2000b, 1935). It is a view that countenances both the possibility of the early evolution of Martian life-forms, perhaps still preserved as fossils in ancient sedimentary layers, and a history more dynamic and unpredictable than uniformitarian views of planetary evolution (Hartmann 2003, 234–49).

Almost invariably, the data from the MGS has led planetary scientists to develop sophisticated models to explain a complex and sporadic hydrological cycle. Victor Baker recently has proposed a "genetic model" to account for both long-term (one hundred million to one billion years) and short-term (ten thousand to one hundred thousand years) aspects of a Martian hydrological cycle. In brief, he suggests that long periods of cold and dry conditions are punctuated by "quasi-stable, short duration warmer (cool) and wetter conditions" (2001, 229), possibly the result of volcanic or magmatic activity that creates landforms associated with such comparatively temperate conditions: glaciated terrain, crater lakes, runoff channels, and teardrop islands. Volcanic activity would warm the planet temporarily through a greenhouse effect triggered by the build-up of carbon dioxide in the atmosphere. Such conditions, however, would be inherently unstable: water evaporating off a transient ocean would be transferred to the uplands (the Tharsis volcanoes and parts of the southern highlands) as glaciers and lost through the "porous . . . martian surface." Much of the carbon dioxide eventually would be locked up beneath a frozen layer of permafrost as carbonate rocks, "gas-charged groundwater," and clathrates; "only occasionally" would carbon be transferred to the atmosphere, briefly warming the planet on a timescale of ten thousand to one million years (228–36). A compatible analysis based on high-resolution images from the MOC indicates that ground ice is trapped in mid-latitude soils (30 degrees to 60 degrees) as well as in near-surface deposits near the poles. This reservoir, researchers estimated, might be enough to cover the surface to a depth of ten to forty centimeters (Mustard, Cooper, and Rifkin 2001, 411–14). These estimates, however, already seem too low. As William Hartmann puts it, "the more [scientists] learn, the wetter Mars seems to be" (2003, 107), and the evidence of water persisting long enough to affect the mineral composition of rocks and soil detected by Spirit at Gusev Crater and Opportunity at Meridiani Planum strengthen this view (Squyres et al. 2004a, 1698-1703; Squyres et al. 2004b, 1709-14; Arvidson et al. 2004, 1730-33; Rieder et al. 2004, 1746-49; Herkenhoff et al. 2004, 1727-33; Christensen et al. 2004, 1733-39; Soderblom et al. 2004, 1723-26; Klingelhöfer et al. 2004, 1740-45).

Yet such a history is based on best-guess scenarios as well as up-to-date data, and, within the overall model of an ice-rich Mars, different, even incompatible, theories try to explain the processes that have shaped its surface. Some scientists even suggest that liquid water might be stable on the Martian surface at certain times and in certain areas. Studying Viking temperature and pressure data as well as recent MOLA data, a team of scientists has concluded that the channeled surface of Mars lies in areas where theoretically liquid water may exist "over large areas on the edge of the northern lowlands and in the Isidis, Argyre, and Hellas plains"; in half of this area, "liquid water may be possible during more than five percent of the martian orbit" (Lobitz et al. 2001, 2133–34). Recently, Gilbert Levin has suggested that the photographs from the Opportunity rover indicate that water activity may be presently occurring on the planet's surface (2004, 126-38). But the complexity of the phenomena and the incompleteness of the data necessitate simplifications in any model of the dynamic processes by which the Martian surface has been shaped, and other scientists have reached very different conclusions. In 2002 Teresa Segura, Owen B. Toon, Anthony Colaprete, and Kevin Zahnle argued that the outflow channels on Mars can be correlated to the intense bombardment of the surface by comets and asteroids during the Noachian period. In their model, different-size impacts ejected differing amounts of rock vapor, water, and dust. Ice and subsurface water were rapidly heated and outgassed, leading to scalding rains, the formation of rivers, occasionally massive floods, and brief periods of moderate temperatures that punctuated the deep freeze of the planet (Segura et al., 2002, 1977–80).[7] Significantly, however, they acknowledge leaving out of their model the very factors that other scientists rely on to explain how a greenhouse effect could have warmed Mars for comparatively long periods of time: a "hydrological cycle in which the rainfall [caused by asteroid impacts] might evaporate back into the atmosphere," "the radiative properties of clouds," the carbon dioxide that might be outgassed into the atmosphere from the soil and the melting of the polar caps, and the evaporation of water into the atmosphere (Segura et al. 2002, 1978). These efforts to factor out complicating or asystemic factors may clarify the theoretical model but have provoked critiques by scientists who think that these complicating factors can have profound effects on the model's validity. As Hartmann noted, at a conference in 2002 Mars researchers debated the two models of the Noachian climate—the warmer and wetter model and the deep-freeze model—with a "slight majority" favoring the former (2003, 111–12). The findings of the Spirit and Opportunity rovers have increased that majority substantially.

In redefining the planet's geological history, the data from the Mars Global Surveyor both complicated and extended the analytical tradition founded on heuristic analogies between Mars and Earth. The fourth planet remains a generic hybrid, an object of analysis understood, in large measure, through its differences from and similarities to Earth. It is a dynamic composite, always in the process of being redefined by complex interactions among technologies of observation, data collection, and competing theories of planetary history. The overall effect of recent missions, including the Mars Odyssey, has been to create "virtual realities," as Oliver Morton calls them, of overlapping, often conflicting, and always incomplete data that allow scientists to establish, refine, and rethink these "imaginary spaces" (2002, 234). These virtual realities locate the surface of Mars within synchronous and diachronic time frames: rather than a fixed object, the planet becomes the dynamic site of interacting geological, hydrological, thermal, and meteorological systems—all dependent on the technologies of representation available to register and transmit data. NASA's vision of multiple orbiting probes in 2010 streaming Martian data back to Earth seizes on the possibility of exploiting new bandwidth capabilities to construct a real-time view of the planet (Morton 2002, 234–35). In one sense, NASA's goal is to move from "virtual realities" to the all-Mars, all-the-time channel, a small flotilla of communication satellites orbiting the planet and relaying data from and commands to increasingly sophisticated rovers on the surface.

Mars Global Surveyor was joined in orbit around the planet by another mission two years after the failure of the Polar Lander. The Mars Odyssey spacecraft reached Mars in October 2001 and continued returning data through 2004. The mission carried three science packages that, in different ways, contribute to an overall strategy centered on searching for subsurface ice and water, identifying possible locations for future paleobiology investigations, and determining radiation levels on the planet's surface.[8] The official NASA web site for the mission (http://mars.jpl.nasa.gov/odyssey/science/index.html) is straightforward about the purpose of the radiation detection experiment: it is to allow "future mission designers . . . to outfit human explorers for their journey to the red planet." The other science packages are intended to uncover the location, history, and geological contexts for the distribution of elements and minerals on Mars. The Thermal Emission Imaging System (THEMIS) detects minerals by identifying their characteristic signatures in the infrared spectrum and studies past liquid environments in the visible spectrum. The composition of rocks provides a means both to identify likely areas for past subaqueous environments (dry lake beds, for example) and to generate data necessary for scientists to understand the geological history of landforms. Using nine spectral bands, THEMIS provides significantly better data than the TES aboard the Mars Global Surveyor, producing color-coded maps of the deposition of minerals. The Gamma Ray Spectrometer (GRS), a suite of three instruments (the Gamma Subsystem, the Neutron Spectrometer, and the High Energy Neutron Detector [HEND]), measures the abundance and distribution of twenty elements on the Martian surface. When exposed to cosmic radiation, chemical elements in soil and rocks emit identifiable signatures in the form of gamma rays, and the intensity of these emissions reveals the chemical composition of the surface. The neutron spectrometer and HEND detect and measure scattered neutrons in different energy bands (thermal, epithermal, and fast), and distinguish neutrons originating from the planet from the background neutron flux. In layperson's terms, the instrument package can infer the presence of hydrogen below the surface, an element that indicates the existence of water or water-ice.

The Mars Odyssey mission quickly delivered a bombshell. In May 2002, NASA announced that the GRS had detected large amounts of hydrogen a meter or less below the surface near the southern polar cap. Because water would not be stable at the low subsurface temperatures, the science team headed by William Boynton concluded that the hydrogen must be bound in the form of ice (Boynton et al. 2002, 81–85). In some areas, ice seemed to comprise as much as 60 percent by volume of the subsurface soil, significantly exceeding the predictions of the models developed by Carr (1995), Baker (2001), and others. Over the following six months, as summer arrived in the northern hemisphere, the GRS tracked the uncovering of water-ice in the north polar cap as carbon dioxide frost sublimed into the atmosphere. The volume of water on Mars, in short, may be greater than Lowell imagined, although debates continue (even after Opportunity's findings at Meridiani Planum) over whether Mars ever experienced long periods when the water would have formed primordial oceans on its surface. The wide distribution of low concentrations of carbonate minerals in the Martian dust implies that much of the surface has been dry in the geologically recent past, as does the lack of weathering of olivine, a common green mineral on Mars that is easily transformed by water (Bandfield, Glotch, and Christensen 2003, 1084–87). But the photographs from the MOC, the detection of abundant subsurface ice, and the geological data returned by Spirit and Opportunity suggest that persistent surface water did exist on Mars and may reappear in the future when Mars enters a new ice age—a period of increased obliquity when the planet warms enough to deplete the polar caps, hydrate the atmosphere, and begin a more robust hydrological cycle, with ice and possibly snow reaching latitudes as low as 30 degrees (Head et al. 2003, 797–802). A new consensus about Mars's climatological past awaits further data, theory building, and testing.

The post-Viking history of Mars exploration underscores the dependence of theory building, as well as data collection, on increasingly sophisticated technologies of mapping, prospecting, and computer modeling. The THEMIS experiments, for example, adapt for interplanetary exploration geological mapping techniques that have been used for decades to identify promising sites for gas and oil drilling. The instruments orbiting Mars, however, must fit into the engineering constraints of size, weight, durability, and software capabilities: experiments on a spacecraft that must function for forty months in the frigid regions of interplanetary space one hundred million miles or more from Earth tax the ingenuity of their designers and the reliability of their subcontracted components. Like the other experiments on MGS and Mars Odyssey, THEMIS involves a variety of specialists; in this regard, the need for more sophisticated technologies has expanded the size of many science teams, even as scientists have refined and improved experiments to provide the geological and atmospheric contexts that Murray and many others had wanted since the 1970s. If different assumptions about Mars (and Martians) informed the design, experimental protocols, and interpretation of the three Viking life-detection experiments, the science packages aboard the Mars Odyssey mission are the products of NASA's scientific consensus and a significant degree of professional coordination. The *Science* article describing the detection of subsurface ice deposits has twenty-five coauthors, including principal investigators and science and engineering teams from the three systems that comprise the GRS package. The lead article in the special section of *Science* in December 2004 devoted to the data returned by Opportunity has fifty coauthors. Such corporate authorship is one measure of the increased specialization—at once rigorously focused and cross-disciplinary—that characterizes NASA's current missions to and mission planning for Mars.

THE METEORITE CONTROVERSY AND EXTREMOPHILE MICROBIOLOGY

Searching for chemical signatures, identifying underground water-ice, and remapping the surface of Mars periodically made headlines between 1997 and 2003, but this data collection and analysis took place within a context of renewed debates about the possibility of Martian life. In 1996, a team of scientists headed by David McKay argued in a controversial article in *Science* that it had identified in a Martian meteorite possible nanofossils of ancient bacteria (McKay et al. 1996, 924–30). In some respects, the controversies about ALH84001 are reminiscent of the disagreements that characterized the Viking life-detection experiments in 1976 and 1977 as scientists tried to develop coherent explanations of often puzzling results: both debates are structured, in part, by assumptions about conditions on Mars, past and present; on the definition of life; on the contexts in which data are interpreted; and on different conceptions about what a biological "theory" is supposed to do—offer probabilistic arguments for ancient microorganisms on Mars or present evidence that meets standards of certainty for terrestrial life-forms. But between 1976 and 1996 developments in extremophile microbiology led to profound changes in traditional ideas of the limit conditions under which life could exist. As early as 1990, Lynn Rothschild argued that terrestrial microbes capable of surviving in extreme environments provided possible analogues for organisms that might exist on Mars (1990, 246–60; see also C. P. McKay et al. 1992, 1234–45). By that time, scientists were studying meteorites that had originated on Mars, such as ALH84001, found in Antarctica by Roberta Score in 1984. Extremophile microbiology offered new modes of analysis that could be brought to bear on the question of exobiology.

ALH84001 (the designation identifies the meteorite as having been found in the Allan Hills of Antarctica in 1984) is one of a class of meteorites called SNC (for Shergotty, Nakhla, and Chassigny, locations where three of the Martians meteorites were found). The ratios of three oxygen isotopes within these meteorites rule out the possibility that they originated on the Earth or Moon; several contain gas globules trapped within these rocks that are identical in composition to the atmosphere of Mars and distinct from any other source of gas in the solar system (Jakosky 1998, 130–31). Initially, McKay and his colleagues were excited by the opportunity to study a meteorite that had been undisturbed on the surface Antarctic ice sheet for thirteen thousand years. ALH84001 offered a kind of time-dilated sample return mission—a rock from Mars that had been ejected into space, probably by the impact of a meteorite some sixteen million years ago (dated by the meteorite's exposure to cosmic rays). Although the rock seemed to be part of the ancient crust of Mars, over four billion years old, the interdisciplinary team of scientists were intrigued by carbonate structures that looked strikingly reminiscent of terrestrial bacteria, and sent samples for chemical analysis by mass laser spectroscopy to Richard Zare at Stanford.[9] Although initially skeptical, Zare and his colleagues found chemical evidence that, in a terrestrial rock, would be indicative of biological activity. However, because ALH84001 came from Mars, the assumptions, methods, and conclusions of McKay, Zare, and their collaborators were scrutinized by a host of often hostile critics.

The initial paper by McKay and his collaborators in *Science* offered four lines of evidence to support a biological interpretation. Although the authors conceded that none of the four in and of itself was conclusive, they suggested that, taken together, the evidence indicated that they had identified and examined fossilized nanobacteria from Mars. First, the meteorite contained polycyclic aromatic hydrocarbons (PAHS). Although PAHS can be formed by inorganic processes, they are often associated on Earth with "decayed organic matter" because "gentle heat can transform dead organisms into PAHS" (Treiman 1999, 54, 56). Second, the sections of the meteorite examined had unusual, small crystals of magnetite (an iron oxide) seemingly identical to crystals that are made only by terrestrial bacteria; on Earth, such perfectly regular crystals would be treated as biogenetic markers. In addition, there was evidence of incompatible minerals existing in close proximity, and again, on Earth, such proximity would suggest organic action. Finally, the cover of *Science* featured one of several photographs of evocative, bacteria-shaped formations in the meteorite; these photographs, first unveiled at the NASA news conference in 1996, looked like segmented biological forms and offered the same kind of challenge that Lowell's first photographs of the "canals" had posed in 1905: to interpret the photographs was to imagine not simply a microorganism or an inorganic mineral formation but an environment in which that interpretation made sense. The McKay team photographs, like the article's larger argument, were context dependent, and those contexts were shaped by scientists' understanding of microbiology, the conditions under which the carbonate formation took place, and the conditions on Mars over three billion years earlier.

Very quickly, criticism of the McKay team's paper emerged, both specific and general. Some scientists maintained that carbonates could be formed only at high temperatures (700 degrees centigrade) and argued that these structures were the result of high-impact shocks (Harvey and McSween 1996, 49–51); when evidence suggested that the carbonates were indeed formed at much lower temperatures, the same researchers suggested an evaporation model to explain the existence of the carbonates (Valley et al. 1997, 1633; McSween and Harvey 1998, 774–83). Other scientists contended that any fossils in ALH84001 were the result of terrestrial contamination, although McKay and his colleagues took pains to argue that any contamination deep in the rock was highly unlikely. Most problematically, for many skeptics, the bacteria-shaped objects in photos seemed too small to contain "the minimum of biological molecules (such as DNA, RNA, and the complex proteins) for the machinery of life as we know it to work" (Treiman 1999, 57). The debates about ALH84001 over the last nine years, consequently, have become what Allan Treiman has termed "a Rorschach test for researchers—[they] see what [they] are trained to see" (55). Training, however, is a vexed term: not only have the debates been about the interpretation of evidence by practitioners in different disciplines, they also have revisited the philosophical and methodological questions that have characterized previous debates about life on Mars.

In a rhetoric that recalled the aftermath of the Viking mission, some members of the scientific community cast the controversy in stark terms by questioning the competence and professionalism of scientists who had called a full-blown press conference to present their research. Malcolm Walter, director of the Australian Centre for Astrobiology at Macquarie University, compared the affair to the cold fusion fiasco (1999, 91–92, 77–84). Edward Anders, writing in *Science*, was merciless: the McKay team's report was "half-baked work that should not have been published. For all these observations, an inorganic explanation is at least equally plausible, and, by Occam's razor, preferable" (1996, 2019–20). Anders, like most Viking scientists before him, assumes that inorganic explanations for ambiguous results are "simpler," and that consequently anything short of cut-and-dried evidence for Martian nanofossils invariably will be less acceptable, less "scientific" than inorganic explanations—even if the inorganic chemistry to explain the nearly perfect structural crystals found in the ALH84001 carbonates has no analogue on Earth. But there are significant differences between the ALH84001 controversy and the debates over the Viking life-detection results. Gilbert Levin was pretty much isolated within the NASA community and had comparatively little support from other team members as he pursued his biological interpretation of the labeled release data in the 1980s and 1990s. In contrast, McKay's team included a variety of researchers both within NASA and without. Its heaviest hitter was Richard Zare, then chairman of the National Science Board, a member of the council of the National Academy of Sciences, the holder of the Marguerite Blake Wilbur Chair in Chemistry at Stanford University, and the recipient of dozens of awards (including the National Medal of Science) for his research in the area of laser spectroscopy and the study of chemical reactions at the molecular level. Zare initially had analyzed the samples sent by Kathie Thomas-Keprta of NASA without knowing that they were sections of a Martian meteorite, and describes his reaction after learning their origin as "interested but cynical" (Markley et al. 2001, "life on mars?" 11). After Zare and his colleagues identified PAHs in the samples, the original coauthors and additional collaborators spent the next five years conducting new tests on sections of the meteorite, corroborating results, and responding to the criticisms leveled against their conclusions. Like the identification of subsurface ice on Mars using the GRS, the analysis of ALH84001 has become a cross-disciplinary endeavor, one that some advocates suggest is redrawing the disciplinary boundaries of cosmo-chemistry.

To understand the implications of the meteorite controversy, one needs to have a grasp of the biochemistry that makes life on Earth possible. Since the 1950s and the discovery of the double helix structure of DNA, scientists have studied the prerequisites for life and speculated about its origins on Earth. Life requires organic molecules, a source of energy to drive disequilibrium processes, and access to biogenic elements: carbon, hydrogen, oxygen, nitrogen, sulfur, and phosphorous. A variety of chemical reactions can drive biological processes. On Earth, the "medium of exchange" for all life-forms is the complex molecule adenosine triphosphate (ATP), which includes two high-energy bonds between phosphorous and oxygen atoms. These bonds can be forged, and new ATP molecules created, by chemical reactions that require external energy sources, or these bonds can be broken, releasing energy to drive other reactions. Because ATP represents an efficient, even elegant, "biochemical solution to the problem of storing energy," many biologists see it as evidence for a single origin of life (Jakosky 1998, 73). The energy required to form ATP molecules involves a redox reaction: the oxidation of hydrogen, for example, and the reduction of oxygen to form a molecule of water. What is reduced in such a reaction is the charge of an atom; oxygen takes on an electron, which has a negative charge, so its overall charge is reduced. Hydrogen gives up an electron in combining with oxygen, and with the loss of a negatively charged electron its charge increases. (In oxidation, the element that relinquishes an electron is said to be oxidized.) In biological systems, any atom or molecule that can act as an electron donor can combine with any other that can act as an electron acceptor. In practice, then, there are many ways of driving biological systems through such reactions: through photosynthesis, oxidation, and fermentation (Jakosky 1998, 73–75). As Jakosky and other scientists note, early Mars seems to have met all of these preconditions for the possible formation of life (Jakosky 1998, 132; Jakosky and Philips 2001, 237–44; DiGregorio 1997, 272–76; Shapiro 1999, 209–11). In theory, as extremophile microbiology has expanded the environmental conditions under which such reactions can be driven, so too has the possibility that Mars currently might support conditions, particularly beneath the surface, for exotic microbial life-forms to exist (Boston, Ivanov, and McKay 1992, 300–308; Taylor 1999).

Norman Horowtiz's findings in the early 1970s that the dry valleys of Antarctica were lifeless proved, to paraphrase Samuel Clemens, greatly exaggerated. By the 1990s, these cold, almost waterless deserts—the closest terrestrial approximation to current conditions on Mars—had become key sites for the study of extremophile microbiology. So, too, had other extreme environments that might reflect conditions early in the planet's geothermal evolution: deep oceanic thermal vents that provide energy sources and microenvironments for bacteria that can tolerate temperatures well above the boiling point of water (130 degrees centigrade); the alkaline, oxygenless mud at the bottom of California's Lake Mono; and basalt deposits more than a mile below the surface of the Columbia River Gorge that harbor microorganisms that survive by metabolizing rock. In each case, extremophile organisms have evolved means to survive in environments far more hostile than Antarctic conditions, redefining scientific assumptions about ecologies, and calling into question traditional cultural constructions of "nature." The bacteria thriving below the Columbia River Gorge, for example, exploit water and rock as sources of chemical energy, using hydrogen to convert dissolved inorganic carbon into more reduced forms of carbon and methane (Jakosky 1998, 77–79). Michael Ray Taylor has made explicit the connections between such "rock-eating bugs" and the Martian nanofossils (Taylor 1999; see also Nisbet and Sleep 2001, 1083–91; Rothschild and Mancinelli 2001, 1092–1101). The physicist and science-fiction novelist, Gregory Benford, bases his novel, *The Martian Race* (1999), on the discovery of sentient (though non-intelligent) colonies of exotic microorganisms beneath the surface of Mars. In the context of an extremophile microbiology—a branch of science that did not exist as such before the work of Carl Woese in the 1980s—Benford's novel seems as much a thought experiment as it does a work of fantasy. Some of the implications of microbiology for the search for life on Mars and Jupiter's moon Europa may be gauged by Zare's comment that "certainly the possibility . . . very much exists" that some form of biology will be discovered eventually on the red planet. If life did emerge early in

Martian history, Zare suggests, it may still exist in microenvironments deep below the surface: "I know how hardy life is, so once started, it's not easily removed" (Markley et al. 2001, "interviews," 5).

Even if Mars did not enjoy a long, temperate period three billion years ago, the presence of water-ice may not be a barrier to environmental niches, such as subsurface aquifers, where microorganisms could exist. Since the 1970s, scientists have known that large subglacial lakes exist below the ice of Antarctica, kept liquid by the pressure of an ice sheet almost 4,000 meters thick and possibly by geothermal heating as well. The largest of these freshwater reservoirs, Lake Vostok, has been mapped by radio-echo sounding: it has a surface area of approximately 14,000 square kilometers and a depth of as much as 670 meters. Geochemical analyses of the accreted ice taken a mere 120 meters above the lake have recovered microorganisms, and researchers surmise that Lake Vostok "and others like it may contain previously undescribed relic populations of microorganisms" that have adapted to an environment that is "low-nutrient, low-biomass, and low-energy flux" (Karl et al. 1999, 2144). The discovery of these microorganisms has made these subglacial lakes important analogues for exobiologists investigating the possibility of life in the liquid ocean presumed to exist below the icy surface of Europa and beneath the surface of Mars (Priscu et al. 1999, 2141–44; Karl et al. 1999, 2144–46). Peter Doran, who studies extremophile biology in Antarctica, argues that "in the Dry Valleys" it becomes possible to do simulated "ecological studies on Mars" (Walker 1999, 53). In December 2002, he announced that 2,800-year-old bacteria and algae had been retrieved and revived from three meters above Lake Vida, a briny body of water (seven times the salt content of ocean water) buried beneath nineteen meters of ice. Speculating that such an ecosystem, trapped beneath the Antarctic ice sheet, offers " 'a good analogue for something in Mars' past,' " Doran is just one of the scientists who employ terrestrial analogies to study what might be called virtual exobiology.[10] In an important sense, the study of sediments within dry paleolakes, microbes in the ice above Lake Vostok, and cyanobacteria revived from above Lake Vida have turned Antarctica into a site for these virtual "ecological studies"—allowing scientists to envision Mars not as a frozen rock field but as the site of a paleoecology where extremophile life could (still) flourish (Doran et al. 1998, 28).[11]

In the context of this work in the 1990s on extremophile microbiology, some members of the original team that worked on ALH84001 and additional collaborators have conducted extensive tests on the disputed nanofossils. Their 2001 paper is the most comprehensive survey of the literature on the Martian meteorite controversy (Gibson et al. 2001, 15–34). While hardly neutral, the paper responds to the initial criticisms leveled against a biological explanation for the carbonate structures, notably the contention that the carbonate structures could have been formed by inorganic processes (Harvey and McSween 1996, 49–51; Bradley, Harvey, and McSween 1997, 454–55). Tests on the isotopic compositions of the three Martian meteorites, ALH84001, Nakhla, and Shergotty, ruled out the possibility that the carbonate globules were the result of terrestrial contamination (Gibson et al. 2001, 29). While Everett Gibson, Kathie Thomas-Keprta, and their colleagues acknowledge that carbon isotopic analyses do not prove that the globules are biomarkers, they argue that this lack of definitive proof is a function of the limitations of scientific knowledge about the conditions under which they were formed; because scientists lack "a complete understanding of the different carbon pools with distinct isotope compositions that reside on Mars," all conclusions have to be conjectural (Gibson et al. 2001, 29). Such an understanding will have to await extensive tests on the surface and perhaps subsurface chemistry of the planet. In the absence of such firsthand knowledge, the two overlapping teams of researchers headed by Gibson and Thomas-Keprta have ensured that the magnetofossils in the ALH84001 have been far more extensively and rigorously studied than any such structures on Earth (Wharton 2002, 232–40).

In their survey of the scientific literature on or related to the Martian meteorite debate, Gibson and his collaborators cite 113 articles, including extensive studies on biomineralization by bacteria. They conclude by asking if the evidence they have documented about ALH84001 dovetails with eight widely accepted criteria for determining "biogenic activity" in terrestrial samples. They define these criteria as follows:

1. The geological context, and its compatibility with life.
2. The age and history of the sample.
3. Evidence of cellular morphology.
4. Evidence of structural remains in the sample.
5. Biominerals that show chemical or mineral disequilibria.
6. Stable isotope patterns unique to biological systems.
7. Organic biomarkers.
8. Indications that any biogenic evidence is indigenous to the sample. (2001, 26)

While scientists still have to make educated guesses about the suitability of ancient Mars for life, the six biomarkers that have been identified for terrestrial magnetofossils (Kirschvink and Vali 1999, 1681) are all met by Martian meteorite ALH84001, as Friedman and his collaborators have argued: crystal size and shape, magnetite chains, elongated crystal structures, anisotopic growth of crystal faces, and chemical purity of the crystals (Friedman et al. 2001, 2176–81; Gibson et al. 2001, 19). In responding to the criticisms of the initial 1996 article, the expanded research teams have done extensive comparisons between the structures in the meteorite and terrestrial organisms. Both the Thomas-Keprta team and Imre Friedman and his collaborators demonstrate that the magnetite crystals in ALH84001 meet the six criteria of the magnetite assay for biogenicity (MAB), and therefore they conclude that the hexa-octahedral magnetite crystals are the product of biological activity (Thomas-Keprta et al. 2002, 3655–56). While the 1996 article in *Science* relied on visual similarities between the alleged nanofossils and terrestrial bacteria, Gibson, Friedman, Thomas-Keprta, and their collaborators have extended this analogical reasoning by comparing the crystal structures in the meteorite to biomarkers that indicate the presence of terrestrial magnetobacteria.

The magnetite crystals found in ALH84001 are "both chemically and physically identical" to "truncated hexa-octahedral magnetites" produced by the terrestrial bacteria strain MV-1 (Thomas-Keprta et al. 2001, 2164). Thomas-Keprta and her colleagues argued that this finding is significant because it provides evidence both for a geological context for the carbonate structures and what may be an evolutionary mechanism to account for these unusual formations. A key aspect of their argument subsequently was challenged by a team of mineralogists headed by D. C. Golden. These researchers generated inorganic crystals morphologically identical to those in ALH84001; having compared almost one thousand magnetite crystals, they argue that small but significant structural differences render both the laboratory and meteorite formations significantly less regular than those produced by MV-1 (Golden et al. 2004, 681–95). Golden and his colleagues suggest that thermal decomposition could have formed the ALH84001 crystals (Treiman 2003, 369–82; Golden et al. 2004, 694).

Morphology, though, constitutes only one aspect of the arguments advanced by Thomas-Keprta, Friedman, and McKay. On Earth, "magnetotactic bacteria" (those that orient themselves by means of a global magnetic field) are "ubiquitous in . . . aquatic environments, and they appear to use the magnetic properties of magnetite" in order to navigate (Thomas-Keprta et al. 2002, 3664). Magnetobacteria produce only one type of magnetite, which is identical (or at least quite similar) to the structures found in ALH84001. Biological crystals are a uniform size; so are those in the meteorite. Magnetic crystals in linear chains occur only in biological configurations on Earth; outside cells, such energetically unfavorable configurations would collapse. While Golden and his colleagues produced linear chains of magnetite crystals inorganically, the photographs they published in their 2001 article reveal that the size of their crystals vary significantly (Golden et al. 2001, 370–75). In contrast, the crystals analyzed in ALH84001 are uniform in size and have an unusual truncated hexa-octahedral form, which features two characteristic width-to-length ratios of the faces; these ratios are found only in biologically produced magnetite on Earth. Inorganic crystals do not exhibit this characteristic hexa-octahedral form (Clemett et al. 2002, 1727–30), although this claim has been challenged by those researchers who note that these crystals have "rounded edges" rather than the "well-developed faces" of MV-1 (Golden et al. 2004, 691). The crystals observed in ALH84001 are defect free; in terrestrial bacteria such chemically pure magnetite crystals (containing only iron and oxygen) facilitate the uptake of iron. Inorganic crystals, the Thomas-Keprta team claims, are not pure and exhibit a variety of structures, another claim challenged by Golden and his colleagues. Friedman, however, has demonstrated that the crystals in ALH84001 are separated by dark areas of inorganic substances, again a characteristic feature of terrestrial bacteria; the inorganic crystals produced by Golden's team in 2001 have no such uniform dark areas to separate them because they are bound magnetically. Furthermore, he argues that biological membranes hold the crystals together in flexible chains; inorganic chains are not flexible because there is no elastic material between structures, no "halo" around them indicative of a biological membrane. Finally, the elongated crystals in ALH84001 are oriented along an axis, again in a form identical to magnetobacteria on Earth; outside the laboratory, inorganic crystals are elongated and lie parallel to each other, forming characteristic bands.

Although Golden and his collaborators maintain that if "an inorganic process can produce magnetite crystals, the claim of a biological origin is greatly weakened" (Golden et al. 2004, 682), their argument does not address the other criteria that Friedman, McKay, and Thomas-Keprta argue are essential to a probable biological origin for the magnetite crystals. Individually and collectively, then, the features that the Thomas-Keprta and Friedman teams identify never have been observed in inorganic magnetite crystals on Earth, with the significant exception of the crystal morphology described by Golden and his collaborators in

2004. Still, the argument for a biological origin for the magnetite grains in ALH84001 rests on the claim that "no inorganic magnetite . . . produced . . . either naturally or synthetically, has simultaneously displayed all six MAB properties" (Thomas-Keprta et al. 2002, 3666). Even after the 2004 challenge by Golden, the Friedman and Thomas Keprta teams seem to be sticking to their guns. The best—that is, the simplest—way to account for the existence of these features in ALH84001, these scientists maintain, is to attribute them to evolutionary design: magnetobacteria on ancient Mars evolved in similar, if not identical ways, to magnetobacteria on Earth, developing the same energy-efficient configurations to navigate in a watery environment.

Friedman, Thomas-Keprta, and their collaborators make it a point to note how their interpretation of the evidence is supported by the findings of the Mars Global Surveyor. One of the initial critiques of the McKay team's findings was that since Mars had no magnetic field, there would be no reason for putative bacteria on the planet to develop mechanisms in order to orient themselves along a magnetic axis (Anders 1996, 2119–21; Goldsmith 1997, 106). However, the evidence of magnetic stripes in the Martian crust indicates that early in its history—about the same time that the carbonate structures in ALH84001 were formed—Mars had a strong magnetic field analogous to Earth's. Magnetic fields in general shield planets from cosmic radiation; consequently, the existence of such a field in the planet's past would suggest that the environment on Mars was more conducive to the evolution and protection of organic molecules than it is today (Thomas-Keprta et al. 2002, 3669).

By embedding their arguments in a plausible history of ancient Mars, Friedman and Thomas-Keprta provide a narrative structure with which to counter other objections made by skeptics to the nanofossil argument. On Earth, the bacteria that create such magnetite chains live in mud and navigate using them to find regions of optimal oxygen. Friedman concludes that the decomposed chains in ALH84001 were initially deposited by carbonate-rich fluids that then evaporated, and the chains were crushed by the meteoric impact or impacts that the Allen Hills meteorite is known to have undergone. There is no chance that these chains are the result of terrestrial contamination because magnetobacteria cannot be formed in or live on ice, and they can hardly navigate through solid rock (the carbonate structures in ALH84001 were found well inside the meteorite). The chemically pure magnetites are embedded in an iron-magnesium carbonate, yet the decomposition of such a carbonate produces "Mg-Fe [magnesium-iron] spinels, not pure magnetite" (Thomas-

Keprta et al. 2002, 3669). For Friedman, the non-Gaussian distribution of the fossilized magnetite chains is telling: the linear fragments themselves in terrestrial samples would be a telling indication of biology. Finally, some researchers have suggested that the Martian meteorites show evidence of possible biofilms, "polysaccharide secretions produced by bacterial colonies in order to make their environments more favorable for survival" (Gibson et al. 2001, 24). Given the weight of this evidence, the teams of researchers headed by Friedman and Thomas-Keprta conclude that unless the magnetite crystals in the Martian meteorite result from "an unknown and unexplained inorganic process on Mars that is conspicuously absent on Earth," the "likely" explanation is that they were "produced by a biological process" (Thomas-Keprta et al. 2001, 2164).

While Golden and others have offered such a mechanism, their view, in turn, has been challenged by researchers working on terrestrial magnetotactic bacteria. Two Australian scientists, who have studied the structure and formation of the gels surrounding organically produced magnetite on Earth, have concluded that Friedman Gibson, Thomas-Keprta, McKay, and others "collectively provide solid evidence for past life on Mars" (Taylor and Barry 2004, 195). In brief, A. P Taylor and J. C. Barry developed a new technique using ultraviolet light, which allowed them to investigate and catalogue structural features inside the gel surrounding the magnetic crystals inside terrestrial bacteria. These features are identical to the structures identified by Friedman, Thomas-Keprta, and their coauthors. By studying the "ultrafine structures" of the minerals of the "magnetosomal matrix" surrounding the chains of crystals in terrestrial bacteria, Taylor and Barry suggest that they have identified "immature" structures in the process of "biomineralization," a form of "pre-magnetite" that will allow scientists to understand the formation of magnetosomes (2004, 180–97). Potentially at least, the notion of morphologically imperfect "pre-magnetite" interacting chemically with its environment provides a means to understand why the biological argument for ALH84001 can survive the counterblasts of Golden and his colleagues.

In invoking Occam's razor, the two sides in the Martian meteorite debate voice different conceptions of scientific "simplicity," and their arguments and counterarguments reframe philosophical (and theoretical) questions about the usefulness of biological—or ecological—analogies between Earth and Mars. Friedman, Thomas-Keprta, Taylor, and Barry argue that cataloguing precise morphological analogies between the carbonate structures in ALH84001 and known magnetotactic bacteria is preferable to invoking inorganic, biomimetic processes that have no known analogues on Earth and that until 2004 could not reproduce experimentally morphologically identical crystals. The arguments against nanofossils in the Martian meteorite, in contrast, pose obstacles for any biological interpretation of the data because they presuppose that any biogenetic system is too complex to qualify as a "simple" explanation for mysterious carbonates that might have been formed inorganically. As in previous debates about life on Mars, from the canals to the Viking life-detection experiments, the skeptics have demanded rigorous criteria of proof before they concede that ALH84001 harbors nanofossils—uncovering the mechanisms for self-replication, fossilized evidence of RNA or even DNA. In contrast to a long tradition of geological analogies between Mars and Earth, the analogical reasoning for biomorphology has been rejected by researchers, such as Anders, McSween, Harvey, and Golden, who assume that any "equally plausible" inorganic explanation invalidates all biological interpretations. To date, no one has attacked the full range of the arguments advanced by advocates of biomineralization in the Martian meteorite. The asymmetry of explanation that characterized NASA's interpretation of the Viking life-detection data—life requires extraordinary proof, inorganic chemistry does not—applies as well to the meteorite debate. Incommensurate narrative and theoretical contexts for analyzing ALH84001, in short, reflect and are informed by different methodological assumptions and values.

The controversy over ALH84001, not surprisingly, has focused attention on the limitations of scientific knowledge about bacterial life on Earth. All of the images of the carbonate structures and magnetite crystals in ALH84001 employ sophisticated electron microscopes, an instrument that has revolutionized the study of microbial life. One of the crucial unanswered questions posed by the suspected nanofossils is whether these structures (under 100 nanometers) are large enough to have contained the genetic material necessary for cell replication. A number of extremophile biologists have argued that microbes can exist at sizes much smaller than the previously accepted limit of 200 nanometers (Folk 1993, 990–99; Taylor 1999, 95–98). Before Robert Folk's studies began to be published in the early 1990s, 200 nanometers seemed to most scientists to be the minimum size required for replication by the universal mechanisms of DNA and RNA. Now that limit has been questioned. Even if the oval and spherical structures between 20 and 200 nanometers identified by Folk and Lynch (1997, 583–89) are inorganic, they could well be "bacterially mediated, mineral precipitates" (Gibson et al. 2001, 22)—that is, by-products that are best explained as biological alterations of the environment or the "templates" for biomineralization that Taylor and Barry suggest (Taylor and Barry 2004, 187–95). If the limit size for microorganisms using perhaps primitive forms of RNA replication is lower than previously accepted, then the case for Martian nanofossils is strengthened significantly. Nonetheless, as Gibson's team admits, fundamental questions about the morphology, environmental interactions, and exotic chemistry of extremophile life need to be addressed because an "exhaustive database . . . does not yet exist of terrestrial fossils and biomarkers, both recent and ancient," that could be used as a basis for informed comparisons between terrestrial organisms and possible exobacteria (Gibson et al. 2001, 30). This database may be a long time coming.

As the microbiologist Norman Pace points out, a terrestrial soil sample contains billions of microorganisms, "so many different types that accurate numbers remain unknown" (1997, 734). Many of these lifeforms were unsuspected as recently as the era of the Viking missions. The revolution in microbial biology in the last twenty years since Carl Woese described the phylogenetic "tree of life" based on small subunit ribosomal Ribo-Nucleic Acid sequences (SSU rRNA) has rendered the plant and animal "kingdoms" a small and late-developing branch of the eukaryotes (organisms with cells defined by a membrane enclosing the genetic material in a nucleus), one of three independent domains (1987, 221–71). The other domains, bacteria and archaea, are prokaryotic (organisms that do not have a nuclear membrane in their cell structure), and their existence has forced scientists to overhaul what had seemed to be rock-solid assumptions about evolution, ecology, and systems of classification.[12] The criteria for classifying species of all three domains are based not on structure and metabolism but on gene sequencing and molecular analysis. Coevolution between species from different domains is the norm rather than the exception, and the symbiotic relationships among different orders of microbes clearly play crucial, if still little understood, roles in the larger world that humans call "nature." To take only one example, methagens (archaea that produce methane as a by-product of metabolism) are responsible for most of the methane in the Earth's atmosphere and in the upper few miles of its crust; they "constitute a large component of global biomass," offering "an inexhaustible source of renewable energy to humankind" (Pace 1997, 736; Cockell 2003). With the confirmation of methane in the Martian atmosphere, some scientists, such as Robert Zubrin, have claimed that a biological explanation is "the most likely explanation for this discovery" and that consequently life currently exists on the fourth planet (2004; Formisano et al. 2004, 1758–

61). In this strange new world of extremophile microbiology, anaerobic species have been identified that could live comfortably in the conditions that exist on Mars, either underground or near the ice-rich poles (Jakosky et al. 2003, 343–50; Kargel 2004; Hanlon 2004).

MARS AND THE IMPLICATIONS OF EXOBIOLOGY

The current debates about life on Mars are no less entangled in philosophical and even theological narratives than they were when Lowell and Wallace squared off a century ago. Almost invariably, specific questions about chirality, biomineralization, and magnetite chains are framed within larger questions about the implications of exobiology, the relationships (if any) between life-forms (if they exist) on Mars and those on Earth, and humankind's place in the cosmos. To some extent, how scientists respond to the meteorite debate depends on their initial assumptions and underlying beliefs about the prevalence of life in the universe, or at least in our solar system. In the absence of anything approaching hard data about the abundance of life elsewhere in the universe, these larger questions are really case studies in the rhetoric of belief rather than in experimental science, as David Grinspoon (2003) suggests. Underlying the meteorite debate, the Beagle 2 life detection experiments, and NASA's plans to search for life on Europa is the widespread assumption—shared by scientists who hold diametrically opposed views on exobiology—that determining whether life exists elsewhere in the universe is of fundamental importance to self-definitions of our species.

This is a huge burden for a few suspected nanofossils to bear, the only plausible candidates at the moment for exobiologists to study. Yet as the cultural critic Molly Rothenberg has pointed out, the presence (or absence) of microbial life on Mars has little bearing on whether there may be intelligent life on planets dozens, hundreds, or thousands of light-years from Earth (Markley et al. 2001, "interviews," 4). The presence of Martian meteorites on Earth indicates that the two planets, particularly early in their histories, may have exchanged all sorts of organic materials in what Jeff Moore, a planetary scientist at NASA-Ames, terms the "day-care theory" of interplanetary evolution: "If one [planet] came down with it, they all came down with it" (Markley et al. 2001, "interviews," 5). Life on Mars, in this case, would indicate that biology may be indigenous to the planets of the inner solar system; if Martian life-forms turn out to be similar to terrestrial microbes, their existence would tell scientists relatively little about the odds of life arising elsewhere in the universe. In this regard, both Rothenberg and Moore actually lower the stakes of the exobiology debate, in contrast to those scientists—from Lowell to Sagan—who suggest that the search for extraterrestrial life has belief-altering implications for human parochialism.

One of the ironies of planetary science is that many of its practitioners justify vigorous exploration of the solar system in metaphysical terms at least a century old. Sagan, for example, unconsciously echoes Lowell in mocking human "posturings" and "imagined self-importance, the delusion that [humans] have some privileged position in the Universe" (1994, 6). The recognition that life on Earth is "inconsequential, a thin film . . . on an obscure and solitary lump of rock and metal" (6) is both a premise and one of the conclusions of the scientific quest for truth and knowledge. In an important sense, Sagan's rhetoric exhibits a familiar dialectical tendency to present the possibility of extraterrestrial life as either millenarian or apocalyptic: Lasswitz's techno-Kantians or Wells's ruthless invaders. Frank Drake, the founder of the SETI program (Search for Extraterrestrial Intelligence) suggests that the benefits of contact with an alien civilization might include "a richness of knowledge, even wisdom, that would help us achieve a higher quality of life, save our resources, and give us significant information . . . within years rather hundreds of years, or thousands" (1997, 88). Alien radio signals, in short, will do for humankind what the benevolent Martians of late-nineteenth and early-twentieth-century science fiction promised to do—provide humankind with technological, political, and moral shortcuts to a utopian future (Dick 1996). In contrast to this view, Sagan places the burden of solving planetwide problems on a deeply flawed modern civilization. Even as science abandons what he calls "the childishness and narcissism of our pre-Copernican notions" (1994, 79) of the universe, human civilization is blindly destroying the very conditions that are essential for its survival. Imagining how the Earth might be perceived from space at a hundred-meter resolution, Sagan suggests that an extraterrestrial intelligence would conclude that "something has unmistakably gone wrong": "The dominant organisms, whoever they are—who have gone to so much trouble to rework the [planet's] surface—are simultaneously destroying their ozone layer and their forests, eroding their topsoil, and performing massive, uncontrolled experiments on their planet's climate" (79). Sagan's imaginary observers are neither space invaders nor interstellar altruists but a projection of an ethical intelligence: the moral and scientific guardians of the ozone layer, forests, and topsoil. They are, in short, both the descendents of the red Martians in Burroughs's *Princess of Mars,* who observe humanity through their telescopes but offer no material or moral guidance, and a future incarnation of the ideal scientist, ethical and dispassionate, who views and is perplexed by "the childishness and narcissism" of late-twentieth-century civilization. For Sagan as for Lowell, human progress and environmental destruction are dialectically bound.

Understanding the environmental consequences of human activities on Earth becomes Sagan's primary justification for going to Mars: "Planetary exploration is of the most practical and urgent utility," he argues, because the program underscores "the fragility of our planetary environment and the common peril and responsibility of all nations and peoples of Earth" (1994, 229, 278). In his mind, even the barren, post-Viking surface of Mars remains an imaginary vantage point for assessing the consequences of human civilization's addiction to fossil fuels, social inequality, and unrestrained profit. At its most extreme, this perspective reproduces the logic of Bradbury's last Martian colonists looking back toward an Earth destroying itself in a thermonuclear holocaust. For scientists on the lunatic fringe of professional astronomy—such as John Brandenburg, who believes that the Face on Mars was carved by an ancient civilization trying to warn future humans about environmental destruction—the two planets become, in effect, mirror images of an identical eco-apocalypse (Brandenburg 1999). What such sensationalism has in common with serious scientific speculation is a tendency (inflated to an obsession by Brandenburg) to see planetary exploration as a means to warn against the "massive, uncontrolled experiments" that a resource-guzzling civilization is currently performing on Earth.

The debates about exobiology suggest a counterintuitive possibility: if proponents of sending humans to Mars temper their rhetoric about the significance of extraterrestrial life, the more appealing the idea of exploring the planet may become to taxpayers and their elected representatives. The most important "reason to go to Mars," according to Kim Stanley Robinson, "is to do comparative planetology" (in Markley et al. 2001, "interviews," 1) because such knowledge is essential to understanding the dynamics of planetary climate change. Much of the data returned by Mars missions since 1997 has suggested new analogies to climactic conditions on Earth, and, probably to a greater extent than it did after the Viking missions, the planet provides an essential point of comparison for understanding aspects of terrestrial ecology. While it may be useful to run computer simulations and construct mathematical models of global climate change, the robotic and perhaps human exploration of Mars remains crucial for many scientists in their efforts to understand short and long-term effects of global warming, ozone depletion, and atmospheric circulation. It is still an open question what exotic microbes or fossilized nanobacteria might contribute to this process of knowledge building. In 2004, it remains difficult to rule out categorically the possibility that life existed or perhaps still exists on Mars. If life did begin on Mars, there seems to be no barrier to imagining life-forms that might have adapted to existence in microecologies. The key word, of course, is "imagining": rather than assigning a specific probability for the survival of communities of exotic extremophiles, even skeptics are left with a kind of Xeno's paradox, where significant extinctions may have occurred for different reasons (loss of water, failure to adapt to ultraviolet radiation, inability of organisms to survive underground, the extinction of its symbiotic life-forms), but the arrow of absolute extinction never reaches zero.[13] The crucial question is what such survival would signify, what could be learned from a remnant ecology on Mars.

FOLLOW THE WATER: SPIRIT AND OPPORTUNITY ON MARS

With the Mars Global Surveyor and Mars Odyssey missions still ongoing, NASA launched its ambitious Mars Exploration Rover mission (MER) in the summer of 2003. Adapting landing techniques that had worked successfully for the Pathfinder mission, NASA sent two rovers to sites on Mars that apparently had been reshaped by water in the distant past. Gusev Crater, ninety-five miles in diameter, was formed by an ancient meteor impact and is intersected by a 550-mile long "meandering valley . . . believed to have been eroded long ago by flowing water" (NMR 2: 200); Meridiani Planum, halfway around the globe near the equator, is among the flattest and smoothest places on Mars, rich in the iron oxide gray hematite that, on Earth, usually forms in the presence of water. In its prelaunch press kit, NASA once again emphasized its strategy of following the water as the most logical means to determine whether life ever existed on Mars; but, the author maintained, "even if life never developed on Mars," an intensive geological investigation of formerly water-rich areas of the planet "can assist in understanding the history and evolution of life" on Earth by "yield[ing] critical information unobtainable by any other means about the prebiotic chemistry that led to [terrestrial] life" (NMR 2: 196). Ironically, given the emphasis that the NASA press kit put on extremophile microbiology and the ALH84001 debate, the $800 million

NASA rovers, christened Spirit and Opportunity, carried no life-detection experiments and engaged in no hunt for biomarkers. That search was ceded to the European Space Agency's Mars Express mission and its shoestring inventiveness.

The MER mission was intended to provide the kind of detailed geological investigations that scientists had called for since the 1970s. Each rover was equipped with an instrument package more sophisticated and robust than the equipment on Sojourner. A panoramic camera would provide high-resolution photographs, and the geological suite of instruments included an alpha particle X-ray spectrometer, a Mössbauer spectrometer, and a mini-thermal emission spectrometer to determine the mineral composition of rocks and soil samples. In addition, a rock abrasion tool, the "equivalent of a geologist's rock hammer" (NMR 2: 199), would allow each rover to dig small circles, five millimeters in depth, to examine the interior composition of rock samples. Another key innovation was a microscopic imager, a combination of microscope and camera, designed to take extreme close-ups of rocks and soils. The imagers could reveal features at a scale of hundreds of microns, too large to detect microbes but fine enough to make detailed mineralogical observations. Each rover was built to last a minimum of three months and travel at least one half mile from its landing site.

In contrast to NASA's approach, the Mars Express mission focused on biology as well as geology and atmospheric studies. The Beagle 2 lander never phoned home after it separated from its orbiter, but it represented an ambitious attempt to pick up where the Viking life-detection experiments had left off. There is no European equivalent of NASA, and the British-built lander was a collaborative effort involving universities, trusts, private firms, monies from the National Lottery, and syndicates. Colin Pillinger, the chief scientist for Beagle 2, oversaw a mission that cost approximately $60 million. The ESA lander had no rover and no back-up systems, yet it vied with the NASA rovers for top billing in the popular press because it was designed to search for unambiguous indications of life (Broad 2003, D1).

Beagle 2 borrowed the Pathfinder landing strategy: cocooned in a hi-tech cushion of airbags, it separated successfully from the Mars Express orbiter and was supposed to drop onto the planet's surface on Christmas Day 2003. Its science package included a stereo camera; a meteorology package; an extendable arm with a grinder to chip away at the surface of rocks to analyze their interior composition; a microscope; a Mössbauer spectrometer that uses gamma rays to study the mineral composition of rocks; an X-ray spectrometer to detect elements; a mass spectrometer to hunt for traces of methane in the atmosphere; and a gas analyzer that employed twelve ovens to search for organic biomarkers. Beagle 2 used a relatively simple test for life. Any carbon dioxide driven off from soil samples placed in the ovens and heated was to be analyzed and the ratio of carbon 12 to carbon 13 measured. Life on Earth shows a distinct preference for the former isotope, and this "isotopic fractionation" experiment on Mars—if it had shown a similar ratio—presumably would have been considered an indication of life. Beagle 2 was designed to dig five feet below the surface and use its arm to burrow horizontally under rocks in order to sample subsurface soils. Its loss meant that scientists in Europe and the United States would have to let inferential biology take a back seat to geology.

By any standard, Spirit and Opportunity have proved among the most successful planetary probes NASA has launched. Spirit touched down in early January of 2004, and Opportunity a few weeks later. By January 9, even before Spirit had rolled off its lander, its mini-thermal emission spectrometer had detected carbonates in Gusev Crater. A few days later, its first color images showed a landing trail made by the airbags that had cushioned its fall; the drag marks indicated that a cohesive agent in the soil had allowed a thin layer of the surface to be "peeled away" like a "crumpled carpet fold" (NMR 2: 281). Months later, David McKay, chief scientist for astrobiology at NASA's Johnson Space Center, suggested that the "cohesiveness" of the surface might be analogous to the properties of terrestrial "microbial mats," the "oldest known ecosystems" (David 2004b), a suggestion that grew out of his continuing work on ALH84001 and terrestrial extremophiles. On January 15, Spirit rolled onto Mars and began to study the soil and nearby rocks. A communications glitch caused by the rover's overloaded memory was resolved over a period of days as Opportunity neared Mars, then landed successfully on January 25. Scientists were delighted to find that the lander had come to rest in a small impact crater on Meridiani Planum; Steve Squyres, the principal investigator for the science instruments on both rovers, called it "a 300-million mile interplanetary hole in one" (NMR 2: 321). The walls of what came to be dubbed Eagle Crater revealed an abundance of rock outcrops that allowed Opportunity to begin to study "a geologist's treasure trove" (321)—a record of eons of Martian geological history. With Spirit's memory problem resolved, both rovers by the beginning of February were returning hundreds of megabytes of information daily.

For the next few months Opportunity circled the inner walls of the crater, studying rock formations and confirming the presence of hematite. This small patch of Mars was littered with odd "tiny spheres" that were dubbed "blueberries," and scientists began to float tentative ideas about their formation: droplets left by volcanic eruptions and "concretations" of minerals formed as water diffused through rocks were two early theories. By mid-February, Opportunity's microscopic imager had taken striking photographs of these tiny spheres embedded in and seemingly emerging from porous rocks, and scientists had used the rovers to dig trenches into the Martian soil and expose subsurface layers for scientific study. As Spirit rolled across the floor of Gusev, Opportunity visited sites within Eagle Crater, took photographs of rocks that showed clear evidence of layering, and detected a sulfur-rich layer on the surface. By early March, NASA's generic reticence about the implications of what Opportunity was finding gave way to a rhetoric that would have warmed Percival Lowell's heart: NASA's press release on March 2 was titled, "Opportunity Rover Finds Strong Evidence that Meridiani Planum Was Wet" (NMR 2: 400). Several of the outcroppings and rocks that Opportunity examined were characterized by sedimentary layers that did not lie in parallel lines but were marked by uneven or intersecting layers, what geologists refer to as crossbedding. Volcanic depositions characteristically lie in parallel layers; sedimentary layers formed in standing water often do not. Moreover, the chemical composition of the layers in Eagle Crater indicated high concentrations of salts apparently left by water as it evaporated. The blueberries were found to be rich in hematite and, by summer, were widely considered by scientists to have been formed by the percolation of water through layers of Martian rock. When Opportunity left Eagle Crater to begin its journey to a larger and deeper crater, Endurance, it found blueberries scattered across Meridiani Planum, solidifying the case for the plain's water-rich past (Bell et al. 2004, 1703–9; Squyres et al. 2004a, 1698–1703; Squyres et al. 2004b, 1709–14; Herkenhoff et al. 2004, 1727–30; Hanlon 2004).

Both Opportunity and Spirit continued to function well beyond the three months of their initial missions, and both continued to return significant data. By August, Spirit had rolled two miles across the floor of Gusev to the Columbia Hills, a range that rises approximately three hundred feet above the datum, and climbed (as of the end of August) forty-three feet up one of the slopes. There it found evidence of significant chemical alterations in a rock dubbed Clovis, suggesting the prolonged action of water. On the plains below the Columbia Hills in April, May, and June, Spirit had documented what appeared to be the effects of small amounts of water on rocks and soils, and scientists theorized that the floor of Gusev Crater may have been covered by a lava flow that buried much of the evidence for the paleolake that they believe once existed within the crater. Clovis may well prove to be evidence of that past. Meanwhile, Opportunity slowly descended into another and deeper crater, Endurance, and examined layers with rock outcroppings that suggested a complex history of flowing water. The strata exposed in this crater recorded geological changes that apparently were much older than those it had explored at Eagle. The middle layers of the lowest accessible outcrop, Axel Heiberg, showed a threefold increase in chlorine. Magnesium and sulfur declined nearly in tandem in older layers, suggesting that those elements could have been dissolved and removed by water. By the end of 2004, the first series of scientific articles published on Opportunity's findings (summarized in Squyres et al. 2004a, 1698-1703) were unambiguous: for significant periods of time—hundreds of thousands if not millions of years—Mars had water in sufficient quantities to alter the chemical composition of at least some areas of its surface.

Millions of people had followed Sojourner's crawl across a small patch of an ancient Martian outflow channel on the Web in 1997, but the number of hits on NASA's MER site came to more than half a billion over the first eight months of the extended Spirit and Opportunity missions. The daily and then biweekly updates and the thousands of photographs from the surface of the planet provided an ongoing narrative—a mission with a beginning, middle, and, if not an Aristotelian end, at least definable goals: Endurance Crater for Opportunity, Columbia Hills for Spirit. Sojourner had traveled a few yards to take readings of the mineral composition of several rocks; the 2004 rovers provided a photographic record that seemingly mimicked the experience of humans walking across a Martian landscape reminiscent of terrestrial deserts. Spirit and Opportunity returned the first vacation photographs from another world. The photographs from Spirit recorded a journey to a once-distant goal; it allowed scientists and web users to live out a version of science fiction's most resilient generic form—the epic trek across previously undiscovered territory or space. The heroic marches across Mars that figure prominently in novels by Ben Bova (1992), Kim Stanley Robinson (1993, 1994), and Geoffrey Landis (2000), among others, acquired an after-the-fact visual and experiential analogue. Previously insignificant craters and outcroppings became concrete goals to measure the success of the mission, and, over weeks and then months, part of the semiotics of human visualization.

While the rovers explored the Martian surface, three teams of scientists, two using sophisticated spectrometers on Earth and the other receiving data from the Planetary Fourier Spectrometer (PFS) on the Mars Express orbiter, independently detected the presence of methane in the Martian atmosphere at the parts per billion level, a potentially significant finding with far-reaching implications (Formisano et al. 2004, 1758–61). Unless it is continually replenished, methane can persist in the Martian atmosphere for only about three hundred years because it quickly oxidizes into carbon dioxide and water vapor. On Earth, methane is a byproduct of biological processes, such as fermentation, although it can also be produced by volcanic activity. The existence of methane on Mars, then, raises two exciting possibilities: it is either a product of some form of life, presumably microbial colonies beneath the surface, or it indicates active volcanic processes (thermal vents, for example, that on Earth harbor rich extremophile ecologies). One of the teams, headed by V. A. Krasnopolsky, provocatively titled its abstract in March 2004 "Detection of Methane in the Martian Atmosphere: Evidence for Life" (http://www.cosis.net/abstracts/EGU04/06169/EGU04-A-06169.pdf), and argued that the best explanation for their results was "methagenesis" by "extremely scarce" subterranean microbes. The principal investigator for the PFS, Vittorio Formisano, and his colleagues were more cautious in their assessment of the source for the methane, suggesting that volcanic activity might be a likely possibility as well (Formisano, et al. 2004, 1758–61). Almost immediately, some scientists began to ask questions about the data. Like the Sinton band controversy in the late 1950s, the initial spectrographic results were met with requests (or demands) for better, even definitive, data. The detection of methane relies on scientists sorting through and interpreting the signatures of chemical components in the Martian atmosphere, a complicated process made all the more difficult by the fact that the methane concentrations reported by Formisano's team (ten or eleven parts per billion) lie at the very limits of the PFS's sensitivity. Michael Mumma, the leader of the NASA team that first detected a possible methane signature on Mars in 2003, expressed caution in August 2004 about taking any of the reports as established fact (David 2004c). Since Formisano and his collaborators detected concentrations of methane as high as thirty parts per billion over some areas of the planet's surface, their interpretation seems likely to be accepted until more sophisticated spectrometers aboard future orbiters return new data.

The detection of methane on Mars falls into what by now is a familiar pattern in the history of areological controversies. The phenomenon in question lies at the very boundaries of detection by state-of-the-art technologies. The readings are open to varying interpretations, and, after an initial dramatic announcement, other scientists urge caution. The implications of the phenomenon potentially are far reaching, and the debate spills over into the popular press. Rather than producing a revolutionary new consensus, the initial reports become the impetus for further speculation and scientific debate and for new technologies or missions to resolve the question. In the case of methane detection, NASA scientists are pushing for a new proposed orbiter, the Mars Volcanic Emission and Life (MARVEL) spacecraft, which would be able to detect methane with one hundred to one thousand times greater sensitivity than the PFS and allow researchers to localize its sources. Paradoxically, as in the case of the meteorite controversy, the value of methane on Mars lies precisely in the fact that the evidence for its sources, if not its existence, is ambiguous, and consequently it provokes dialectical responses—skepticism and belief—that in turn will promote new research programs, new technological innovations, greater economic investments in interplanetary exploration, and new incentives for the continuing exploration of Mars.

MARS DIRECT AND THE HIGH FRONTIER

Throughout this chapter and chapter 6, I have concentrated on NASA's missions to Mars because these government-funded efforts have had a near monopoly on the production of scientific data about the planet, with the exception of the Soviet Union's handful of Martian probes. Although NASA's rover missions in 2004–5 have upstaged the contributions of the ESA's Mars Express orbiter, the ESA has announced plans under its Aurora program to launch another Mars mission in 2009 and is taking exploratory bids from hi-tech companies to build a lander for a future sample return mission. Nonetheless NASA remains the most expensive game in town, an agency with the resources to fund multiple, ongoing programs in planetary exploration, however often they may be delayed or deferred.

Even as NASA and ESA continue to justify their planetary exploration programs in terms of a grand search for exobiology, other justifications for exploring Mars are motivating a small but vocal group that claims it has the means, the motivation, and (soon) the political muscle to promote human exploration and colonization of the red planet, far in advance of the 2004 presidential commission's timeline recommending leisurely progress to landing humans on Mars. NASA itself is not a closed institution, and some of its leading planetary scientists have become affiliated with the Mars Society, a nongovernmental organization that includes prominent scientists, enthusiasts for colonizing space, a few venture capitalists, a noted film director (James Cameron), and some offbeat types who are ready to homestead the planet. In some respects, the Mars Society is an advocacy group; in others, it represents scientific and engineering alternatives to what it perceives as the bureaucratic inertia of the federal government. The centerpiece of the society is its faith in the financial and scientific feasibility of the Mars Direct plan developed over a decade ago by Robert Zubrin, then an engineer at Martin Marietta and now (unelected) president of the Mars Society.

Mars Direct is the antithesis of the $450 billion plan that NASA drew up in response to George H. W. Bush's call in 1989 for putting humans on Mars in the twenty-first century. This plan for a massive spaceship constructed in low-Earth orbit incorporated a wish list of engineering and scientific projects, many of which harked back to Werner von Braun's vision in *The Mars Project* and the exploration by flotilla scenarios of science-fiction novels of the 1950s and 1960s. Zubrin derided NASA's plan for being too "focused on realizing the science fiction vision of the giant interplanetary spaceship rather than actually getting to Mars or doing anything useful on arrival" (quoted in Bowden 1997, 104), and set about exploring alternatives for low-cost ways of getting humans to and sustaining them while they were on the planet. In January 1990 Zubrin, working with another engineer, Ben Clark, developed an outline of the Mars Direct plan; the next month Martin Marietta, where Zubrin and Clark both worked, developed a twelve-person scenario development team to come up with "broad new strategies" for the human exploration of space (Zubrin 1996b, 51–52). In contrast to NASA's "Battlestar Galactica," Mars Direct depended on both engineering ingenuity and an unabashed use of what Zubrin perceived as the triumphal experience of the American frontier: "Living off the land," he decided, was the way to go: the "intelligent use of local resources" was "the way the West was won . . . and it's also the way Mars can be won" (1996b, 2). As it evolved, Mars Direct elaborated on scientific speculation in the 1980s and 1990s that human explorers on Mars could make use of indigenous resources for fuel and food (Meyer and McKay 1996, 393–442). Rather than the endless string of zeroes on NASA's price tag for a Mars mission, Zubrin calculated that Yankee ingenuity could drive the cost down to somewhere between $20 billion and $50 billion.

Mars Direct envisions using NASA's off-the-shelf technology to minimize research and development expenditures (Turner 2004, 207–26). A heavy-lift booster (like the Saturn V rocket used for the Apollo missions or the Delta rocket used for the Space Shuttle) launches an unmanned, forty-ton payload to Mars—an Earth return vehicle (ERV) with its two methane/oxygen-driven propulsion stages unfueled. The lander also includes six tons of liquid hydrogen, a one-hundred-kilowatt nuclear reactor mounted in the back of a methane/oxygen-propelled light truck, a chemical processing unit, and small compressors. This payload represents a small chemical factory that can produce rocket fuel and water for future human explorers. The chemistry itself is straightforward: the hydrogen can be reacted with the carbon dioxide in the Martian atmosphere to produce methane and water. The methane is then liquefied and stored to serve as rocket fuel for the ERV, which will carry astronauts back to Earth, and the water is electrolyzed to produce hydrogen (which is recycled through the methanator) and oxygen, which also is stored. The chemical reactions, methanation and water electrolysis, produce twenty-four tons of methane and forty-eight tons of oxygen over ten to eleven months. Additional oxygen necessary to burn the methane at the optimal mixture ratio for rocket fuel (thirty-six tons) can be dissociated directly from the atmosphere. The 108 tons of propellant are enough to fuel the return vehicle and support the use of long-range methane/oxygen vehicles to explore the surface of Mars for a period of more than a year.

Twenty-six months later, two craft are launched to Mars; one is another fuel processing plant, the other is a habitation module, carrying four astronauts, dehydrated provisions, and another rover. Before the astronauts leave Earth, Mission Control can confirm that the rocket fuel and water manufactured by the first mission are waiting for them. The human mission lands at the site of the original factory, with a fully fueled ERV ready for the voyage back to Earth at the next launch window eighteen months later; if all goes well, the second craft lands several hundred kilometers away and begins processing fuel for a future mission. Should the astronauts land more than a few hundred kilometers from the original factory site (the distance that can be traversed by the rover), the second automated factory can have a second ERV ready within less than a year, in plenty of time for the launch window. With two missions every two years, Zubrin asserts, humans can quickly build up a network of stations and return vehicles to support an ambitious program for the continuing exploration and eventual colonization of Mars (Zubrin 1996b, 79–85; Zubrin 1999, 102–5).[14]

Rather than trying to transport all the food, water, and air necessary to sustain a mission to Mars, astronauts from the beginning of their stay can treat the planet as a stockpile of resources to exploit. Countering critiques that a human mission to Mars would be undone by radiation hazards, six months in zero gravity, dust storms, back contamination from unknown Martian organisms, and the psychological stresses of three years in close proximity, Zubrin has argued since the mid-1990s that humans could be colonizing the red planet within a decade (Zubrin 1996b, 114–35; Dreifus 1999, D3). Not surprisingly, most science-fiction novelists in the 1990s have jumped on the Mars Direct bandwagon: novels by Benford (1999), Hartmann (1997), Geoffrey Landis (2000), and Zubrin himself (2001) all envision future missions along the lines developed in the Mars Direct plan. More significantly, a number of scientists both in and outside of NASA have accepted the premises, technologies, and mission scenarios that Zubrin has outlined, and a detailed secondary literature has begun to be developed (Turner 2004).

Zubrin put an initial price tag on each set of missions of $20 billion to $30 billion; NASA subsequently produced a scaled-up version of the Mars Direct plan with twice as many astronauts at an estimated cost of $50 billion (Zubrin 1999, 104). NASA's incorporation of a version of Mars Direct into its long-range planning constitutes a de facto acceptance of Zubrin's logic of "living off the land." A NASA-ESA study (available online at http://www.marssociety.org) in 2004 compared scenarios for two versions of Mars Direct: using different assumptions and parameters, Charles D. Hunt (NASA) and Michael O. van Pelt (ESA) provided cost estimates for the first and second operational phases (from initial planning through the first two missions) of the ESA and NASA plans: ESA, $26.6 billion and $5.2 billion; NASA $39.4 billion and $7.05 billion. In constant 2002 dollars, estimates of ten missions over ten years range from Zubrin's $55 billion to NASA's $98 billion, much less than the costs for the Apollo program ($120 billion) that put astronauts on the Moon. However rough these estimates are, they are substantially less than the cost estimates implicitly attached to the 2004 presidential commission report.

The drawing board success of Mars Direct and its offshoots can be attributed to the coherent mission strategy—the plausible narrative—it projects into the near future. This future can be seen as both a logical extension of NASA's previous and current investments in the technologies of human spaceflight and interplanetary robotic exploration. As a thought experiment, Mars Direct invokes the generic appeal of Fredric Jameson's "nostalgia for the present" (1989, 517–37), a nostalgia for the shortcuts and idealized problem solving of science fiction that provides an alternative to the delays, budget cuts, and lost spacecraft that have slowed and threatened to derail plans for the human exploration of Mars. In this regard, the Mars Direct plan serves as a bridge between the alternative histories of science fiction, novels that envisioned humans on the red planet before the end of the twentieth century (Stephen Baxter's *Voyage*, 1999), and current nongovernmental efforts, notably by the Mars Society, to fast-track efforts to explore the planet.

Even among advocates of a vigorous space program, however, Zubrin's vision has come under attack for what some consider its technological sleights-of-hand as well as its ideological assumptions and values (Park 2000, 82–87). Some of these critiques call attention to Zubrin's triumphalist reading of American history: in his eagerness to sell a larger public on Mars Direct, Zubrin describes an unproblematic narrative of technological and cultural progress that becomes both the means and ends of humankind's opening a new frontier on Mars. In large measure, his vision of the American frontier seems lifted from a 1960s high school textbook. Celebrating (and simplifying) Frederick Jackson Turner's frontier thesis, he offers an idealized American past as a template for Martian exploration.[15] Like "the Great Frontier that shattered the static, stultifying, irrational, dogmatic, and completely stratified world of medieval Christendom," Zubrin asserts, "the colonization of Mars will usher in a new stage of social progress and technological innovation" (1996a, 15). In harking back to Locke's invocation of a "vacant" continent, and in proposing solar system exploration as the solution to twenty-first-century problems of scarcity, poverty, and social disorder, he depicts Mars as a reservoir of "infinite" use-value: colonists will find on the planet "all the metals, silicon, sulfur, phosphorous, inert gases, and other raw materials needed to create not only life but an advanced technological civilization" (16). Zubrin's verb "create" is revealing; rather than describing Martian exploration in terms of the scientific unknowns involved in altering a planetary environment, his rhetoric recalls nineteenth-century apologies for the United States' westward expansion. The alternative to the exploration and exploitation of Mars, he continues, is "technological stagnation"—a failure of will and know-how that is provoked by the widespread awareness that civilization has reached its limits: "Unless people can see broad vistas of unused resources," Zubrin argues, "the belief in limited resources tends to follow as a matter of course." Trapped on Earth, refusing to search for their manifest destiny among the stars, people revert to Hobbesian despair, fighting for diminishing shares of scarce resources. Without a far-reaching program of interplanetary colonization, he concludes, "humanity will create hell for itself in the 21st century" (19, 22). Not surprisingly, in Zubrin's starkly contrasted futures, a colonized Mars becomes the vehicle for humanity to relive an idealized vision of the past, correcting the mistakes that have led to technological and cultural stagnation: Mars "is far enough away [from Earth] to free its colonists from intellectual, legal, or cultural domination by the old world, and rich enough in resources to give birth to a new" (16). His projection of the American frontier onto Mars, in effect, appropriates the imagined futures of science fiction for an interplanetary vision of manifest destiny.

While Zubrin has used his bully pulpit as president of the Mars Society to advance his views, others within the organization have debated fiercely the ethics of cannibalizing Martian resources for human settlements, leading (at least imaginatively) to the prospect of terraforming an alien world. The controversies about how humans might use or abuse the Martian environment reflect competing visions of the utopian possibilities of starting over on a new planet: the nuclear-powered, aggressive colonization scheme that Zubrin describes versus an effort to redefine humankind's relationship to the complex environmental systems on both Earth and Mars. The society itself is, in part, an outgrowth of the Mars Underground that took shape in Boulder, Colorado, in the 1970s and 1980s. To some extent, elements within the Mars Society maintain allegiances to counterculture or New Age values, even as they help promote the idea of space exploration. Originally a loosely organized collection of like-minded researchers and graduate students, including Chris McKay, Penelope Boston, Tom Meyer, Carol Stoker, and Carter Emmart, among others, the Mars Underground has been described as a "hippie band of Colorado visionaries" who in their current incarnations as (mostly) NASA scientists may get the chance to spearhead a "ruthlessly efficient, eminently doable" mission to Mars (Bowden 1997, 62, 64). Yet although many members of the Mars Society are planetary scientists, biologists, engineers, meteorologists, and geologists, the society has been derided in the popular media as "a model rocket club without the rockets" (McNichol 2001, 142) and lumped together with faith healers and advocates of pyramid power as yet another cult of adherents to "voodoo science" (Park 2000, 82–87). Even less-dismissive reporting ignores ideological and scientific disputes among the members. In a 1998 article in *New Times,* Glenn Gaslin describes the Mars Society as a "vocal, media-savvy coalition of Mars fanatics" intent on transforming the red planet "into another San Fernando Valley" (1998, 11, 12). Much of the article rides roughshod over significant differences among planetologists about the feasibility of terraforming; Gaslin transforms speculative papers about terraforming into NASA's "blueprint . . . to turn Mars into an entirely habitable, Earthlike little planet" (16). Such reporting is suggestive of the all-or-nothing coverage that space exploration receives: cynicism, enthusiasm, or, in Park's case, professional disdain.

Yet the eclectic backgrounds of its membership indicate something of the internal divisions at play in the Mars Society and its vision of a spacefaring civilization. At its founding convention in 1998, Chris McKay chaired a panel discussion of terraforming that had would-be planetary engineers, including Zubrin, backpedaling before hostile questions from the audience about various scenarios to terraform the planet, particularly those based on nuclear-powered reactors, launch vehicles, and terraforming plants (Blakeslee 1998). In the years since then, annual meetings of the Mars Society routinely have pitted proponents of nuclear power against the advocates of environmentalism transported to Mars, what some terraforming proponents, like the poet Frederick Turner, have termed ecopoesis (Turner 1988; Turner 1989, 33–34). As in NASA's official press kits since the 1960s, the rhetoric on both sides in such debates is filtered through a science-fiction tradition that promotes space exploration as a visionary goal. What unites most members of the Mars Society is the conviction that colonizing Mars represents a crucial step for humankind in escaping, forestalling, or preventing a grim future of environmental catastrophe, warfare, and social chaos on twenty-first-century Earth.

The Founding Declaration of the Mars Society in 1998 buries many of these tensions under a rhetoric of true belief. Signed by seven hundred conference attendees in 1998, the document remains available for additional signatures on the society's web site (www.marssociety.org). The declaration outlines three primary goals to promote the human exploration of Mars: conducting broad public outreach programs; lobbying for increased spending by the United States and other nations on planetary exploration; and finding ways to fund such exploration on a private or commercial basis. To this end, the society has lobbied for political candidates committed to space exploration; sought to promote its vision of international efforts to colonize the red planet as a means to reinvigorate western civilization, politically, spiritually, and scientifically; and established scientific research stations, on Devon Island in the Canadian artic and in the Utah desert, dedicated to developing techniques for supporting a long-term human presence on the red planet. These stations are staffed by volunteers, but almost all are well-trained (and in many cases well-published) planetologists, geologists, and engineers. Of the seven justifications the Mars Society offers for exploring and colonizing the red planet, however, only two are scientific: the discovery of Martian fossils or nanofossils would demonstrate that terrestrial life is "not unique" and "by implication, reveal a universe that is filled with life and probably intelligence."[16] Equally significant in the declaration is the scientific payoff promised by "comparative planetology"—a knowledge that is figured in apocalyptic terms; just as "Venusian atmospheric studies" revealed the threats posed by "global warming" and the degradation of the ozone layer, missions to Mars "could be key to [human] survival." The rest of the document echoes the rhetoric of Zubrin's *The Case for Mars,* invoking the challenge of a new frontier, the benefits that would accrue from firing the imaginations of young people dedicated to a noble scientific enterprise, the spreading of life to the universe, and the creation of a new home for humanity. This "great enterprise" is shaped both by scientific plausibility and the values and assumptions of Stoker's "*Star Trek* civilization" (Markley et al. 2001, "interviews," 1).

In brief, the Mars Society seems a latter-day incarnation of the Royal Society of London founded in Restoration England, as much a social club of enthusiasts as a professional scientific organization. Yet with Mars Direct as a blueprint and the concept of *in situ* resource utilization as a widely accepted principle, the Buck Rogers meets Tom Swift vision of the Mars Society is one shared, in some form, by many planetary scientists. In large measure, Mars Direct is a contemporary instance of the speculative projection that structures scientific and technological thinking. Generically, there is not really another option: full-fledged mission scenarios do not spring full-blown into existence without having incorporated and relabeled parts of their cultural as well as scientific inheritance. In this regard, Zubrin's plan may be to the next fifty years what Werner von Braun's Mars Project was to the Mariner, Viking, and *Star Trek* era—a grand vision that helped to generate a conceptual and narrative framework that allows scientists and engineers to see themselves contributing to an overall strategy capable of shaping the technoscientific future. Mars Direct offers what many commentators feel the Space Shuttle and the International Space Station lack—a goal to galvanize public support for space exploration. In the possible futures envisioned by science fiction, Mars Direct dominates the colonization of the red planet because it represents not only a chance to revisit, for good or ill, the ethos of the

American frontier, but an opportunity to explore the moral, philosophical, and ecological implications of cannibalizing the resources of another world for humankind's benefit. In the best of these novels, Kim Stanley Robinson's *Red Mars, Green Mars,* and *Blue Mars,* Mars once again becomes a grand thought experiment to explore the political, social, and moral problems confronting an overcrowded and polluted Earth.

NINE

Falling into Theory: Terraformation and Eco-Economics in Kim Stanley Robinson's Martian Trilogy

The world has become a giant science fiction novel which we're all coauthoring.—KIM STANLEY ROBINSON, 1998

• • ● • •

UTOPIA REVISITED

In the introduction to *Future Primitive: The New Ecotopias,* an anthology of "green" science-fiction stories, Kim Stanley Robinson defines science fiction as "a collection of thought experiments that propose scenarios of the future. . . . They are historical simulations . . . images, endlessly reiterated, [that] have come to form in our imagination a kind of consensus vision of the future" (1994a, 3). In his own novels, as well as in this collection, Robinson presents alternatives to the genre's "consensus vision" of humankind "as the last organic units in [the] denatured, metallic, clean, and artificial world" of a cyber-engineered future. In place of this denatured vision, Robinson urges his readers to explore the utopian possibilities of "cobbl[ing] together aspects of the postmodern and the paleolithic" in a "future primitive" that might best be described as an ecocentric turn toward holism. This "future primitive," he implies, can serve as a powerful analytic to reveal—and indeed to gesture beyond—the forms of alienation that structure and are structured by the deep-seated antiecological values and assumptions characteristic of western thought (9–10). In this context, Robinson's Martian trilogy, *Red Mars* (1993), *Green Mars* (1994), and *Blue Mars* (1996), offers a sustained, theoretically sophisticated attempt to conjure into being a future that resists the romantic dystopianism of cyberpunk, the antitechnological bias of much "green" literature, and the blanket denunciations of capitalist technoscience that have become popular in some left-wing circles. Taken together, though, the three novels demonstrate just how complex a notion "utopia" can become. The "future primitive" that Robinson envisions both exploits and critiques what Don Ihde has called the "doubled desire" of technology. Historically, technology presents itself as the essential means for humankind to adapt or to control nature and thereby improve the quality of life; however, the promises that technoscience makes—pleasure, plenty, and self-actualization—ironically seek to render it transparent. Massive investments of labor, capital, and resources, in other words, offer us enhanced versions of a "natural," pretechnological existence (Ihde 1990, 75–76). In his Martian trilogy, the politics of this doubled desire lead Robinson to explore the consequences of people struggling "to yoke together impossible opposites" (1994b, 229): mind and body, spirit and matter, nature and culture, and biosphere and technoscience. In the process, his novels call into question two of the constitutive fictions of modernity: the separation of nature from culture and the consequent privileging of contemporary technoculture at the expense of a devalued, technologically primitive past (Latour 1993, 99–100).

Robinson's introduction to *Future Primitive* emphasizes that science fiction is a genre of ideas: like Samuel Delany, Robinson argues that science fiction does not represent historical experience but generates simulations of what that experience may become. This distinction between representation and simulation is crucial to understanding his Martian trilogy as a theoretical intervention in late-twentieth-century debates about ecology, economics, and technology, and it is worth recalling the argument that Steve Shaviro makes: if representation, as Lacan suggests, is predicated on a fundamental lack, if it entails "the murder of the thing," simulation, he argues, "precedes its object: it doesn't imitate or stand in for a given thing, but provides a program for generating it" (1996, 17).[1] In Shaviro's sense, the "utopian" possibilities of science fiction occupy a register of simulation: they give imaginative form to the desire to think beyond the contradictions of historical existence, and, as the etymology of the word suggests, beyond our location in time, culture, and geography. Brought into the regime of representation, utopian schemes are always in the process, as Robinson suggests in a chapter title in *Red Mars,* of "falling into history," undone by the distance between the idealized operations of a frictionless system and the wear and tear of embodied, historical existence (Serres 1982, 10–12). Utopias can best be understood, then, as expressions of their creators' (and their cultures') desires to imagine conditions that would allow humankind to transcend its *originary* alienation—an alienation, at once, ecological, political, and psychological that severs us from nature, from others, and from ourselves. Robinson's phrase "historical simulations" thus suggests his interest in reprogramming the mindset that divides nature from culture; rather than utopian longings, his trilogy offers a carefully nuanced thought experiment in the greening of science, economics, and politics.

As the titles *Red Mars, Green Mars,* and *Blue Mars* suggest, Robinson's future history focuses on the simulated science of terraformation—a science that (a half century after Arthur C. Clarke's *The Sands of Mars*) exists only as a thought experiment, as the uncertain and arbitrary simulations designed to engineer a biosphere, sufficient at least for plant life, on Mars. Beginning with two short stories, "Exploring Fossil Canyon" (1982) and "Green Mars" (1985), and continuing after the trilogy in his collection of stories, sketches, and poems, *The Martians* (1999), Robinson uses the terraforming of Mars to rethink the complex relationships between planetary ecology, the interlocking systems that create and sustain the tenuous, seemingly miraculous conditions that allow life to flourish, and political economy, the distribution of scarce resources among competing populations and interests. At the conceptual center of this thought experiment lies what Robinson calls "eco-economics," his challenge to the default assumption that economics means the exploitation, degradation, and eventual exhaustion of natural resources—and the subsequent single-minded pursuit of more resources to exploit. On a world where the biosphere itself is being manufactured, notions of value make sense only to the extent that they erase distinctions between quantitative measures of labor and capital and qualitative contributions to social and ecological balance.

In this regard, Robinson radically revalues the science-fiction tradition of Burroughs, Brackett, and Bradbury; his literary and political touchstones become the utopian tradition represented by works such as Bogdanov's *Red Star* and Kornbluth's and Merril's *Outpost Mars.* Yet one of the strengths of the trilogy is its depiction of Mars as "a giant mountainous wilderness," in the experience of "red rock red dust the bare / mineral of here and now" (Markley et al. 2001, "interviews," 2; Robinson 1999, 385). In "Fossil Canyon" a tourist hiking through the canyon systems of the Valles Marineris finds lava pellets that he initially mistakes for fossils. After the guide, Roger Clayborne (who reappears in two subsequent short stories), correctly identifies these "pseudofossils" as pellets from the eruption of Olympus Mons, Eileen Monday feels "a loss larger than she ever would have guessed. She wanted life out there as badly as . . . the rest of them did" (1999, 52). Roger and Eileen voice what become in the trilogy the "red" and "green" positions on colonizing Mars: the reds want to leave Mars in a nearly pristine—and lifeless—condition; the greens seek to terraform the planet to make it habitable by humans. The greens represent a spectrum of technological and political positions, articulating what are ultimately competing versions of planetary inhabitation: the utopias that hark back to the science-fiction paradises of the 1890s, a vast mining colony, a tourist haven, or even a new world that will supercede a worn-out Earth.

But in order to explore humanity's desires to conserve, exploit, and redefine the utopian possibilities for a spacefaring society, Robinson must distinguish his work scientifically, fictionally, and politically from the science-fiction traditions he has inherited. Whereas Larry Niven in *Rainbow Mars* (1999) populates the fourth planet with a century of imagined Martians, from Wells's octopoid cannibals to Burroughs's giant green warriors, Robinson emphasizes a radical break with such traditions. Standing on the surface of Mars, Eileen Monday recognizes that the experience of "red rock red dust" lies outside the literary and conceptual territory of twentieth-century planetology:

> All the so-called discoveries, all the Martians in her book—they were all part of a simple case of projection, nothing more. Humans wanted Martians, that was all there was to it. But there were not, and never had been, any canal builders; no lamppost creatures with heat-beam eyes, no brilliant lizards or grasshoppers, no manta ray intelligences, no angels and no devils; there were no four-armed races battling in blue jungles, no big-headed skinny thirsty folk, no sloe-eyed dusky beauties dying for Terran sperm, no wise little Bleekmen wandering stunned in the desert, no golden-eyed golden-skinned telepaths, no doppleganger race—not a funhouse mirror-image of any kind; there weren't any ruined adobe palaces, no dried oases castles, no mysterious cliff dwellings packed like a museum, no hologrammatic towers waiting to drive humans mad, no intricate canal systems with their locks all filled with sand, no, not a single canal; there were not even any mosses creeping down from the polar caps every summer, nor any rabbitlike animals living far underground; no plastic windmill-creatures, no lichen capable of casting dangerous electrical fields, no lichen of any kind;

> no algae in the hot springs, no microbes in the soil, no micro-bacteria in the regolith, no stromatolites, no nanobacteria in the deep bedrock . . . no primeval soup. (1999, 53)

The litany of science-fiction creatures and doppleganger races depopulates twenty-first-century Mars, from Burroughs and Dick down to the eco-niches that many scientists suspect may still harbor relics of the planet's Noachian past.[2] This depopulation is necessary to imagine how humankind might respond to an unmediated "nature," a pristine condition that would allow humanity to calculate—and take responsibility for—all the caloric, biogenetic, and chemical interventions in a dynamic environment. Eileen's Mars is a thought experiment, an experiment that never can be performed on Earth but that is a necessary starting point for considering the engineering and ethical implications of transplanting humans to a new world. In this respect, Mars becomes the site for a philosophical and political rethinking of the values and assumptions that underlie the discourses of ecology and, more generally, planetology.

In "Green Mars," Roger and Eileen meet two hundred years later (having benefited from the longevity treatments that play an important role in the trilogy) on a climbing expedition up the escarpment of Olympus Mons. Mars has been terraformed, and Roger treats the loss of the "red rock red dust" as "the visible sign of a history of exploitation," the reshaping of the planet to conform to human "history" rather than "topography" (192). In contrast, Eileen invokes Heidegger's "distinction between *earth* and *world*" in order to suggest that *all* experience is mediated: " '*Earth* is that blank materiality of nature that exists before us and more or less sets the parameters of what we can do. . . . *World* then is the human realm, the social and historical realm that gives earth its meaning' " (144). "Green Mars" fictionalizes the dynamic accommodation that must exist between these theoretical postulates. The ongoing deconstruction of this distinction between "earth" and "world" becomes, in effect, Robinson's contribution to a rethinking of the values and assumptions that define human representations of the natural environment. As he writes in the poem "Canyon Colour," "There, on a wet red beach— / Green moss, green sedge. Green. / Not nature, not culture: just Mars" (364).

If "earth" is accessible only through the mediation of senses—not to mention the clothing, oxygen masks, ice boots, tents, and stoves that are necessary for mountaineering on any planet—this recognition does not mean a surrender to the imposition of human desires on a "blank" landscape, to the "projection" that Eileen had analyzed in "Fossil Canyon." At nineteen, on a Mars just beginning to be terraformed, Roger had experienced on "the great northern desert of Vastitas Borealis" an epiphanic moment in the wilderness:

> Light leaked over the horizon to the southeast and began to bring out the sand's dull ochre, flecked with dark red. When the sun cracked the horizon the light bounced off the short steep faces of the dunes and filled everything. He breathed the gold air, and something in him bloomed, he became a flower in a garden of rock, the sole consciousness of the desert, its focus, its soul. Nothing he had ever felt before came close to matching this exaltation, the awareness of brilliant light, of illimitable expanse, of the glossy, intense *presence* of material things. (1999, 145)

As in the trilogy, such passages do not represent moments of psychological self-awareness but the collective experience of an environment that dissolves barriers between "self" and "nature." Yet as the hiss of the oxygen regulator reminds him, there is no idealized escape "back" to a garden, rock or otherwise, no choice to make that could sever earth from world. There is only an ethics of responsibility, of the values that the characters bring to the ascent of Olympus Mons. Roger's experience of "what it *feels* like to be in such wilderness" (206) is recaptured at the end of the climb when they reach the caldera of the tallest volcano in the solar system. It is this lived experience of finding oneself " 'in the middle of such an heartless immensity' " that offers a means to think beyond the received opposition of humankind and nature. Robinson's invoking of Melville is suggestive: where Pip is driven mad by being left alone on the sea, Roger and Eileen represent the ongoing negotiation of "self" and "wilderness" as an opportunity to be explored rather than a battle to be fought or a horror to be avoided. Rather than the hostile Martian nature envisioned by Schuyler Miller in the 1930s and Ludek Pesek in the early 1970s, Robinson offers "the most textured and varied evocations of a mapped Mars that literature has to offer" (Morton 2002, 168–78).

Robinson's trilogy explores the possibilities of science fiction as a political thought experiment—notes toward a utopian future that have proved both a critical and popular success.[3] Having established themselves on the fourth planet in *Red Mars*, the First One Hundred, the initial party of scientists sent to colonize the planet, fragment politically, socially, and geographically. After several years, a scientific team led by Vlad Taneev and Marina Tokareva develop a process to retard the onset of aging, then turn their attention to eco-economics as a means to integrate ecology and "its deformed offshoot economics" (1993, 297), to develop a system of value, in other words, that recognizes the feedback loops between the large-scale development of Martian resources and terraforming. In contrast to economics, "people arbitrarily . . . assigning numerical values to non-numerical things," as Vlad puts it, eco-economics defines "efficiency [as] the calories you put out, divided by the calories you take in." An ethical imperative follows: "Everyone can increase their ecological efficiency by efforts to reduce how many kilocalories they use" (1993, 297, 298). Restricting consumption becomes a far more effective means to increase one's value to the system than accelerating production because production invariably strains scarce resources. Eco-economics, in this regard, calls into question the logic of capitalist production and, more generally, the ongoing exploitation of nature as the means to generate value. It acts as a historical simulation to suggest alternatives to the ever-increasing cycles of intensification and environmental degradation that Marvin Harris describes and that, as I have argued in previous chapters, form the backdrop for the future histories written by Cyril Kornbluth, Judith Merril, Ray Bradbury, and others (Harris 1977, 5). As Robinson writes in a poem in *The Martians,* "in the / Attempt to imagine Mars I came to see / Earth more clearly than ever before" (382).

Robinson's fragile ecology-in-the-making on Mars thus serves as a fictional projection of late-twentieth-century eco-economic crises, a virtual space in which to imagine a society struggling through and toward "some kind of universal catastrophe rescue operation, or, in other words, the first phase of the postcapitalist era" (Robinson 1996, 63). His trilogy works on a variety of levels to imagine the conditions under which capitalism will evolve—haltingly, violently, uncertainly—toward an eco-economic future. What distinguishes these novels from other recent thought experiments about humanity's future on Mars is precisely Robinson's recognition that the unending profits envisioned by late (and future) capitalism require infinitely exploitable resources to escape the diminishing returns and declining living standards of intensification. As one of his characters, William Fort, the head of a metanational corporation that eventually metamorphoses into an umbrella of semiautonomous collectives, declares, "Capital is a quantity of input, and efficiency is a ratio of output to input. No matter how efficient capital is, it can't make something out of nothing" (1994b, 81). In this regard, the novels can be read as simulations that paradoxically remain open to stochastic self-organization, thought experiments that engage the contingencies on which most utopian aspirations founder: social unrest, economic competition, psychic crises, national rivalries, racial hatreds, official violence, greed, stupidity, and environmental degradation. As a richly imagined geophysical and political landscape, Robinson's Mars blurs distinctions between fictional and scientific simulations of terraforming, even as it allows readers to question the values on which current justifications for planetary engineering rest.

SCIENTIFIC SPECULATION

Robinson's revaluation of the tradition of Martian science fiction is linked to his complex relationship to the scientific literature on terraforming that became an important subgenre of plans for the human exploration of Mars in the 1980s (Oberg 1982; Fogg 1995; McKay, Toon, and Kasting 1991, 489–95; Shirley 2004). He exploits the work of a number of scientists who have speculated about terraforming, but rejects the ideology of the frontier that Zubrin invokes (1996a, 13–24; 1996b). Terraforming, according to Martyn Fogg, one of its leading advocates, is the hypothetical "process of planetary engineering, specifically directed at enhancing the capacity of an extraterrestrial planetary environment to support life," and perhaps ultimately "to recreate an unconstrained planetary biosphere emulating all the functions of the biosphere of the Earth" (Fogg 1995, 9).[4] In the case of Mars, the consensus candidate in the solar system for such planetary engineering, scientists have their work cut out for them. Mars's atmosphere is 95 percent carbon dioxide; its atmospheric pressure is about six millibars (little more than 1 percent of Earth's), its mean surface temperature is -56 degrees centigrade, its water reserves are bound in subsurface ice, and because it has only trace amounts of oxygen, it has no ozone layer, so the surface is bathed in ultraviolet radiation (Fogg 1995, 202–3). Nonetheless, since the 1970s would-be planetary engineers have proposed a variety of strategies to create a biosphere on this forbidding terrain.

The minimalist approach, termed ecopoeisis by its proponents, stops short of full-scale terraformation; by augmenting the planet's greenhouse effect some scientists believe that they can raise planetary temperatures, thicken and hydrate the atmosphere, and then seed the planet with genetically engineered anaerobic microorganisms to begin a long evolutionary process. But, as Fogg maintains, investing billions to provide a habitat for lichen makes little economic or scientific sense. Scenarios for no-holds-barred terraformation include creating a runaway greenhouse effect by introducing chloroflurocarbons into the atmosphere (the same pollutants that are compromising the ozone layer on Earth), thereby melting the polar ice and carbon dioxide caps, heating the planet, and outgassing carbon dioxide trapped in the regolith; "freeing" the water and ice that exist in subsurface deposits and in the polar caps by detonating thermonuclear explosions; placing giant mirrors in stationary orbits near Mars to increase insolation and warm the surface; and "harvesting" ice-rich asteroids from their orbits and propelling them onto collision courses with the planet, instantly thickening the atmosphere and providing water for plants to survive. "Oxygen availability does not limit our ability to terraform Mars," Owen B. Toon, a leading expert on planetary atmospheres, argues because it is "plentiful in the soils" (1997, 56). Citing "compelling evidence that Mars has a permafrost that is rich in water," Thomas Meyer and Chris McKay suggest that "it is possible to prepare breathable air, water, rocket propellant, fertilizer, and other useful compounds and feedstocks" from gases in the Martian atmosphere (Meyer and McKay 1996, 403, 399). Such *in situ* resource utilization (ISRU) would allow future colonists to relax "the need for tight closure, total recycling and complex toxicogenic filtering of the air supply . . . , allowing the use of simpler semi-closed life support systems where losses could be continuously made up from freshly produced air supplies" (Meyer and McKay 1996, 395). The overriding goal of these terraforming strategies is to create a biosphere by warming and hydrating the planet so that the same evolutionary processes that took place on precambrian Earth can occur—in exponentially accelerated fashion—on Mars.[5]

Significantly, one of the crucial texts on terraformation, cited almost reverently in the scientific literature, is a novel, *The Greening of Mars* (1984), coauthored by two prominent scientists, James Lovelock, the originator of the Gaia Hypothesis, and Michael Allaby. Lovelock and Allaby turn twentieth-century swords (ICBMS) into plowshares (vehicles to carry CFCS to Mars to augment the greenhouse effect), creating a carbon dioxide–rich atmosphere that sustains wide varieties of plant life. In converting the nightmarish excesses and deadly by-products of industrial civilization to benevolent uses, the novel provides the mythic origins for an imagined future, a parable of ecological restitution on a planetary scale: the authors' terraformed Mars exports the Gaia Hypothesis to the red planet, universalizing the balances and feedback loops of Earth's self-sustaining biosphere. As a thought experiment, the novel constitutes the ideational ground—the values, assumptions, and theories—on which the emerging discipline of planetary engineering rests. In one respect, the simulated biosphere of Lovelock's and Allaby's twenty-third-century

Mars reinscribes, in scientific terms, late-nineteenth-century views of the red planet as a bucolic paradise. More generally, the novel suggests the extent to which the science of terraformation relies on mythic archetypes of resurrection for its rationale.[6]

Even as such thought experiments—both literary and scientific—envision massive technological interventions to terraform Mars, their rhetoric invokes seemingly antithetical myths of humankind's relation to a terrestrial nature: the idealized visions of restoration ecology and the endless generation of wealth through exploitation. Mediating this contradiction, I shall argue, is both the strategy and the rationale of Robinson's eco-economics. For its ecologically minded proponents, terraformation is not the imposition of humankind's will on an alien environment but a heroic project to re-create conditions that existed three to four billion years ago when massive floods scoured the surface of a warmer and wetter Mars (Squyres and Kasting 1994, 744–49; Kasting 1997, 1245; McKay and Stoker 1989, 189; Baker 2001, 228–36). As Chris McKay recently put it, "Mars lived fast, died young, and left a beautiful body—the Sylvia Plath approach to planetary science. We could play Ted and just ignore it, or we could do something better and bring it back to life" (Shirley 2004). This postmodernist vision of the Sleeping Beauty myth makes humankind's technological mastery of planetary engineering the equivalent of a magical kiss, and McKay advocates restoring to Mars its hypothesized indigenous biological, geochemical, and hydrological cycles, with Martian microorganisms interacting with "restored" versions of its ancient atmospheric and surface environments (Shirley 2004). In Frederick Turner's ten-thousand-line epic poem *Genesis,* ecopoeisis on Mars is cast in an allusive language that blends epic conventions and Gaian ecology. Before terraformation, humankind encounters Mars in the twenty-first century as the abode of "a stunted and abortive chemistry, / A backward travesty of life." Terraformed by dedicated science and mystical incantation, Mars becomes a self-sustaining biosphere, "an arch-oeconomy / Dynamically balanced by the pull / Of matched antagonists, controlled and led / By a fine dance of feedbacks, asymptotic, / Cyclical, damping, even catastrophic" (1988, 7). As Fogg's enthusiastic praise of Turner's poem suggests, the mythos of biogenic resurrection plays a crucial role in constituting ecopoeisis as the fulfillment of our doubled desires for technology: planetary engineering creates a self-regulating biosphere in which humanity, and lower forms of life—"beetles and bacteria / And molds and saprophytes," as Turner says—can start anew (Fogg 1995, 22–24). The myth of ecopoeisis as resurrection, though, might be called more accurately a form of ideological displacement: the terraformed Mars of the imagined future has not been restored to its pristine state but has become the vehicle to give scientific and poetic shape to fantasies of a prelapsarian terrestrial ecology.

If Mars terraformed becomes the scientific "confirmation" as well as the spiritual projection of Lovelock's Gaia Hypothesis, it is also the imaginary space of a new frontier, a technologized site for an updated manifest destiny. In promoting his Mars Direct scenario (discussed in chapter 8), Robert Zubrin forges explicit connections between the frontier thesis of Frederick Jackson Turner and the ideology of American-led terraformation (Zubrin 1996b, 295–306; see also Zubrin 1999, 101–27). "Without a frontier to grow in," Zubrin asserts, "not only American society, but the entire global civilization based upon Western enlightenment values of humanism, reason, science and progress will ultimately die" (1996b, 303). Only Mars "has what it takes" to continue the march of progress toward a humanist salvation: "It's far enough away to free its colonists from intellectual, legal, or cultural domination by the old world, and rich enough in resources to give birth to the new" (298). Zubrin's libertarian rhetoric of self-actualization thus depends on the economics of resource appropriation, even as it evokes, as fellow enthusiast Turner puts it, "a project that will allow us to pursue beauty and truth on a grand scale" (1989, 33–34). Terraformation is the signifier for the aesthetics as well as economics of plenitude.

Zubrin's romantic vision of the American frontier, however, is founded on dubious or simplified readings of American history that repress both the human and ecological consequences of conquest and colonization.[7] Liberty becomes a function of an idealized "New World" open to seemingly limitless exploitation of its resources. Projected into the future, this romanticized view of the frontier elevates the engineering strategies behind Mars Direct into a metaphysics of unlimited freedom founded on the exploitation of resources: "If the idea is accepted that the world's resources are fixed, then each person is ultimately the enemy of every other person, and each race or nation is the enemy of every other race or nation. Only in a universe of unlimited resources can all men be brothers" (1996b, 304). Zubrin's rhetorical movement from "the world's resources" to "a universe of unlimited resources" enacts the logic of a fantastic political economy in which terraformation—and the harvesting of resources from other worlds—becomes economically, socially, and politically essential to infinite growth and infinite freedom. For Zubrin, "Mars beckons" because capitalist and democratic values were "born in expansion, grew in expansion, and can only exist in a dynamic expansion." Whether one is a proponent of the Gaia Hypothesis or an investment banker, the effect of his arguments is to reinforce the belief that humanity's only hope for the future is to repeat on Mars the cycles of spewing CFCs into the atmosphere, mining, harvesting crops and timber, and devastating wildlife that have compromised the earth's environment. The logic of terraformation, not surprisingly, thus requires new frontiers beyond the red planet. "The universe," Zubrin declares, "is vast. Its resources, if we can access them, are truly infinite" (1996b, 304). Terraforming Mars becomes only the initial impetus to ratchet upward the "two key technologies of power and propulsion" so that humankind can exploit the "infinite" resources of the outer solar system and beyond (1999, 127–222). Ironically, the logic of endless terraformation dictates that without the mind-boggling investments in technology to make accessible the "infinite" resources of new frontiers, humanity lapses into its default condition—the Hobbesian war of all against all. In Zubrin's mind, to terraform Mars—to render it both a biosphere and a commodity—is to reinvigorate ourselves psychically and to reverse the downward spiral of civilization.

The logic of the frontier that Zubrin sketches is founded on the anti-ecological assumption that "natural" resources are always and already marked as objects of exploitation and exchange. In this respect, he displaces onto Mars a vision of infinite resources that has led civilizations from one crisis of intensification to another until a dry, frigid, and almost oxygenless planet seems to many humanity's last best hope for survival. Not all commentators and scientists share Zubrin's enthusiasm for terraforming or his view of the American frontier. Jeremiah Creedon classifies the prospect of terraforming Mars as "grandiose" fantasy, like the Strategic Defense Initiative, that "belong[s] to a special kind of American virtual reality—a make-believe world full of things the public pays for but never sees." Terraforming, he adds, "falls short as science, [but] it does make great myth" (1994, 36). Planetary engineering has critics in the scientific community, particularly when it is presented as a near-future project or a necessary consequence of colonization. The planetary astronomer David Grinspoon suggests that "terraforming will remain a purely intellectual exercise for the forseeable future," and therefore "anyone who suggests seriously that we embark on any of these [terraforming] schemes anytime soon should be institutionalized or forced to teach Freshman Astronomy at a large public university" (1997, 337). More recently, he has conceded that as an intellectual and ethical exercise, "terraforming Mars is very good for us, and is maybe a first step towards attaining the kind of wisdom" that a century or so from now will be necessary to consider seriously planetary engineering (Shirley 2004). Projecting terraforming into the future makes it humanity's exit exam for graduating to the first phase of its new existence as a *Star Trek* civilization.

More like Grinspoon than Zubrin, Robinson does not define the "virtual reality" of terraformation in terms of new frontiers or simplistic solutions to problems of overpopulation and environmental degradation on Earth. If Zubrin projects an idealized past into the future, Robinson calls into question the values and assumptions that have motivated previous colonialist enterprises. It is precisely this fantasy of the "mastery" of nature which eco-economics seeks to counter. At a crisis point in *Red Mars,* Frank Chalmers, the codirector of the mission to Mars and an inveterate politician, explains to the idealistic John Boone the logic of interplanetary colonization:

> "Why were we sent here in the first place, Frank?"
>
> "Because Russia and our United States of America were desperate, that's why. Decrepit, outmoded industrial dinosaurs, that's what we were, about to get eaten up by Japan and Europe and all the little tigers popping up in Asia. And we had all this space experience going to waste, and a couple of huge and unnecessary aerospace industries, and so we pooled them and came here on the chance that we'd find something worthwhile, and it paid off! . . . And now even though we got a head start up here, there are a lot of new tigers down there who are better at things than we are, and they all want a piece of the action. There's a lot of countries down there with no room and no resources, ten billion people standing in their own shit." (352–53)

In Frank's mind, terraforming Mars is a fortuitous gamble, born of desperation, overpopulation, and the exhaustion of resources. His cynicism echoes throughout *Red Mars* as a counterpoint to both debates about the ethics of ecopoeisis and revolutionary struggles to determine who owns Mars and its resources. In one sense, the "utopian" project of Robinson's trilogy is to render such cynicism, as far as possible, a historical artifact, to replace the politics of desperation with a simulated future in which hard-won forms of cooperation, synthesis, and the dialogic unity of eco-economics wins out over a coercive political economy based on the control of scarce resources. In another sense, Mars, as it undergoes its sea change from red to green to blue, offers its citizens (and the novels' readers) a means to rethink the individualistic and opportunistic values of the frontier.

THE MARTIAN LANDSCAPE

But it's a rough road to utopia. Robinson's trilogy is structured ideationally as a series of conflicts between competing visions of terraforming Mars and, therefore, opposing views of politics, economics, and social organization. During the course of two centuries (the longevity treatment developed by Vlad and his cohorts allows some characters, middle-aged in 2027 when *Red Mars* begins, to survive into the twenty-third century), the conflicts over terraformation mutate, grow sclerotic, and explode in revolutionary upheaval, anarchy, civil war, and repression. In one respect, his meticulous attention to political detail, to investigating the psychological changes that come over characters during decades of infighting, argument, and frustration, make Robinson seem more akin to Anthony Trollope than to, say, Ben Bova. The hero of his novels, nonetheless, remains Mars itself, particularly if one remains alert to the myriad ways in which humans—immigrants and then the native-born—shape and are shaped by its outgassing regolith, thickening atmosphere, proliferating plant and animal life, and expanding oceans. In this respect, the evolving biosphere is not a backdrop for a tale of social evolution but an integral part of the complex workings of eco-economics.

The political, ecological, and philosophical conflicts in all three novels pit the opponents of terraformation, the Reds, against the champions of ecopoeisis, the Greens. In one sense, these struggles project into the future a debate already taking shape: the philosophical implications voiced by Chris McKay in advocating a minimalist ecopoeisis versus the enthusiasm some proponents exhibit for an all-out assault on lifelessness. "On earth," McKay notes, "the notion of life and the notion of nature are inseparable. But on Mars and in the rest of the solar system, life and nature are two different things. Mars appears to be a dead planet, yet it is undeniably a beautiful, valuable planet" (quoted in Kluger 1992, 74–75). In another, they reverse the valence of contemporary ecological discourse by making the most ardent proponents of planetary engineering the avatars of the mystical energy—"viriditas"—of life itself. In *Red Mars,* the key advocates of Red and Green philosophies, the geologist Ann Clayborne and the polymorphous scientific genius Sax Russell, articulate their positions while the course of terraformation remains uncertain, "too big," as Sax says, with "too many factors, many of them unknown" to "model adequately" (171). In an effort to halt the terraformation of Mars, Ann sends private messages back to Earth, then must face her co-colonists, most of them terraforming enthusiasts. Her "tirade" consistently casts her opponents as careless children:

> "Here you sit in your little holes running your little experiments, making things like kids with a chemistry set in the basement, while the whole time an entire world sits outside your door. A world where the landforms are a hundred times larger than their counterparts on Earth, and a thousand times older, with evidence concerning the beginning of the solar system scattered all over, as well as the whole history of a planet, scarcely changed in the last billion years. And you're going to wreck it all. . . . You want to do that [the "mass alteration of the environment"] because you think you can. You want to try it out and see—as if this were some big playground sandbox for you to build castles in. A big Mars jar! You find your justifications where you can, but it's bad faith, and it's not science." (176–77)

At stake in Ann's comments is the moral relationship of humankind to the land. For her, the Martian landscape itself challenges anthropocentric and biogenic justifications for terraforming the planet; creating the conditions for life is purposeless because the geology of the planet is inherently valuable as a "historical record" of planetary and solar system history that dwarfs human technologies, intentions, and desires. If Red Mars is "a beautiful pure landscape," however, its purity can be appreciated only through human perceptions and values, through an aesthetic appreciation of its beauty and an intellectual, and even spiritual, recognition of the knowledge it offers.

In response to Ann, Sax emphasizes our inability to imagine beauty, knowledge, or usefulness without giving in to a mystical anthropocentrism. His scientific defense of rapid terraformation heroicizes the irrevocable imposition by humans of a metaphysics of order on physical reality: " 'The beauty of Mars exists in the human mind,' " [Sax] said in that dry factual tone, and everyone stared at him amazed. 'Without the human presence it is just a collection of atoms, no different than any other random speck of matter in the universe. It's we who understand it, and we who give it meaning' " (177). Sax's pronouncements suggest something of the attraction and limitations of his traditional scientific outlook, a worldview that evolves throughout *Green Mars* and *Blue Mars*. If Ann's defense of a "pure" Mars provokes a questioning of biocentrism,

Sax identifies knowledge rather than the exploitation of resources as the ultimate rationale for terraformation. In this regard, his response to Ann becomes a kind of philosophical one-upsmanship; it is precisely human intervention that produces the "meaning" that structures even her celebration of an aesthetics and science of "pure" observation, an ideal of nonintervention. Yet Sax's insistence on the anthropocentric nature of meaning in the universe ironically reveals the accuracy of Ann's criticism: the basis of terraformation, of Baconian science itself, is an adolescent faith in human significance, a will to play (and play God) with the universe. For Sax, at least in *Red Mars*, science may be unpredictable and modeling techniques limited, but the mind remains capable of constructing knowledge by the inductive method, of developing experimental programs and then using the results to generate rather than simply recognize meaning in the cosmos.

These Red and Green philosophical positions—reiterated, modified, and contested during the course of the planet's transformation—mutate in response to the historical experiences of terraformation. The conceptual, political, and spiritual development of the trilogy, in this regard, may be described as the movement of Reds and Greens toward reconciliation; antagonists throughout the three novels, Ann and Sax become romantically linked at the end of *Blue Mars*, a measure of the operations of viriditas on both. The alchemical sublimate for the emergence of a blue Mars on which humans can walk, glide, and sail is the philosophy of Hiroko Ai, "the Japanese prodigy of biosphere design" (32), who articulates and embodies the holistic imperatives of a twenty-first-century ecotheology. As the First Hundred branch out from their scientific station at Underhill and other settlers arrive from Earth, Hiroko and her followers leave for the southern hemisphere to further the ecopoeisis of Mars and to begin a communal existence that resists and transcends the antiecological, hierarchical efforts of transnational corporations to treat the planet as a vast mining camp. The isolation of this "Lost Colony" allows its members to survive the civil war of 2061, when corporate forces brutally quash attempts to establish an independent Mars, killing thousands, including many of the First Hundred.

Green Mars, which spans the decades after the war, might be seen as Hiroko's book because it is the moral force of her lived-philosophy of viriditas that brings together the scattered groups of the underground in a loose confederation and that eventually provides the rationale and moral authority for Martian independence. In the process, the nature of politics itself is transformed. At the beginning of this novel, Hiroko and her followers, including a generation of genetically engineered "ectogenes," have created a small utopia, Zygote, in an ice dome under the south pole. As the spiritual leader of this society, she gives voice to a philosophy that seeks to unify microcosm and macrocosm and prepares members of the underground for their eventual reemergence as a political as well as moral force:

> "Look at the pattern this seashell makes. The dappled whorl, curving inward to infinity. That's the shape of the universe itself. There's a constant pressure, pushing toward pattern. A tendency in matter to evolve into ever more complex forms. It's a kind of pattern gravity, a holy greening power we call viriditas, and it is the driving force in the cosmos. Life, you see. . . . And because we are alive, the universe must be said to be alive. We are its consciousness as well as our own. We rise out of the cosmos and we see its mesh of patterns, and it strikes us as beautiful. And that feeling is the most important thing in all the universe—its culmination, like the color of the flower at first bloom on a wet morning. It's a holy feeling, and our task in this world is to do everything we can to foster it." (19)

Hiroko's celebration of viriditas inscribes the abstract principles of a scientific will to meaning (the Artificial Intelligence punned on in her name: Ai) on sensory experience. The greening power she invokes gestures toward a union of spirit and matter, a synthesis of organic complexity and the spiritual growth that attends the processes of technologically, genetically fostering ecopoeisis as the "supreme act of love" (19). As lifeforms spread across Mars, this moral and aesthetic imperative to create beauty becomes a sociopolitical complement to the attempts of the underground to move stochastically, intermittently toward a roughhewn, evolving eco-economics. Viriditas, then, is not a simulation or thought experiment imposed on Mars but the embodied experience of participating in the evolution of green life on a red planet.

Throughout the trilogy, there are anticipations of the eventual reconciliation of Red and Green, of the alien landscape and the unforeseeable consequences of terraformation. Such anticipations, though, are scripted upon bodies and organisms, inscribed genetically, rather than articulated as abstract programs. This is the process of "areoformation," "an endeavor driven at a level below intention." Conscious political intentions and philosophical positions are acted upon and sublimated by the land itself, fostering complex processes of ideational as well as genetic evolution. The opening of *Green Mars* reads: "The point is not to make another

Earth. . . . The point is to make something new and strange, something Martian. . . . All the genetic templates for [the] new biota are Terran; the minds designing them are Terran; but the terrain is Martian. And terrain is a powerful genetic engineer, determining what flourishes and what doesn't, pushing along progressive differentiation, and thus the evolution of new species" (13). In Robinson's descriptions of the landscape, Mars is sensed and felt as well as seen. The terrain acts upon its human visitors, beginning a process of conceptual change before the effects of significant terraformation—heat, construction, and engineered life-forms—take hold.

On an early expedition to the north pole led by Ann, Nadia, a Russian engineer, experiences the alien beauty of Mars; Robinson's description extends the strategies of aesthetic and psychological inquiry that had characterized Roger's epiphany in "Fossil Canyon":

> The sun touched the horizon, and the dune crests faded to shadow. The little button sun sank under the black line to the west. Now the sky was a maroon dome, the high clouds the pink of moss campion. Stars were popping out everywhere, and the maroon sky shifted to a vivid dark violet, an electric color that was picked up by the dune crests, so that it seemed crescents of liquid twilight lay across the black plain. Suddenly Nadia felt a breeze swirl through her nervous system, running up her spine and out into her skin; her cheeks tingled, and she could feel her spinal cord thrum. Beauty could make you shiver! It was a shock to feel such a physical response to beauty, a thrill like some kind of sex. And this beauty was so strange, so alien. . . . She had been enjoying her life as if it were a Siberia made right, so that really she had been living in a huge analogy, understanding everything in terms of her past. But now she stood under a tall violet sky on the surface of a petrified black ocean, all new, all strange; it was absolutely impossible to compare it to anything she had seen before. (141–42)

For Robinson, who has digested seemingly all of the information available from the Viking missions, and then imagined the sensory overload of experiencing the planet's unearthly colors, massive dimensions, and weak gravity, beauty is both physical and geophysical, the product of the sublime engagement of human physiology and Martian landforms. Nadia's response to the alien beauty of violet skies and frozen silicate oceans is emblematic of the changes that Mars works on its colonists. The terrain itself suggests the inadequacy of frontier metaphors and economic rationalizations to describe areoformation, the changes wrought by the planet on humans as well by humans on the planet. Descriptions such as this one thus have a maieutic function: the impossibility of fitting Mars into paradigms imported from Earth forces characters to move beyond false historical analogies and, consequently, to take moral responsibility for the complex changes—socioeconomic as well as biospheric—initiated by terraformation. This responsibility is what ultimately distinguishes viriditas from corporatist models of terraformation as business investment and the passive worship of a romanticized nature. Areoformation, another name for this responsibility, resists the acts of simplification and demonization that construct Mars—or the Earth—as a storehouse of materials and energies waiting to be extracted, priced, and marketed. In this light, the ebb and flow between Red and Green areophanies reveals the paradox that there is value in both the pristine terrain of Mars and in life spreading across and irrevocably altering the planet's surface and atmosphere. If viriditas in the abstract tends toward a kind of ecofeminist mysticism, it is constrained as practice by the geology itself, by what Sax refers to repeatedly as the "thisness" of specific forms of biospheric alchemy, of life evolving on and transforming the planet.

RETHINKING HISTORY, RETHINKING ECONOMICS

Terraformation provokes numerous reflections in the three novels on what settling Mars means in historical terms. These reflections extend the ethical and political dilemmas that Robinson explores in his earlier fiction, notably the short stories "The Lucky Strike" (1984), "A Sensitive Dependence on Initial Conditions" (1991), and the novel *Icehenge* (1984). To some extent, *Icehenge* anticipates both the political questions and literary strategies of the Mars trilogy; it marks as well Robinson's initial fascination with the fourth planet as a means to think through humanity's possible futures. *Icehenge* offers three linked narratives that deal with the consequences of a democratic uprising on twenty-third-century Mars: it begins with the story of Emma Weil, a systems ecogeneticist, who returns to devastated colonies on Mars in 2248 rather than join the remnants of the defeated rebels on a desperate venture to become the first humans to venture beyond the solar system; the second narrative follows Hjalmar Nederland, an archaeologist, who in 2547 sets out to prove that the Martian rebellion was more than the anarchic rioting claimed by the colonial authorities; and the third section is narrated by Nederland's great grandson, Edmond Doya, who devotes his life to proving that Icehenge, a Stonehenge-like megalith found on Pluto, was not erected by the rebels on their way out of the solar system. Thematically, *Icehenge* examines the problems of memory, history, and autobiography in an age when people routinely live to be five hundred years old. Their attempts to authenticate Weil's journal—the only firsthand evidence that a social-democratic rebellion did occur—lead Nederland and later Doya to fantasize about meeting Emma, who becomes an imaginative projection of an ultimate truth or knowledge. The gap between such an idealized knowledge and the methodological problems of historical inquiry—shaped by the limitations of memory and the experience of trauma—fictionalizes the alienation of intellectual labor in the late twentieth century. As a child, Nederland had survived the destruction of a rebellious Martian city: his path to self-realization lies in validating the ethical and political values of the rebellion. In contrast, Doya is a rootless and marginalized part-time academic, who seeks to debunk the very politics of memory that motivate his great grandfather.

In this context, Icehenge is a monument to competing reconstructions of the failed rebellion and, by implication, of the historical traumas of the twentieth century. On one level, the novel intimates that Icehenge was built by Emma (who apparently has reinvented herself and used her expertise to become a reclusive and mysterious billionaire) to commemorate the rebels and the democratic-socialist values that their quixotic voyage symbolizes. But this "revelation" is advanced at the end of the novel as only one of many reconstructions of the past. While Nederland believes that "history is made, because facts are not things" (88), his trust in the self-sufficiency of "things," of archaeological artifacts, is challenged by the controversies that swirl around Icehenge. In part, Nederland's dream is to recover an authentic history by excavating "one of the lost Martian cities" that he links to "all those cities that had been razed and abandoned by conquerors, Troy, Carthage, Palmyra, Tenochtitlan, [now] all resurrected by scientists and their work" (71). But this project of recovery gains a political authenticity only if this otherwise forgotten genocide can serve as a call to political action in the present. His interpretation of Icehenge as a testament to the legacy that "once all the Martians revolted together, and broke spontaneously toward utopia" (138) anchors his belief that such a revolution could occur again. "To love the past," Nederland contends, "is to become fully human" (165), but his great-grandson ends the novel by quoting a sensationalist author who claims that Icehenge was built by aliens: "In the beginning was the dream, and the work of disenchantment never ends" (262). The danger is that by adding the destroyed New Houston to such a list of lost cities, the significance of its destruction becomes a target for "the work of disenchantment," and a cynical and ultimately neoconservative resolution to the uncertainties of history.

To give into "disenchantment," however, is to risk precisely the senses—both political and ecological—of connectedness that animate the utopian rebels in *Icehenge* and, in *Green Mars,* Hiroko's settlement under the south polar icecap. In the earlier novel, politics and ecology are linked by the crew's efforts to create a closed, regenerative ecosystem onboard the jerry-rigged starship. "We worked," writes Emma, "for hours and hours, mutating and testing bacteria, juggling the physiochemical processes, trying to make a tail-in-mouth snake that would roll across the galaxy" (41). The biochemical difficulty of creating such a closed system is a metaphor for the fate of the rebels on their voyage toward utopia, destruction, or a continuing historical struggle. In the middle section of *Icehenge* and more extensively in the trilogy, political conflict is implicated irrevocably in the terraformation of Mars. As Robinson has indicated, the fictional process of terraformation is not a blueprint for the future but a way to think about the interanimating logics of economics, ecology, and political power as they currently exist on Earth (Markley et al. 2001, "interviews," 2, 4, 5).

In *Red Mars, Green Mars,* and *Blue Mars,* various characters search fruitlessly for past analogues to explain present circumstances, and history itself comes to obsess many of them as they search for definitions, patterns, and meaning in human experience. On the initial voyage to Mars, John Boone and Phyllis Boyle, a true believer, debate the theological implications of history (1993, 52–54); later in the novel, while traveling around Mars seeking a consensus on what form a new Martian society might take, John defines history as "what happened when you weren't looking—an unknowable infinity of events . . . a nightmare, a compendium of examples to be avoided" (283–84). Decades later, Sax searches for a "science of history" to explain to himself the illogic of social stratification, but must relinquish his inquiries, concluding that history is "non-repeatable and contingent" (1994b 205–6). In the 2170s, Charlotte Dorsa Brevia, a product of an autonomous matriarchal commune, publishes a "metahistory," a "kind of master narrative," to explain the emergence of a "democratic Martian society" from the wreckage of the "dominance hierarchies" characteristic of both feudalism and capitalism (1996, 393, 392). Her analysis of history tracks "a fundamental shift in systems" from the feudal-capitalist coercion of labor and monopolizing of profits to a "cooperative democratic economy" in which "everyone saw the stakes were high; everyone felt responsible for their collective fate; and everyone benefited from the frenetic burst of coordinated construction that was going on everywhere in the solar system" (393). Although Robinson's description of a democratic economy (like Zubrin's arguments for funding Mars Direct) requires access to additional resources and expansion throughout the solar system, he critiques theocentric and naively empiricist accounts of history for ignoring or marginalizing the complex effects of human needs, desires, and conflicts—and the spiraling cycles of intensification and the painful adjustments they dictate. Abstract systems and disembodied beliefs, in other words, represent an anthropocentric, masculinist belief in the superiority of ideas to the lived experience of history; such models invite disillusionment when they lead inevitably to violence, stagnation, and environmental degradation, leaving "ten billion people standing in their own shit."

Robinson's future history in the trilogy begins with an act of near-biblical betrayal: Frank Chalmers suborns the murder of his erstwhile friend and romantic rival, John Boone, by misrepresenting John's desire for a democratic Mars as a threat to the social and theological practices of a radical Arab faction. John, the first man to land on Mars, is an idealist, and his efforts to forge "a scientific system [of social organization] designed for Mars, designed to [the settlers'] specifications, fair and just and rational and all those good things" make him the ethical touchstone for the as yet unfocused attempts in *Red Mars* to "point the way to a new Mars" (283). Frank's motives for the murder remain, to some extent, unclear even to him. Frank fears being cut out of the negotiations with Earth to revise the treaty that governs interplanetary relations; he finds John's plans for Mars unrealistic, insufficiently attuned to "the ethnic hatreds, the religious manias" (16) that characterize an expanding, multiethnic Martian society; and he is jealous of John's continuing relationship with his former lover, Maya Toitovna, the leader of the Russian contingent of the First Hundred. Frank does not want authoritarian power but the authority to negotiate for Mars in its unending squabbles with Earth. His resort to murder—"diplomacy by other means" (17)—testifies to the ethical confusion inherent in a politics that simply imposes terran values and assumptions on Mars. Frank becomes a crisis manager without a vision, "empty, and cold in the chest" (400), bickering with Maya and constantly placating contending factions on Earth and Mars, trying to unify Mars by self-defeating strategies of playing one group against another. He dies without having confessed to John's murder, but engaged in one of his few uncalculated, unselfish acts—saving Maya, Ann, Sax, and other refugees from the violence of 2061 during the massive floods triggered by the revolutionaries' sabotage of subsurface aquifers. Frank's death, then, coincides with the catastrophic reconfiguring of the landscape, the floods that alter "every single feature of the primal Mars," signaling irrevocably that "Red Mars was gone" (550). As he is swept away by the flood, the conventional notions that Martian politics can be micromanaged by terran realpolitik—expediency, arm-twisting, and violence—are swept away as well.

The survivors of 2061 who continue the struggle toward eco-economics fall, then, not only into history but into theory—that is, into meta-explanations of the ongoing processes of areoformation. In *Green Mars,* Sax emerges as the hero of this quest to understand the complex transformations occurring on Mars. Part 4 of the novel, by far the longest, is entitled "The Scientist as Hero," and tracks Sax's progress from the anthropocentric views of terraformation he voiced in *Red Mars* to his efforts to further the greening of the planet and its inhabitants. During the course of *Green Mars,* Sax is given a new face and new identity so that he can work above ground as a plant geneticist; is seduced by Phyllis, who represents the unholy alliance of Christian apologetics and capitalist ruthlessness, is unmasked by her, tortured and mind-probed to reveal what he knows about the underground, rescued by Maya and others, and then forced to struggle during a long rehabilitation to overcome the effects of a torture-induced stroke and relearn the intricacies of putting thoughts into words. Sax's efforts to regain his speech metaphorically underscore his emergence as a symbol and practitioner of a science committed to the ethical imperatives of viriditas and eco-economics. During his rehabilitation, Sax goes through extensive conversations with Michel Duval, the psychologist sent with the First Hundred, who had saved himself from despair by joining Hiroko's group. For Michel, the scientist's job is "to explore everything. No matter the difficulties! To stay open, to accept ambiguity. To attempt to fuse with the object of knowledge. To admit that there are values shot through the whole enterprise. To love it. To work toward discovering the values by which we should live. To work to enact those values in the world. To explore—and more than that—to create!" 373). Sax's response, "I'll have to think about that," testifies to his persistent professional dispassion even as he puts many of Michel's injunctions into practice. In the second half of *Green Mars* and in *Blue Mars,* Sax becomes a key figure in the development of a democratic Martian society, whether destroying Deimos so that it cannot be used as a base to attack the rebels during the second revolution against metanational authority, seeking to reconcile Ann and other Reds to the effects of terraformation, or developing an antidote for the memory losses that increasingly plague the aged survivors of two centuries of Martian history. Science, for Sax, loses none of its commitment to exploring the cosmos but, transformed and embodied, redefines the relationship between objective values and ethical commitment. Science creates rather than simply describes.

This reimagining of science necessarily informs and is informed by a rethinking of both conventional and revolutionary politics.[8] Even as it intervenes to ensure that the "whole enterprise" of settling Mars is "shot through" with egalitarian values, science provides a partial model for recasting politics so that decisions about immigration from Earth, resource management, and governance reflect its commitments to the "truths" of eco-economics. Few novels devote as much painstaking attention as *Green Mars* and *Blue Mars* to the complications and frustrations of political debate and compromise. In the former, the underground gathers at Dorsa Brevia to hash out a statement of principles that becomes the basis for Martian independence; in the latter, Reds, Greens, anarchistic collectives, and a range of ethnic and religious communities struggle to write a constitution that encodes the fundamental assumptions and values of eco-economics.[9] These political meetings are foreshadowed, in some respects, by the scientific conference on the progress of terraformation that Sax attends, in his new identity as Stephen Lindholm, in *Green Mars*. Initially eager to catch up on developments that have occurred during his years in the underground, Sax becomes increasingly dismayed by the politicization of science as different speakers plug the latest schemes of the corporations who fund their research: a "degraded dark zone invade[s] the heretofore neutral terrain of [the] conference" (199). This blasted ideal of disinterested scientific knowledge, though, reemerges as the animating force behind the efforts of Maya, Nadia, Sax, and others to broker an ecologically sensitive politics for Blue Mars.

What finally succeeds at the constitutional conference is the process of compromise itself, a kind of utopia by committee. The realities of governing by eco-economic principles are fraught with conflict, but a free Mars evolves to meet crisis after crisis in the years following independence. Such agreements, though, are unthinkable without the terraformation of Mars: in 2061 the revolution fails because the rebels, in their domed structures, are easy prey to devastating attacks from space. At the end of

Green Mars, when Reds mine the dikes that hold back one of Mars's new oceans and, in the confusion of a transnational counterattack, detonate the explosives and send a flood racing toward the rebel stronghold of Burroughs, the entire population dons masks to filter the carbon dioxide remaining in the atmosphere and walks in the cold but thickened and oxygen-rich atmosphere seventy kilometers to safety. In the course of the three novels, the idealists, dreamers, and politicians are killed off: John, Frank, Arkady Bogdanov (Nadia's anarchist lover), and Phyllis. Hiroko disappears in a metanational attack at the end of *Green Mars.* The scientists—Vlad, Sax, and Nadia, who becomes the reluctant first president of Mars—and the nomads, notably the stowaway, Coyote, who survives for two centuries as trickster, jack-of-all-trades, roving ambassador to underground settlements, revolutionary, and party-goer—press on.

The ideal that survives revolutions, floods, conflicts, and conferences is eco-economics, the effort to find a means to live in concert with the realities of areoformation. At the constitutional convention in *Blue Mars,* Vlad defends eco-economics as "more democratic, more just" (119) than efforts by some of the younger generation to institute on Mars the verities of capitalist acquisition and ownership:

> "If democracy and self-rule are fundamentals, then why should people give up these rights when they enter the workplace? In politics we fight like tigers for freedom, for the right to elect our leaders, for freedom of movement, choice of residence, choice of what work to pursue—control of our lives, in short. And then we wake up in the morning and go to work, and all those rights disappear. We no longer insist on them. And so for most of the day we return to feudalism. That is what capitalism is—a version of feudalism in which capital replaces land, and business leaders replace kings. But the hierarchy remains. . . . There is no reason why a tiny nobility should own the capital, and everyone else therefore be in service to them. There is no reason they should give us a living wage and take all the rest that we produce. No! The system called capitalist democracy was not really democratic at all. . . . History has shown us which values were real in that system." (116–17)

Eco-economics rewrites the rules governing investment, capital, and labor. For economists, the conflation of feudalism and capitalism makes little historical sense, but Robinson insists on this identification at several points in the three novels (1994b, 85; 1996, 392–93). In what amounts to an authorial endorsement of eco-economics, the narrator describes the effects of Vlad's speech in the rhetoric of Old Testament prophetic fury: "One of the ancient radicals had gotten mad and risen up to smite one of the neoconservative young power mongers" (120). Vlad emphasizes that ownership has been the guiding force of economic history—ownership defined as the unchecked and scientifically unsound privilege to treat common resources as private property. In contrast, the Dorsa Brevia accord recognizes "an economics based on ecological science. The goal of Martian economics," the document continues, "is not 'sustainable development' but the sustainable prosperity for its entire biosphere" (1994b, 358). To charges that he is a utopian dreamer or the avatar of terran socialism returned, Vlad reiterates the ecocentric principles of the Dorsa Brevia agreement: "The land, air, and water of Mars belong to no one, . . . we are the stewards of it for all future generations" (1996, 119). This concept of stewardship challenges the logic of property common to both feudalism and capitalism; its refusal to commodify the resources that terraforming has produced means that the control of capital remains in the hands of those who produce it. As Vlad puts it, "in our system workers will hire capital rather than the other way around" (119). In one respect, the ideal of a self-regulating biosphere advanced by Lovelock and Allaby in *The Greening of Mars* is extrapolated in Robinson's eco-economics to the realm of sociopolitical organization. On Blue Mars, people not only try to live in harmony with a newly created biosphere, but participate in an open, evolving system of elaborate feedback loops, checks and balances, and safeguards to ensure that there are no threatening accumulations of capital by a "tiny nobility," a prospect as dangerous, Robinson implies, as a deadly buildup of atmospheric pollutants.

TERRAFORMING AND ITS LIMIT CONDITIONS

In all three novels, major characters—John, Frank, and Maya in *Red Mars;* Sax, Maya, Nadia, and Nirgal (one of Hiroko's sons) in *Green Mars;* and Nirgal, Sax, Maya, Ann, Jackie (John's daughter), and Zo (Jackie's daughter) in *Blue Mars*—wander the planet, at times almost aimlessly, working on various projects, meeting new settlers and old friends, and taking stock of the infinite changes being wrought on the planet and its inhabitants. In some respects, this rootlessness seems a necessary antidote to the bureaucracy, interference, and tyranny of metanational capitalism; in others, however, this nomadic existence testifies to the redefinition of notions of identity that the terraformation of Mars fosters.[10] In his travels, John comes to recognize that he was "probably wrong" to assume that "if he only saw more of the planet, visited one more settlement, talked to one more person, that he would somehow . . . get it—and that this holistic understanding would then flow back from him to everybody else." The implicit politics of representation, of John's efforts to become the "articulator of all [the settlers'] hopes and desires" (1993, 284), fails, in part, because it reinforces the alienation of social, ethnic, economic, and psychological descriptions of identity from the processes of areoformation, from the terrain itself. As John comes to recognize, Red Mars is already being transformed and transforming its inhabitants; character is interpenetrated by a sense of place, of geography, as well as by historical experiences and psychic traumas (Jameson 2001, 208–32). Nirgal, who (rather than his sometime lover, Jackie Boone) inherits John's role as the ethical consciousness of his generation, finds himself, more than a century later, rootless in the aftermath of independence. His disorientation marks both his recognition of and his resistance to the mutual inflections of identity, vocation, and place:

> All his life he had wandered Mars talking to people about a free Mars, about inhabitation rather than colonization, about becoming indigenous to the land. Now that task was ended. . . . It was hard to give up being a revolutionary. Nothing seemed to follow from it, either logically or emotionally. . . . On the one hand he wanted to stay a wanderer, to fly and walk and sail all over the world, a nomad forever, wandering ceaselessly until he knew Mars better than anyone else. Ah yes; it was a familiar euphoria. On the other hand, it was familiar, he had done that all his life. It would be the form, of his previous life, without the content. And he already knew the loneliness of that life, the rootlessness that made him feel so detached. . . . Coming from everywhere, he came from nowhere. He had no home. And so now he wanted that home, as much as freedom or more. A home. He wanted to pick a place and stay there, to learn it completely. (1996, 301)

Nirgal's efforts to turn farmer, however, are devastated by a dust storm, and he returns to a nomadic existence, for a time joining a tribe of hunter-gatherers, future primitives who roam Mars living off the terraformed land. His dilemma, in one respect, is that his "home"—Mars itself—is constantly undergoing alchemical transmutations: ancient craters fill with water and become seas, the population expands into previously pristine areas, the atmosphere thickens enough so that, with some genetic adjustments, humans can breathe the air without masks, and the sky evolves to various shades of an oxygen-rich reddish-blue. On Blue Mars, "home," like one's sense of self, is subject to areoformation. Nirgal, therefore, cannot be described as a "postmodern" self in the usual sense of that term; his subjectivity is a function of his political-spiritual calling as a spokesperson and exemplar of viriditas. His rootlessness is neither a sign of neurosis nor a hopeless quest for an absent ideal but the natural condition for settlers whose lives extend to centuries and for whom the "pharonic" projects (1994b, 438) of creating a new biosphere define, in effect, where they will be and who they will become. For this ectogenic *homo martialis,* one does not practice eco-economics so much as one becomes a function of its aerophonic energies. The generation of Martian natives that he represents marks the end of the classically conceived *homo economicus,* that phantom of endless self-aggrandizement, who must be banished for any ecotopia to thrive.

By the end of *Blue Mars,* Ann and Sax are lovers, the opposition of Reds and Greens subsumed by aerophonic blue. Mars has avoided a third interplanetary war and offered itself as a model for Earth as the mother planet struggles through crises of overpopulation, the result of the longevity treatment, and ecological devastation caused by the flooding of coastal regions when the Ross Ice Shelf in Antarctica is melted by volcanic eruptions. The trilogy ends on a beach with children eating ice cream and Ann willing herself to survive a bout of arythmia. The technologies of terraformation offer, ultimately, a vision of small-town life, or such a life experienced in an ecologically pristine equivalent of Santa Barbara: scenic beauty, good restaurants. Robinson returns his readers to the doubled desires of the technologies of terraformation—the utopian possibility of a future primitive beyond the ecological degradation and economic injustices of the late twentieth century. With Mars terraformed, the massive projects needed to transform the planet give way to a self-regulating biosphere; the true ecopoeisis becomes the creation of new forms of social as well as biological life. Robinson's 1,700-page thought experiment finally presents itself less as a utopian dream than a falling into ecotopian theory—both a policy statement and a hard-won course charted to an imagined holism.

And yet terraformation remains a sequence of dynamic and unpredictable interactions between human intentions and irreducibly complex environmental changes, adaptations, and reconfigurations. Although the First One Hundred imagine Mars as "a blank red slate" on which, according to Arkady, they can "transform . . . ourselves and our social reality," the planet ultimately proves recalcitrant. At the end of *The Martians,*

Robinson returns to the romance of Eileen Monday and Roger Clayborne on a Mars on which terraforming has begun to fail: the planet is locked into an ice age that may require a re-engineering of the planetary environment. Blue Mars has given way to a deep freeze that redefines the limits and possibilities of the human experience on the planet. "Winterkill is winterkill," says Eileen, "but this is ridiculous. The whole world is dying" (349). Their ice boat trip across the frozen oceans of the north leads them back to the problems posed by terraforming that had sparked debates between the Reds and Greens throughout the trilogy. Hans Boethe, an areologist who had ascended Olympus Mons with them centuries earlier, offers a litany of ways that the ice age might be reversed: "Bombs below the regolith. . . . A flying lens to focus some of the mirrors' light, heat the surface with focused sunlight. Then bring in some nitrogen from Titan. Direct a few comets to unpopulated areas, or aerobrake them so that they burn up in the atmosphere. That would thicken things up fast. And more halocarbon factories" (352). These "industrial" solutions are countered by Roger who reinvokes "ecopoeisis" as offering "less violence to the landscape" (352). This debate about reterraforming Mars is not resolved, and "it begins to seem as if they are on an all-ice world, like Calisto or Europa"—or in Antarctica, the setting for Robinson's 1997 novel.

The deep freeze on Mars both closes the trilogy and offers a meditation about the limitations of human intentions in environmental engineering or, more broadly, about humans' ability to transform nature into habitat. Even at the end of *The Gold Coast,* the second volume of his California trilogy, Robinson counters the dystopian vision of a landscape of triple-tiered highways and endemic pollution with a climb into the Sierra Nevada mountains that presents his hero the potential for both escape and renewal; as elsewhere in his fiction, mountain environments represent the possibility of a human relationship with the Earth that resists greed and ecological degradation. The view from the mountains provides the glimmer of a utopian—or at least a different—future from the one force-fed to us by late capitalism. The reversal of terraforming in this final story, "A Martian Romance," however, is not experienced by the younger characters as a tragedy. In contrast to "the despair of the [environmental] crash" perceived by the older generation and the prospect that "warm[ing] things up again . . . could take thousands of years," the young Jean-Claude "shrugs": " 'It's the work that matters, not the end of the work.' " The story ends with his affirmation that even if " 'everything alive now will die, [and even if] the planet will stay frozen for thousands of years, . . . there *will* be life on Mars' (360). This is not the affirmation of a "red" or "green" philosophy so much as it is a meditation on the bioexpansionism that is, after all, one of the generic bases of science fiction. Rather than a theological "destiny of man" that Walter A. Miller evokes (1980, 68) in his 1950 vision of terraforming Mars, human love is not directed toward a transcendence of suffering but refigured as an ethical commitment to the dynamic relationship between life and environment, a relationship that is transforming humans as humans transform the land. In this regard, while the experience of love, friendship, and dialogue are crucial to Robinson's achievement in the Mars trilogy, such experience is never divorced from the politics and ethics of being "visitors on this planet" (1999, 385), whether Earth or Mars. Love is finally defined by human efforts "to do something good something useful," by the complex relationships between "red rock red dust the bare / mineral here of now / and we the animals standing in it" (385). The ultimate challenge posed by planetary transformation is ultimately as much ethical as it is scientific.

EPILOGUE: 2005

The success of the rover missions in 2004 clearly has given the exploration of Mars a scientific and public relations boost, even as the analysis of gigabytes of data continues. The next NASA mission, the Mars Reconnaissance Orbiter (MRO), is scheduled to go into orbit in 2006, and features the high resolution imaging experiment (HIRISE), a camera that will be able to take photographs at a variety of scales down to one meter: from orbit, HIRISE will reveal objects the size of basketballs on the surface of Mars. For the first time, an interplanetary spacecraft will be able to produce images that rival the photographic resolution of cameras on surface landers. A web site now exists for members of the public to suggest targets to photograph as scientists plan to zero in on a variety of enticing targets: possible sources of methane and water vapor detected by Mars Express, layered terrain in ancient flow channels, the gullies photographed by the MGS, and many others. Although the MRO mission is being defined by NASA primarily as a series of geological and climatological investigations, the spacecraft will also scout landing sites for the next generation of surface rovers.

Significantly, the success of Spirit and Opportunity makes it very, very difficult to imagine another twenty-year gap (the hiatus between Viking and Pathfinder) before new landers explore the red planet's surface. NASA's current Mars plans include a series of missions, including the Phoenix lander (projected launch date: 2007), which will include updated versions of some of the experiments lost when the Mars Polar Lander crashed in 1999. A new generation of rovers (projected launch date: 2009) will be the first interplanetary SUVs, with the capability to travel up to ten miles over rough terrain and be instrumented with a more complex array of spectrometers and microscopic imagers than those on the 2004 rovers. Scientists envision landing these rovers by a crane mechanism so that they touch down, ready to roll, directly on the surface. The rovers will then be able to travel from relatively safe landing areas to more geologically interesting and irregular terrain. Also by 2010, NASA and ESA hope to have the beginnings of an integrated communications system of orbiters around Mars, facilitating the transmission of radio commands to vehicles on the surface and the uploading of data from the surface to these satellites and then relaying gigabyte after gigabyte to Earth.

By 2011, NASA hopes to be ready to launch a sample return mission, designed to load Mars rocks from the surface onto a launch vehicle that then would rendezvous with a spacecraft designed to return the rocks to scientists on Earth. A decade before such samples might arrive on Earth, there is already a vigorous scientific debate about the possibility of unknown Martian microbes wreaking havoc with human health and the environment. Writing on the op-ed page of the *New York Times* in 2004, Olivia Judson, an evolutionary biologist at Imperial College London, warned that "our ignorance of life" on Earth makes it unlikely that scientists could recognize health and environmental threats posed by Martian microbes, if they exist (Judson 2004). The ghost of H. G. Wells looms over such debates as scientists try to assess—without any data—the likelihood that hypothetical micro-Martians could do to earthlings what Wells's terrestrial germs did to his fictional invaders more than a century ago. But back contamination may be the least of NASA's problems in the first and second decades of the twenty-first century. If NASA, ESA, and Japanese missions meet with more successes like those achieved by Spirit and Opportunity, then the key question will become the one that has been deferred since the 1980s: Will technological and scientific breakthroughs provide enough leverage to move the powers that be in Washington, Berlin, London, Tokyo, Paris, Moscow, and Beijing to begin the decades-long process of funding necessary to put humans on Mars? For Zubrin's vision of a new Martian frontier to become more than science fiction, robotic exploration and thought experiments will have to give way to large-scale efforts to design, build, test, and operate the spacecraft, habitats, life support systems, and propulsion technologies necessary to send humans to the red planet. In the 1950s, Ray Bradbury imagined the first expeditions leaving for Mars in 1999; four decades later Kim Stanley Robinson envisioned human exploration before 2020. Zubrin has said repeatedly that a mission using the Mars Direct approach could be launched within ten years. But, given NASA's current mission plan, even the launch windows in 2018 and 2020 seem far too soon. Is 2030 possible? Or 2050, a date hinted at in the presidential commission's 2004 Moon to Mars report?

Science-fiction novels have offered many rationales for eventual exploration—a new space race for Martian resources, a make-work project for giant aerospace industries, a desperate quest to find new resources to substitute for Earth's depleted oil, coal, and minerals. A best-guess scenario for optimists would be that NASA and members of the Mars Society can sell lawmakers on the idea that a humans-to-Mars mission before, say, 2030 would generate better payoffs for contractors, workers, and state budgets than current pork-barrel fiascos like the International Space Station and the Star Wars missile defense system. The wild card in any scenario for future exploration would be the discovery of ironclad evidence of life on Mars, perhaps detected by a drilling apparatus of the sort that Beagle 2 was supposed to deploy. Perhaps, too, an inventory of Martian minerals will reveal something that looks so obscenely profitable that billions for new propulsion technologies and assembly techniques will seem like shrewd and far-sighted business investments. Whatever NASA's budget looks like in 2020, the analogies between the third and fourth planets will persist: as projects like HIRISE get underway, Mars is likely to remain the great laboratory for understanding the dynamics of climate change on a world where "ten billion people" may well be "standing in their own shit" (Robinson 1993, 353) by mid-century.

Killer bacteria from outer space? Disturbing evidence that, as Lowell said, "the earth . . . is going the way of Mars?" (1909, 122). The next giant leap to Gerard O'Neill's High Frontier? Even agnostics recognize that Mars beckons, and Mars still haunts our dreams.

NOTES

INTRODUCTION

1 I am indebted in particular to the valuable histories and scientific studies of Mars and its explorers that are offered in Cooper 1980; Ezell and Ezell 1984; Wilford 1990; Sheehan 1996; Raeburn 1998; Moore 1999; Strauss 2001; Bergreen 2000; Morton 2002; and Hartmann 2003, among others.

2 On the "overwhelming fascination exerted by Mars" in science fiction, see Guthke 1990, 367 n.54.

3 Albedo is an index of reflexivity.

4 The issues that are raised by the cultural study of science are too complex to discuss adequately in a single note or, for that matter, a single chapter. I deal with some of them in previous works (Markley 1993; Markley 1999a; Markley 1999c). For important contributions to the cultural study of science, see particularly Serres 1982; Woolgar 1988; Haraway 1991; Hayles 1991, 76–85; Latour 1987, 1993; Rotman 1993; Shapin 1994, 1999; Pickering 1995; Barbara Herrnstein Smith 1997, 243–66; and Plotnitsky 2002.

5 See Crumley 1994, 1–11. See also Winterhalder 1994 and Ingerson 1994. On the idea that socionatural "totalities" are continually degrading the very conditions on which their survival depends, see Lewontin and Levins 1985, 133–42, 272–285.

1. MARS AND THE LIMITS OF ANALOGY

1 This section extends material found in Markley et al. 2001, "early views." In this section of *Red Planet,* animations, photographs, and graphics illustrate essential scientific concepts for planetary astronomy: opposition, conjunction, good seeing, retrograde motion, and so on. Good introductions to the science of areography can be found in Sheehan 1988, 1996; Burgess 1978, 1990; Moore 1977, 1999; Murray, Malin, and Greeley 1981; Wilford 1990; and Raeburn 1998. Post-Viking discussions of the suitability of Mars for some form of life can be found in

Chandler 1979; Cooper 1980; Goldsmith 1997; DiGregorio 1997; Walter 1999; Bergreen 2000; Morton 2002; Boyce 2003; and Hartmann 2003.

2 An arc second is a unit of angular measure, ¹⁄₃₆₀₀ degree, used to express the apparent diameter or separation of astronomical objects. Because there are 360 degrees in a circle, an arc second describes a measure of ¹⁄₂₁₆,₀₀₀ of a hypothetical circle with the viewer at its center.

3 The six outer planets all appear to exhibit retrograde motion, but the orbits of Jupiter and Saturn are so distant from Earth and they move comparatively so slowly around the sun that their retrograde course is not nearly as dramatic as that of Mars. Uranus (1781), Neptune (1849), and Pluto (1930) were not discovered until after the invention of high-powered telescopes.

4 Kepler's Laws are as follows: (1) planets travel in elliptical orbits with the sun as one foci of the ellipse; (2) planets sweep out equal areas under the ellipse in equal periods; and (3) the ratio of the periodic times of any two planets is the ratio of their mean distances raised to the $3/2$ power ($P = a^{3/2}$).

5 On Hooke's report, see Flammarion 1981 (43–47 [28–30]). English translations from Flammarion's *La planète Mars et ses conditions d'habibilité* (Paris: Gauthier-Vilars, 1892) are taken from the typescript *The Planet Mars*, by Patrick Moore in the Lowell Observatory Library. Page numbers from the French edition are given in brackets.

6 On the development of modern notions of scientific objectivity, see Shapin 1994 and Markley 1993.

7 On planetary observation and telescopes, see Brush 1996; Schindler 1998; Sheehan 1988; and Smith 1997, 49–77.

8 Fontenelle, like Cassini and other continental astronomers of the late seventeenth century, was a Cartesian materialist, and he described gravity as the effect of vortices that required no notion of action at a distance. As Stephen Dick argues, Descartes's attempts to explain gravitational effects in terms of vortices played an important role in buttressing the notion of a plurality of worlds by "extending the idea of a plurality of Earthlike planets to that of a plurality of solar systems" (1982, 126).

9 *Cosmothereos* went through five editions in English by the end of the eighteenth century.

10 See Markley 1999b, 831. If the blank spaces on terrestrial maps provoked dreams of infinite profits, the Moon and the five other planets offered entire worlds on which humankind could project idealized visions of social and political values. As early as 1638, John Wilkins (later president of the Royal Society in London) brought the hypothetical inhabitants of the Moon into the framework of seventeenth-century political economy: " 'Tis possible," he suggested in his *Discovery of a New World in the Moone*, "for some of our posteritie, to find out a conveyance to this other world, and if there be inhabitants there, to have commerce with them" (1640, 241).

11 In addition to Guthke 1990, 105–11; see also Alkon 1987, 45–85; Suvin 1979, 103–8;

Philmus 1970, 13; and Dick 1982. As David Grinspoon suggests, "Planet Earth was also discovered by telescope" (1997, 37). On Kepler's *Somnium*, see Paxson 1999, 105–23.

12 On Fontenelle, see Marsak 1959, 14; Rendall 1971, 496–508; Niderst 1972; and Gelbart, in Fontenelle 1990, vii. For a different approach to Behn's translation of *Entretiens*, see Cottegnies 2003, 23–28.

13 In her hilarious farce, *Emperor of the Moon* (1687), Behn satirizes the deluded Baliardo, who, according to his daughter, "religiously believes there is a World [in the Moon] [and] discourses as gravely of the People, their Government, Institutions, Laws, Manners, Religion and Constitution, as if he had been bred a *Machiavel* there." His mind has been turned, she tells the audience, "with reading foolish Books," including those by Godwin and Wilkins, and "a thousand other ridiculous Volumes too hard to name" (1996, I i 86–97). Galileo and Kepler become characters in a farce-within-a-farce intended to cure Baliardo of his delusions.

14 On *Cheats of the Pagan Priests*, see Gelbart in Fontenelle 1990, xv.

15 On telescopes in the nineteenth century, see Sheehan 1988.

16 Dawes presented his findings to the Royal Astronomical Society on June 9, 1865 (Flammarion 1981, 288 [204]; Sheehan 1996, 53–55; Dawes 1865, 225–68). He died in 1867 before Proctor's map was published.

17 Flammarion, respected both as a scientist and a spiritualist, was an extraordinarily popular writer. Two of his bestselling novels, *Urania: A Romance* (1889) and *Omega: The Last Days of the World* (1893), combine planetary astronomy with the belief that the dead are reincarnated on other planets, including Mars, in other bodies and other sexes. In his 1862 work *La pluralité des mondes habités* he declared that the "striking points of resemblance" between Earth and Mars "lead us to believe that both planets are inhabited by beings whose organization is of similar character" (cited in Flammarion 1981, 214 [142–43]). This updating of the plurality of worlds idea went through thirty-seven editions by 1892.

2. LOWELL AND THE CANAL CONTROVERSY

1 William Sheehan, for example, compares Lowell's books about Mars to William Paley's *Natural Theology* (1802) and concludes that both men exhibit "a primitive, childlike tendency to construct artificialist explanations for natural phenomena" (1988, 3). For other harsh treatments of Lowell, see Hetherington 1976, 303–8; 1981, 159–61; and 1988, 49–64. In contrast, William C. Heffernan argues that Lowell's scientific methods and reasoning were comparable to those of his critics (1981, 527–30). This view has been seconded by Strauss (2001). Norman Horowitz, one of the principal investigators on the Viking life detection experiments, describes Lowell as "probably the best-informed Mars-watcher of his day" (1986, 79).

2 As Wily Ley and Werner von Braun put it forty years after Lowell's death: "To a

well-read and open-minded man at the turn of the century this discovery [of canals], first reported from Italy and then confirmed by observers in the United States as well as France, was really just a final proof for a philosophy that had grown strong on the discoveries of nineteenth-century science" (1956, 11).

3 On Lowell's use of the nebular hypothesis, see Strauss 2001, 152–54; on Newton's voluntarism, see Markley 1993, 131–77.

4 Astronomers from Johannes Kepler to Isaac Newton to William Whewell maintained that the beauty and order of the physical world could be guaranteed only by positing a divine agency and purpose rather than accepting the spontaneous arising of order, what the Newtonian Colin Maclaurin derided as "a lucky hit in a blind uproar" (1748, 4; cited in Markley 1993, 176). On the argument from design, see Markley 1993, 56–57, 122–24.

5 For a psychoanalytical interpretation of Lowell's career, see Lears 1981, 198. Hoyt 1976 and Strauss 2001 emphasize Lowell's rebellion in choosing his career paths as a diplomat in Japan and an astronomer and lecturer.

6 Lowell was particularly interested in the ability of life-forms to adapt to thin air and colder temperatures and thus in early studies of mountain ecology. In 1907, he studied tree heights and density on the Kaibab plateau to support his contention that a plateau, even at great elevation (and therefore presumably analogous to the Martian surface), retains heat and provides more favorable conditions for life than mountain peaks (Lowell 1909, 94–103).

7 The excellent conditions for observing bodies in the solar system allowed the Lowell Observatory to remain an important center for mapping the lunar surface into the 1960s (Morton 2002, 43–44).

8 Lowell argued that "so much light is grasped by any glass over 12 inches aperture that workers on planetary detail have to employ appliances to get rid of it, appliances which so far as light-getting power is concerned, are a dead loss." Covering the eyepiece with a filter of colored glass, he contended, "has frequently made all the difference between good definition and poor"; routinely, he would use a diaphragm to reduce glare and distortion (1905a, 14, 16–17).

9 On Lowell and popular conceptions of astronomy, see Strauss 1993, 164; Lankford 1997, 79, 197; and, more generally, Dick 1996 and Crowe 1986. Shlovskii and Sagan (1966, 276) are among the few commentators to have recognized the role that Lowell's rhetorical strategies played in popularizing his theory.

10 See Lowell 1895, 66–67. Recently, Kevin Zahnle and William K. Hartmann have suggested a possible correspondence between the major Martian canals with geological features: the most frequently seen "canal," Agathodaemon, corresponds to the Valles Marineris; Daemon, the second most frequently seen, to the chaos at the origin of the Valles Marineris; and Gigas, the most prominent "double" canal, to the "parallel rifts aligned with the great Tharsis volcanoes" (Zahnle 2001, 211). Hartmann notes the correspondences between Schiaparelli's and Lowell's sketches and the geological features in the Xanthe region (Hartmann 2003, 43–49).

11 On ideology, see the discussion in the introduction, and Lewontin 1991, 240–53; Kingsland 1985, 206–10; Woolgar, 1988; Hayles 1991, 76–85; Rouse 1991, 141–69; Latour 1993; Shapin 1994, 1999, 1–14; and Markley 1999a, 47–70.

12 On Serviss's fiction, see chapter 3.

13 The clipping books at the Lowell Observatory in Flagstaff include both positive and negative pronouncements; the majority opinion seems to have accepted the logical narrative as described by Lowell.

14 Lowell's opponents were frequently ridiculed in Boston papers. In a story headlined "Prof. Lowell Attacked," the *Boston Advertiser* in November 1907 mocked the criticism of Professor Harold Jacoby in a lecture at Columbia University by noting that "in the same lecture, [he] attacked Newton and declared that 'all the intricacies of motion can be explained just as well by Ptolmey's theory,' as by Newton's" (1907, LOA).

15 On displays of Lowell's drawings and photographs see Hoyt 1976, 179–86.

16 Some of the 1907 photographs of Mars as well as Lowell's and Slipher's drawings are reproduced, courtesy of the Lowell Observatory, in Markley et al. 2001, "canals" 12, 13, 15.

17 See Sagan 1971, 511–14; Sagan 1973, 513–14; Burns and Harwit 1973, 126–30; Sagan, Toon, and Gierasch 1973, 1045–49; and Pollack 1979, 479–553.

18 On air pollution and environmental degradation during the nineteenth and early twentieth centuries, see McNeill 2000.

19 Pickering's defense of the canals in the 1920s is discussed in chapter 4 below.

20 For a different view of Wallace, see Raby 2001. See also Heffernan who maintains that "Wallace scrupulously separated his scientific and religious beliefs" (1978, 99).

21 On Haeckel and evolutionary biology, see Bramwell 1989, 39–44.

22 See Morse (1907, 102, 128–29) for an extended criticism of Maunder, who admits to bad seeing during the opposition of 1892 but then ridicules the idea of canals.

23 Arizona's constitution included "equal suffrage for both sexes," "legalization of the boycott," "adoption of the initiative, the referendum, and the recall," and "abolition of the injunction against labor unions" (Lowell 1910, 6–7). It passed by a vote of forty-three to eleven at the territorial convention in 1910.

24 As Jim Moore argues, Social Darwinism "inexorably structure[s] social expectations and invest[s] the interpreters of biological language with Nature's authority when they pronounce on historical contingencies" (1986, 74). On Social Darwinism in the nineteenth and twentieth centuries, see also Young 1985.

25 Ward's article appeared in the Brown University alumni magazine, and Ward later corresponded with Lowell. See Hoyt 1976, 217–18.

3. MARS IN SCIENCE FICTION, 1880–1913

1 On Wells's scientific training (he earned a PH.D. in biology late in life), see Haynes 1980, 40–46, who argues that his scientific contemporaries treated Wells as a

professional. Wells discusses his early interests in science and the influence of his study with T. H. Huxley in his *Experiment in Autobiography* (1934, 159–236).

2 On efforts by Wallace and others to define evolution theologically, see Brush 1987, 245–78, and Raby 2001.

3 Markley 1997, 77–103.

4 See Harris 1977, and, for two different views, Sanborn 1998 and Yue 1999.

5 On Wells and contemporary views of nature, see Wagar 1983, 165–66, and Russell 1990, 145–52. On the pollution of London, see McNeill 2000, 57–58.

6 See the discussion of Derrida on hauntology in the introduction.

7 One of the key technological problems for science-fiction writers before the age of airplanes is conceiving how flying machines could take off from and land on a rotating Earth. Since the spacecraft from Greg through Bogdanov operate on antigravity principles, they take off and land vertically, like fictionalized precursors of helicopters. Consequently, space stations over the North Pole become familiar devices to explain how the Martians get to and from Earth.

8 On late-nineteenth- and early-twentieth-century German attitudes toward coal production and water and air pollution, see McNeill 2000, 86–87.

9 The details of Bogdanov's life are taken from Stites 1984, 1–16.

10 On relations between Lenin and Bogdanov, see Service 2000, 181–94.

11 Kendall Bailes, "Alexei Gastev and the Soviet Controversy over Taylorism, 1918–24," cited in Stites 1984, 8.

12 Lenin, for one, was not impressed, and wrote a stinging critique of Bogdanov's "empiriomonism" (a version of tectology), and later Russian science-fiction novelists, such as Yevgeny Zamyatin in *We* (1921), offered powerful dystopian visions of socialism corrupted into fantasies of absolute control.

13 Rather confusingly, different characters in *Engineer Menni* have the same names as prominent figures in *Red Star.*

14 A search of the Web in December 1999 turned up eight hundrred (mostly Russian) sites dedicated to *Aelita,* a testament to Tolstoy's successful combination of the romance of space travel, adventure, and transcendent love. My thanks to Lisa Zunshine who conducted the search.

15 Adapted for the screen by Fyodor Otzep and Alexei Faiko, the 1924 *Aelita* starred Nikolai Tseretelli, Igor Ilinski, and, as the Martian princess, Yulia Solntseva. Short segments of the film, now available on video, are reproduced in Markley et al. 2001 "dying planet," 8. On the response to *Aelita,* see Youngblood 1985, 30–32.

4. LICHENS ON MARS

1 In addition to works discussed below, see Horowitz 1986, 79–82; Dick 1996; Raeburn 1998; Rovin 1978; Wilford 1990; Crowe 1986; Moore 1999; and Sheehan 1996 for different approaches to the persistence of the canal theory in the middle of the twentieth century.

2 On the difficulties of the observatory during this period, see Hoyt 1976; Doel 1996; and Tatarewicz 1990.

3 During the opposition of 1918, for example, Slipher's notes night-in and night-out record minute observed changes on the surface of Mars. On Wednesday, April 3, 1918, a detailed drawing of Mars is accompanied by notations about the seeing conditions, and the times of observations are precisely noted. Between 11:15 and 11:45 that night the seeing was rated at 5½ and the tip of Syrtis Major passed the center plane. "Two bright patches at south [pole]—frost or clouds—one is the Hellas." A line from the dark region north of the south polar region leads to this text: "The south sea a marked chocolate brown color all above this line (dotted) anyway—" The seeing deteriorates during this half hour to "4½–3" ; at 11:43 it is "not as good as first" and no other observations of the north polar region are made. Slipher continued to draw canals and to refute the anticanalists. This sketch is reproduced in Markley et al. 2001, "canals" 15), courtesy of the Lowell Observatory. The sketch and accompanying notes are found in Earl C. Slipher Working Papers, Mars Drawings 1918, Box 2, Folder 1A, LOA.

4 While on a lecture tour in England, Lowell dropped in unannounced on a meeting of the BAA on March 30, 1910. The exchange was cordial but no opinions were changed (see Crowe 1986, 537–38).

5 On the context of science and new data, see Tatarewicz 1990; Doel 1996; and Shapin 1994.

6 Dick 1996, 105–6, treats Garbedian's article briefly, but it has been ignored by other historians.

7 V. M. Slipher was awarded the Gold Medal by the Royal Astronomical Society in 1933 (Doel 1996, 38–39).

8 Chesley Bonestell's illustration of von Braun's 1953 article in *Collier's,* a popular redaction of *The Mars Project,* is reproduced in Markley et al. 2001, "red and dead," 10.

9 The map is reproduced in Markley et al. 2001, "red and dead," 15.

10 This edition was intended "for authorized government use." An identical civilian edition, edited by John C. Hall, under the title *The Photographic Story of Mars,* was issued by Northland Press in Flagstaff and Sky Press in Cambridge, Massachusetts, the same year. The inside cover reads, in part, " 'Photographic History of Mars 1905–1961' was produced under Air Force Cambridge Research Laboratory Contract No. AF 19(604)-5874 and publication was sponsored by the USAF Aeronautical Chart and Information Center under Contract No. AF 23(601)-3602. Coordination with the scientific community was established through Commission No. 16 'Physical Study of Planets and Satellites' of the International Astronomical Union."

11 In his notes on the images reproduced in figure 3, Slipher offers the following analysis: "Comparison of the Antoniadi drawing , no. 5, with my photograph of the same region, no. 6 (magnification 1700 diameters) reveals that all the nu-

merous details in the photograph to the left and above the Thoth, namely in the Aethiopis and Elysium regions, are not recorded in his drawing. However, his representations of the aspect of the Thoth itself is not supported by the photograph. Examples are the short double canal running downward from the base of the Thoth, and also the dark core-like medial line through the center of the Thoth, flanked by diffuse irregular shadings on each side" (Slipher 1962, 142).

5. MARS AT THE LIMITS OF IMAGINATION

1 See the discussion of science fiction in the introduction above, particularly Suvin 1979, 9–10; Jameson 1989, 517–37; and Delany 1994, 187–94.

2 See Holtsmark 1986, 32; Wilson 1992; Mullen 1995, 144–56; and Brown 1993, 129–63.

3 On cannibalism, see Harris 1977.

4 Paul Carter suggests that "The Cave" was "little noted at the time it appeared" (1977, 266), but became a classic that influenced Heinlein's description of the water-sharing ceremonies in *Stranger in a Strange Land* and in Frank Herbert's *Dune.*

5 Koch recounts his six-day drafting of the script and the subsequent broadcast in Koch 1970.

6 Although Gaudet and Herzog are not credited with coauthorship of the study, Cantril acknowledges that Gaudet "was in charge of the actual administration of the investigation [and] not only made most of the tabulations based on the interviews, but [contributed] many of the ideas reflected in the tabulations and the text" (1940, xiii). Herzog had made the first study of the broadcast's effects independently, then joined forces with Gaudet and Cantril to author *The Invasion from Mars.*

7 On the radio broadcasts in South America and Buffalo, see Holmsten and Lubertozzi 2001, 74–78. In Quito, the broadcast terrorized thousands who fled into the streets. When the hoax was revealed, an angry mob burned the building that housed the radio station, killing fifteen employees.

8 The standard psychoanalytic description of paranoia can be found in Swanson, Bohnert, and Smith 1970. For the implications of this analysis, see McElroy 1985: "The sound mental health recognized by society is a construct of rationalizations and evasions designed to shield us from an intolerable reality that the paranoiac, in the honesty of his desperation, accepts as given" (219).

9 On Palmer's significance as an editor of science fiction, see the entry in Clute and Nicholls 1995, 905.

10 Brackett cowrote the 1946 screenplay for Howard Hawks's *The Big Sleep* with William Faulkner and Jules Furthman; she wrote the 1973 Robert Altman remake of *The Long Goodbye* as well as the script for *The Empire Strikes Back* (with Laurence Kasdan [1979]) and the screenplays for the John Wayne films *Rio Bravo* (1949, with Furthman) and *Hatari* (1962).

11 The Russian astrophysicist I. S. Shlovskii floated this idea seriously in the Russian work that Sagan had translated and subsequently expanded as *Intelligent Life in the Universe* (Shlovskii and Sagan 1966, 373; Bryld and Lykke 1999, 102).

12 In the 1970s, the original stock footage of a model rocketship against a cheesy backdrop was replaced by three minutes of new special effects, including the descent of the rocket to a cratered, post-Mariner surface of Mars. This updating of a cult classic suggests the extent to which the popular appeal of science-fiction films depends on more than special effects and rubber-suited monsters.

6. THE MISSIONS TO MARS

1 The conference proceedings were edited by Colin Pittendrigh, Wolf Vishniac, and J. P. T. Pearman as *Biology and the Exploration of Mars* (1966). On this institute, see Ezell and Ezell 1984, 61.

2 On the Mariner 2 mission to Venus in 1962 and its results, see Grinspoon 1997; Shklovskii and Sagan 1966; and Tatarewicz 1990.

3 NASA focused on the moon at its inception in 1958 and into the early 1960s (see Wilhelms 1993; and Tatarewicz 1990, 33).

4 The Mariner 4 photographs and those of subsequent missions have been archived by NASA on its Web site and can be accessed at http://www.nasa.gov/mars. Many of these photographs have been reproduced in NASA publications Glasstone 1968, as well as in Markley et al. 2001, "missions to mars" 1–2, and elsewhere.

5 On the scientific debates and bureaucratic infighting that characterized the Voyager program, see Ezell and Ezell 1984, 83–120. This canceled Mars mission is not to be confused with the interplanetary probe sent to the outer planets a decade later.

6 Mariner 6 took fifty far-encounter and twenty-six near-encounter photographs; Mariner 7 took ninety-three far-encounter and thirty-three near-encounter photos. See Ezell and Ezell 1984, 175–77, for a discussion of the scientific results from Mariners 6 and 7.

7 Although information about the aims of all the Russian missions remains sketchy, the huge size of the Soviet landers in the 1970s indicates that they were intended to conduct life detection experiments. Unlike Mariner 9, Mars 2 and Mars 3 did not have the capability to extend their missions by delaying the launch of their landing vehicles. Both orbiters returned some data, but the landers apparently succumbed to the dust. In 1973, Mars 4 and Mars 5 returned some sixty photos of Mars, but Mars 6 missed its orbital insertion and the Mars 7 lander stopped transmitting data after a few seconds on the surface (Murray 1989, 56–58). Phobos 2 in 1989 returned photographs of the planet and useful data on the Martian moon of that name. At the time, the photographs of a thousand-mile swath of the surface provided the most detailed thermal analysis of surface materials and geological activity; evidence of a solar wind ripping away the Martian atmosphere at the rate of ninety tons per day seemed to confirm the belief that

without a stronger magnetic field in the past, the atmosphere would have been lost more quickly. Either outgassing from volcanic activity or the hydrological cycle of the polar caps were considered likely sources of volatiles to replenish the atmosphere (Burgess 1990, 126–27). Russian spacecraft did land on Venus in 1973 and 1976 and transmitted the first pictures from the surface of that planet before succumbing to the tremendous heat (over 800 degrees Fahrenheit) and atmospheric pressure (ninety times that of Earth).

8 See the photographs in Markley et al. 2001, "missions to mars" 5–7.

9 Since the advent of spaceflight, studies in planetary astronomy peak at and after interplanetary probes (see Tatarewicz 1990, 120).

10 On the obliquity of Mars and its significance for the planet's climatological history, see Hartmann 2003, 296–97; and Markley et al. 2001, "ancient floods" 3–5. These screens include an animation and a video description by Jeff Moore (NASA-Ames).

11 Tragically, on a trip to Antarctica to conduct a version of Levin's experiment in the dry valleys, Wolf Vishniac fell to his death (Ezell and Ezell 1984, 235–36).

12 On the emergence of the dominant view of the Viking experiments, see Ezell and Ezell 1984; Cooper 1980; Burgess 1990, 99–104; and Raeburn 1998, 100–7.

13 On the radical transformations in molecular evolution instigated by Carl Woese's description of the phylogenetic tree, see Pace 1997, 734–40.

14 See the discussion of the meteorite controversy in chapter 8.

15 On the listserv for the Washington, D.C., chapter of the Mars Society in 1999–2001, about a dozen regular participants in e-mail discussions about the Viking life-detection experiments had scientific affiliations (including NASA); during one six-month period in 2000, e-mails ran about two to one in favor of Levin's interpretation.

16 Although Levin had his champions such as Barry DiGregorio, whose popular study (1997) both raised serious questions about the interpretation of Viking data and accused NASA of engaging in an orchestrated coverup of evidence for Martian organisms, the Web site of Biospherics (now Spherix) has become the primary means for Levin to promulgate his views. See http://biospherics.com/mars/index.html for on-line versions of Levin's Mars articles.

17 Levin argues that a reexamination of the data from the pyrolytic release experiment actually confirms the presence of organic compounds on Mars. During the developmental stage of the PR instrument, an ultraviolet filter had been developed to guard against false positives. The initial positive on the PR experiment was short of the premission standard set to register as a positive signal for life. Arguing that the UV filter "did not completely eliminate the formation of organic matter," Levin suggests that PR registered the formation of organic matter within the instrument itself (Levin 1997, 4). More recently, the researchers in the Atacama Desert ruled out superoxides and peroxides as the oxidizing agent in the soil (Navarro-Gonzalez et al. 2003, 1020–21).

18 These experiments are discussed in chapter 8.

19 Levin has also argued about the colors of rocks and soils in photographs from the surface of Mars, first by the Viking landers and, in 1997, by Pathfinder. Where others saw grayish or rust-colored dust, Levin maintained that they had a greenish, and therefore presumably biological, tint. His insistence on this point has contributed to the cool reception his arguments have received from many planetologists.

7. TRANSFORMING MARS

1 Unlike Martians earlier in the century who barely had begun to colonize the solar system, the aliens of post–World War II science fiction traverse galaxies at tachyonic, post-Einsteinian speeds. After the Mariner missions, Leigh Brackett recast her series of Eric John Stark novels, originally set on Mars, and resituated them on the far-off planet of Skaith (see Clute and Nicholls 1995, s.v. Brackett).

2 In John Carpenter's film *Ghosts of Mars* (2001), demonic Martians possess human colonists to exact revenge on the terrestrial invaders of their planet. Although the Martians have been unleashed from a massive stone fortress, they have been transformed from presumably material beings to dustborne microscopic agents of infection.

3 Dick's Mars is known to most of his readers as the site of Paul Verhoeven's 1989 film *Total Recall.* Ironically, however, the short story that serves as a springboard for the film, "We Can Remember It for You Wholesale" (*Fantasy and Science Fiction,* April 1966), is set on Earth. Because the hero Quaile (not Quaid as in the film) never makes it to Mars, his ultimate repressed memory is not of skullduggery on the red planet but of saving Earth from an interstellar invasion. *Total Recall* is discussed below.

4 In his introduction to a special issue of *Science-Fiction Studies* on Dick, Carl Freedman calls Dick "the most accomplished, interesting, and significant American novelist to have emerged since the Second World War" (1988, 121); similarly, the novelist and critic Norman Spinrad declares that "Dick is arguably the greatest science fiction writer who ever lived and certainly a central figure in the literary history of the field" (1970, 198). Other critics of the genre often second such enthusiasm (Paul Williams 1986; Mackey 1988) or, more critically, register the significance of Dick for a critique of popular literature and pulp fiction. See Rabkin 1988, 161–72; Hoberek 1997, 374–404; and Hayles 1999.

5 Blish's afterword, dated July 30, 1965, notes the correspondence between his fictional Mars and the Mariner 4 photographs, although he also quotes an unnamed NASA scientist who remarks that "the probe took its pictures 'in the wrong places, at the wrong season' " (Blish 1967, 156) to see evidence of the canals.

6 In his afterword to *Return to Mars,* Bova states explicitly that he does not believe the fictional scenario he has concocted—an intelligent race of Martians were wiped out seventy million years ago in the same asteroid storm that killed off dinosaurs on Earth—yet the discovery of ancient cliff dwellings and pictographs

revives the tradition of the archaeological science fiction of H. Beam Piper ("Omnilingual" 1957) and other science-fiction writers of a half century earlier. Allen Steele prefaces his lost-civilization novel *Labyrinth of Night* with a disclaimer about Hoagland's argument for " 'the Face on Mars' ": "The truth probably lies somewhere between the opposite poles of fact and fantasy" (1992, viii).

7 A 1976 NASA report extended Sagan's model to suggest other ways of heating Mars and thickening its atmosphere (see Averner and MacElroy 1976). On early terraforming schemes, see Markley et al. 2001, "dreams of terraforming" 1–6; Fogg 1995, 226–31; Boston 1996, 327–61; Meyer and McKay 1996, 393–442; Oberg 1982; and the discussion in Shirley 2004.

8 I am grateful to Professor Miklitsch for sending me a copy of the page proofs of his article before it was published. I am very grateful as well to Professor Molly Rothenberg for her work on *Total Recall* and for her permission to use some of her insights about the film in the discussion which follows.

8. MARS AT THE TURN OF A NEW CENTURY

1 The articles collected in *Mars* (Kieffer et al. 1992) represent a valuable cross-section of scientific interpretations of the Viking data.

2 This figure is taken from a survey of articles appearing in the following journals: *Science, Nature, Icarus,* and *Journal of Geophysical Research.*

3 On Earth, geologists found an analogue for these outflow channels in the scablands of eastern Washington, an area scoured by floods when ice dams broke thirteen thousand years ago, sending catastrophic floods rushing down the Columbia basin. To some field geologists who studied the region, the terrain began to seem eerily Mars-like.

4 Under certain circumstances, water-ice can trap molecules of gas to form gas hydrates—clathrates. Temperature and pressure changes can then cause the gas to expand rapidly. Nicholas Hoffman, an Australian geologist, has suggested that the gullies and valleys of Mars are the result of clathrates—either as flowing liquid carbon dioxide or pressurized gas—that in some circumstances on Earth (in the deep ocean, for example) mimic the geomorphic forms caused by water runoff. Hoffman and a few other geologists argue that all the water on Mars remains frozen beneath the surface (Kerr 2003, 1037–38; Bandfield, Glotch, and Christensen 2003, 1084–87). On clathrates, see Hoffman 2000, 326–42; Morton 2002, 206–12; and Hartmann 2003, 237.

5 See Mutch et al. 1976; Kieffer et al. 1992; McKay and Stoker, 1989, 189–92; Sheehan and O'Meara 2001; Morton 2002; Walter 1999; and Hartmann 2003.

6 See the photograph reproduced in Markley et al. 2001, "canals," 27.

7 The morphology of some craters indicates that rain must have fallen on Mars at some point in its remote past (Craddock, Maxwell, and Howard 1997, 321–40).

8 The following synopsis of the instruments on the Mars Odyssey mission are

drawn from the relevant NASA web sites and related links: http://mars.jpl.nasa.gov/odyssey/science/index.html; http://mars.jpl.nasa.gov/odyssey/technology/themis.html; http://mars.jpl.nasa.gov/odyssey/technology/grs.html; http://grs.lpl.arizona.edu/instruments/gamma/; http://grs.lpl.arizona.edu/instruments/ns.

9 On Zare and mass laser spectroscopy, see Goldsmith 1997, 82–94. Zare describes the operation of mass laser spectroscopy and the initial testing of the samples from ALH84001 in Markley et al. 2001, "life" 11.

10 Doran is quoted in http://www.spacedaily.com/news/antarctic-02t.html.

11 On Imre Friedman's and Chris McKay's investigations on extremophile life in Antarctica, going back to 1980, see Morton 2002, 242–43.

12 The relationships among the three domains are determined by tracing a molecular sequence–based phylogeny that describes evolutionary relationships (see Pace 1997, 734–40, and the notes therein; Woese 1987, 221–71, 2002, 8742–47).

13 On the difficulty of imagining a planetwide extinction of life on Earth, see Cockell 2003.

14 Zubrin discusses the Mars Direct Plan in Markley et al. 2001, "interviews" 2.

15 Robinson, Rothenberg, and Moore offer sociopolitical critiques of Zubrin's use of the frontier metaphor in Markley et al. 2001, "interviews" 4.

16 Quotations from the Mars Society declaration are taken from the society's web site, www.marssociety.org.

9. FALLING INTO THEORY

1 On the logic of simulation and mathematics, see Markley 1996, 55–77; and Rotman 1993.

2 In Robinson's short story "Discovering Life" (2000), the discovery of Martian microbes by astronauts on the planet short-circuits NASA's plans to begin terraforming Mars. " 'Well, shit,' " one of the project scientists says at the end of the story, " 'I guess we'll just have to terraform Earth instead' " (Robinson 2002, 153–64; quotation from 164).

3 *Red Mars* won the 1993 Nebula Award, *Green Mars* the 1994 Hugo Award. The Mars Society has adopted a red, green, and blue flag as a visual symbol of its members' hopes for the colonization of the planet, even though Robinson remains skeptical of the frontier mentality voiced by some promoters of Martian colonization. See Markley et al. 2001,"interviews," 3, 4.

4 Fogg's work (1995) remains the standard text in a growing body of "serious" scientific literature on terraforming; in his opening chapter Fogg surveys the scholarship on terraforming through 1994. For important overviews of the technological and ecological problems involved in terraformation, see Averner and MacElroy 1976; Oberg 1982; McKay, Toon, and Kasting 1991, 489–95; Haynes 1990, 161–83; Birch 1992, 331–40; Zubrin 1996, 172–210; Jakosky 1998, 160–65; Pollack and Sagan 1993, 921–50; the essays collected from *Analog* in Schmidt and Zubrin

1996; Gerstell et al. 2001, 2154–57; and the discussion in Shirley 2004. Fogg has edited four special issues of the *Journal of the British Interplanetary Society* devoted to terraforming: vol. 42, no. 12 (1989); vol. 44, no. 4 (1991); vol. 45, no. 8 (1992); and vol. 46, no. 8 (1993). Robinson thanks Fogg and McKay in his acknowledgments in both *Green Mars* and *Blue Mars.* Other novelists have drawn extensively on this literature. To take only one example, Robert L. Forward's *Martian Rainbow* (1991) includes a ten-item bibliography, including works by Carl Sagan, Chris McKay, Michael Carr, and James Oberg.

5 Fogg (1995) provides a comprehensive survey of various proposals and offers his own "synergic" approach that applies all of these methods to create a carbon dioxide-rich but significantly terraformed planet within two hundred years.

6 In his 1999 novel *White Mars,* Brian Aldiss, in collaboration with the noted mathematical physicist Roger Penrose, offers a utopian vision of human settlement on the red planet but resists the idea that terraforming will be an inevitable consequence of colonization. At the end of the novel, Aldiss adds a note from APIUM (the Association for the Protection and Integrity of an Unspoilt Mars). This statement contests many of the key points of the Founding Declaration of the Mars Society; attempts to turn the red planet "into something resembling a colony, an inferior Earth, . . . would extend prevailing dystopian tendencies into the [twenty-first] century." Even though he disagrees vehemently with the terraforming dreams of Lovelock and Allaby, then, Aldiss shares an assumption that science fiction has a crucial role to play in shaping the values and assumptions that guide science itself. Inveighing against the "rape and ruination" of Mars, Aldiss envisions a treaty to protect the planet "as unspolit white wilderness . . . a kind of Ayers Rock in the sky" (Aldiss and Penrose 1999, 323).

7 For ecological histories that challenge the traditional account of the frontier, see Crosby 1986; Davis 2001; Crumley 1994, 1–16.

8 In a question-and-answer session at the Modern Language Association conference on December 28, 1998, Robinson reflected on his experiments in utopian fiction. In the utopian conclusion to his California trilogy, *Pacific Edge,* Robinson engaged in what he called a "sleight of hand" by putting the actual transformation from late capitalism to a democratic, environmentally rigorous and socially just society in the twenty-first century off stage. Rejecting a romanticized revolutionary call to action, Robinson characterizes himself as "a radical not really in favor of [armed] revolution." He describes his narrative technique in portraying the successful revolution in *Green Mars* as the application of the adage, "on thin ice skate fast." For his depiction of the incipient Martian utopia in *Blue Mars,* Robinson drew on scant historical models for alternative economies, including Lewis Hyde's *The Gift;* in 1998, he expressed the wish that he had used new work on "participatory economies" such as Herman Daly's *On Growth* and *Steady State Economics* and Lester Brown's *Full House.* See also Jameson 2001, 208–32.

9 In *Antarctica* (1997), Robinson continues to explore the utopian structure of scientific practices, methods, and beliefs. The novel works toward the redefini-

tion of a science in tune with the natural world and used against the excesses of capitalism, social injustice, and environmental degradation. As one of Robinson's characters puts it, "social justice is a necessary part of any working environmental program" (1997, 383). See particularly 322–27; 395–97.

10 On the decentered notions of subjectivity in Robinson's fiction, see Franko 1994, 191–211; and Jameson 2001, 208–32.

WORKS CITED

ABBREVIATIONS

LOA: Lowell Observatory Archives

NMR: NASA Mission Reports

Adams, W. S., and T. Dunham. 1934. "The B Band of Oxygen in the Spectrum of Mars." *Astrophysical Journal* 79: 308–16.

——. 1937. "Water-Vapor Lines in the Spectrum of Mars." *Publications of the Astronomical Society of the Pacific* 49: 209–11.

Aldiss, Brian. 1973. *Billion Year Spree: The True History of Science Fiction.* New York: Doubleday.

Aldiss, Brian, in collaboration with Roger Penrose. 1999. *White Mars, or, The Mind Set Free: A Twentieth-First Century Utopia.* New York: St. Martin's Press.

Alkon, Paul. 1987. *Origins of Futuristic Fiction.* Athens: University of Georgia Press.

——. 1994. *Science Fiction before 1900: Imagination Discovers Technology.* New York: Twayne.

Anders, Edward. 1996. "Evaluating the Evidence for Past Life on Mars." *Science* 274: 2119–20.

Anderson, Kevin J. 1994. *Climbing Olympus.* New York: Warner.

Antoniadi, E. M. 1910. "On Some Objections to the Reality of Professor Lowell's Canal System of Mars." *Journal of the British Astronomical Association* 20.

——. 1975. *The Planet Mars.* Trans. Patrick Moore. Devon, U.K.: Keith Reid. (Orig. pub. 1929.)

Arnold, Edwin. 1969. *Gulliver of Mars.* New York: Ace Books. (Orig. title: *Lieutenant Gulliver Jones*, 1905.)

Arrhenius, Gustav, and Stephen Mojzsis. 1996. "Life on Mars—Then and Now." *Current Biology* 6; 1213–16.

Arvidson, R. E., et al. 2004. "Localization and Physical Property Experiments Conducted by Opportunity at Meridiani Planum." *Science* 306, 1730–33.

Ashley, Mike, ed. 2002. *The Mammoth Book of Science Fiction.* London: Constable and Robinson.

Ashworth, William. 1995. *The Economy of Nature: Rethinking the Connections between Ecology and Economics.* Boston: Houghton Mifflin.

Asimov, Isaac. 1966. "The Martian Way" (1952). In *The Martian Way and Other Stories*. Greenwich: Fawcett.

Austin, Mary. 1974. *The Land of Little Rain*. Albuquerque: University of New Mexico Press. (Orig. pub. 1903.)

Averner, Maurice, and Robert MacElroy, eds. 1976. *On the Habitability of Mars: An Approach to Planetary Ecosynthesis*. NASA SP-414.

Bailey, K. V. 1990. "H. G. Wells and C. S. Lewis: Two Sides of a Visionary Coin." In *H. G. Wells under Revision*, ed. Patrick Parrinder and Christopher Rolfe. London: Associated University Presses, 226–36.

Baker, Victor. 1981. *The Channels of Mars*. Austin: University of Texas Press.

——. 2001. "Water and the Martian Landscape." *Nature* 412: 228–36.

Balée, William, ed. 1998. *Advances in Historical Ecology*. New York: Columbia University Press.

Balfour, Bruce. 2002. *The Forge of Mars*. New York: Ace Books.

Bandfield, Joshua, Victor E. Hamilton, and Philip R. Christensen. 2000. "A Global View of Martian Surface Compositions from MGS-TES." *Science* 287: 1626–30.

Bandfield, Joshua L., Timothy D. Glotch, and Philip R. Christensen. 2003. "Spectroscopic Identification of Carbonate Minerals in the Martian Dust." *Science* 301: 1084–87.

Barbour, Michael G. 1996. "Ecological Fragmentation in the Fifties." In *Uncommon Ground: Rethinking the Human Place in Nature*, ed. William Cronon. New York: Norton, 233–55.

Barnard, E. E. 1892. "Preliminary Remarks on the Observation of Mars 1892, with the 12-in. and 36-in. Refractor of the Lick Observatory." *Astronomy and Astro-Physics* 11: 680–84.

Bartram, C. O. 1908. "The Possibility of Life on Mars." *Nature* 77: 392.

Baxter, Stephen. 1997. *Voyage*. New York: HarperPrism.

Bear, Greg. 1993. *Moving Mars*. New York: Ace Books.

Bederman, Gail. 1995. *Manliness and Civilization: A Cultural History of Gender and Race in the United States, 1880–1917*. Chicago: University of Chicago Press.

Behn, Aphra. 1992. *A Discovery of New Worlds* (1688). Trans. of Bernard le Bovier de Fontenelle, *Entretiens sur la pluralité des mondes* (1686). In *The Works of Aphra Behn*, vol. 4, ed. Janet Todd. Columbus: Ohio State University Press.

——. 1996. *Emperor of the Moon* (1687). In *The Works of Aphra Behn*, vol. 7, ed. Janet Todd. Columbus: Ohio State University Press.

Belcher, Donald, Joseph Veverka, and Carl Sagan. 1971. "Mariner Photography of Mars and Aerial Photography of Earth: Some Analogies." *Icarus* 15: 241–52.

Bell, J. F., III, et al. 2004. "Pancam Multispectral Imaging Results from the Opportunity Rover at Meridiani Planum." *Science* 306, 1703–9.

Bell, Jim. 2003. "Mineral Mysteries and Planetary Paradoxes." *Sky and Telescope* 106, no. 6: 34–41.

Benford, Gregory. 1999. *The Martian Race*. New York: Warner Books.

Bent, Silas. 1924. "Mars Invites Mankind to Reveal His Secret." *New York Times.* August 17, sec. 8, p. 6.

Berendzen, Richard. 1974. "Origins of the American Astronomical Society." *Physics Today* 27: 32–39.

Bergreen, Laurence. 2000. *The Quest for Mars:* NASA Scientists and Their Search for Life Beyond Earth. London: HarperCollins.

Biemann, K., J. Oro, P. Toulmin III, L. E. Orgel, A. O. Nier, D. M. Anderson, P. G. Simmonds, D. Flory, A. V. Diaz, D. R. Rushneck, J. A. Biller. 1976. "Search for Organic and Volatile Inorganic Compounds in Two Surface Samples from the Chryse Planitia Region of Mars." *Science* 194: 72–76.

Biemann, Klaus, et al. 1977. "The Search for Organic Substances and Inorganic Volatile Compounds in the Surface of Mars." *Journal of Geophysical Research* 82: 4641–62.

Birch, Paul. 1992. "Terraforming Mars Quickly." *Journal of the British Interplanetary Society* 45: 331–40.

Bisson, Terry. 1990. *Voyage to the Red Planet.* New York: Avon.

Blakeslee, Sandra. 1998. "Society Plans to Promote Idea of Humans on Mars." *New York Times.* August 18.

Bleiler, Everett F. 1990. *Science Fiction: The Early Years.* Kent, Ohio: Kent State University Press.

Blish, James. 1967. *Welcome to Mars.* New York: Avon.

Bogdanov, Alexander. 1984. *Red Star.* Trans. Charles Rougle, ed. Loren Graham and Richard Stites. Bloomington: Indiana University Press.

Bono, James. 1990. "Science, Discourse, and Literature: The Role/Rule of Metaphor in Science." In *Literature and Science: Theory and Practice*, ed. Stuart Peterfreund. Boston: Northeastern University Press, 59–90.

Boston, Penelope, ed. 1984. *The Case for Mars.* American Astronomical Society Science and Technology Series, vol. 57. San Diego: Univelt.

——. 1996. "Moving in on Mars: The Hitchhikers' Guide to Martian Life Support." In *Strategies for Mars: A Guide to Human Exploration*, ed. Carol Stoker and Carter Emmart. American Astronautical Society Science and Technology Series, vol 86, 327–61.

Boston, P. J., M. V. Ivanov, and C. P. McKay. 1992. "On the Possibility of Chemosynthetic Ecosystems in Subsurface Habitats on Mars." *Icarus* 95: 300–8.

Bova, Ben. 1992. *Mars.* New York: Bantam.

——. 1999. *Return to Mars.* New York: Avon Books.

Bowden, Mark. 1997. "Houston, We Have Landed on Mars: A Fanatic Band of Hippie Scientists Has Sold NASA on a Manned Mission to the Red Planet." *Playboy.* August, pp. 62–64, 84, 104, 108.

Boyce, Joseph S. 2002. *The Smithsonian Book of Mars*. Washington: Smithsonian Institute Press.

Boynton, W. V., et al. 2002. "Distribution of Hydrogen in the Near Surface of Mars: Evidence for Subsurface Ice Deposits." *Science* 297: 81–85.

Brackett, Leigh. 1953. *The Sword of Rhiannon.* New York: Ace Books.
——. 1961. *The Nemesis from Terra.* New York: Ace Books.
——. 1964. *People of the Talisman.* New York: Ace Books.
——. 1967. *The Coming of the Terrans.* New York: Ace Books.
Bradbury, Ray. 1950. *The Martian Chronicles.* New York: Doubleday.
Bradbury, Ray, Arthur C. Clarke, Bruce Murray, Carl Sagan, and Walter Sullivan. 1973. *Mars and the Mind of Man.* New York: Harper and Row.
Bradley, J. P., R. P. Harvey, and H. Y. McSween Jr. 1997. "No 'Nanofossils' in Martian Meteorite." *Nature* 390: 454–55.
Bramwell, Anna. 1989. *Ecology in the Twentieth Century: A History.* New Haven: Yale University Press.
Brande, David. 2006. *Fictions of Postmodernity: Ideology and Desire in Contemporary Literature and Science.* Ann Arbor: University of Michigan Press.
Brandenburg, John E., and Monica Rix Paxson. 1999. *Dead Mars, Dying Earth.* Shaftesbury, U.K.: Element.
Broad, William J. 2003. "Soon, Three New Travelers to Mars." *New York Times.* May 27, sec. D, pp. 1, 4.
Brosnan, John. 1978. *Future Tense: The Cinema of Science Fiction.* New York: St. Martin's Press.
Brown, Bill. 1993. "Science Fiction, the World's Fair, and the Prosthetics of Empire, 1910–1915." In *Cultures of U.S. Imperialism*, ed. Amy Kaplan and Donald Pease. Durham: Duke University Press, 129–63.
Brunner, John. 1967. *Born under Mars.* New York: Ace Books.
Brush, Stephen G. 1978. "Planetary Science: From Underground to Underdog." *Scientia* 113: 771–87.
——. 1987. "The Nebular Hypothesis and the Evolutionary Worldview." *History of Science* 25: 245–78.
——. 1996. *A History of Modern Planetary Physics.* 3 vols. Cambridge: Cambridge University Press.
Bryld, Mette, and Nina Lykke. 1999. *Cosmodolphins: Feminist Cultural Studies of Technology, Animals and the Sacred.* London: Zed Books.
Bukeavich, Neal. 2002. " 'Are We Adopting the Right Measures to Cope?' Ecocrisis in John Brunner's *Stand on Zanzibar.*" *Science-Fiction Studies* 29: 53–70.
Burnett, John. 1979. "British Studies of Mars: 1877–1914." *Journal of the British Astronomical Association* 89: 136–45.
Burns, Joseph A., and Martin Harwit. 1973. "Towards a More Habitable Mars—or—The Coming Martian Spring." Icarus 19: 126–30.
Burgess, Eric. 1978. *To the Red Planet.* New York: Columbia University Press.
——. 1990. *Return to the Red Planet.* New York: Columbia University Press.
Burroughs, Edgar Rice. 1965. *John Carter of Mars.* New York: Ballantine.
——. 1969. *Swords of Mars.* New York: Ballantine. (Orig. pub. 1936.)
——. 1975a. *The Chessmen of Mars.* New York: Ballantine. (Orig. pub. 1922.)
——. 1975b. *A Fighting Man of Mars.* New York: Ballantine. (Orig. pub. 1930.)

——. 1975c. *LLana of Gathol.* New York: Ballantine. (Orig. pub. 1948.)
——. 1979a. *A Princess of Mars.* New York: Ballantine. (Orig. pub. 1912.)
——. 1979b. *The Gods of Mars.* New York: Ballantine. (Orig. pub. 1913.)
——. 1979c. *Thuvia, Maid of Mars.* New York: Ballantine. (Orig. pub. 1916.)
——. 1979d. *The Master Mind of Mars.* New York: Ballantine. (Orig. pub. 1927.)
——. 1979e. *Synthetic Men of Mars.* New York: Ballantine. (Orig. pub. 1939.)
——. 1980. *The Warlord of Mars.* New York: Ballantine. (Orig. pub. 1914.)
Caidan, Martin. 1972. *Destination Mars.* New York: Doubleday.
Caidan, Martin, and Jay Barbee, with Susan Wright. 1999. *Destination Mars: In Art, Myth, and Science.* New York: Penguin Studio.
Campbell, Mary Baine. 1999. *Wonder and Science: Imagining Worlds in Early Modern Europe.* Ithaca: Cornell University Press.
"Canals of Mars Photographed." 1905. *Scientific American.* June.
"Canals on Mars: Further Proof that They Have Actual Material Formation." 1906. *New York Times,* January 2.
Canfield, J. Douglas. 2001. *Mavericks on the Border: The Early Southwest in Historical Fiction and Film.* Lexington: University of Kentucky Press.
Cantril, Hadley. 1940. *The Invasion from Mars: A Study in the Psychology of Panic, with the Complete Script of the Famous Orson Welles Broadcast.* Princeton: Princeton University Press.
Carr, Michael H. 1981. *The Surface of Mars.* New Haven: Yale University Press.
——. 1995. *Water on Mars.* New York: Oxford University Press.
Carr, Michael H., and James Garvin. 2001. "Mars Exploration." *Nature* 412: 250–53.
Carter, Dale. 1988. *The Final Frontier: The Rise and Fall of the American Rocket State.* London: Verso.
Carter, Lin. 1973. *The Man Who Loved Mars.* New York: Ace Books.
Carter, Paul A. 1977. *The Creation of Tomorrow: Fifty Years of Magazine Science Fiction.* New York: Columbia University Press.
Chandler, David. 1979. *Life on Mars.* New York: Dutton.
Christensen, P. R., et al. 2004. "Mineralogy at Meridiani Planum from the Mini-TES Experiment on the Opportunity Rover." *Science* 306, 1733–39.
Clareson, Thomas D. 1985. *Some Kind of Paradise: The Emergence of American Science Fiction.* Westport: Greenwood.
Clarke, Arthur C. 1974. *The Sands of Mars.* New York: New American Library. (Orig. pub. 1952.)
——. 1995. *The Snows of Olympus: A Garden on Mars.* New York: Norton.
Clarke, Bruce. 2001. *Energy Forms: Allegory and Science in the Era of Classical Thermodynamics.* Ann Arbor: University of Michigan Press.
Clemett, S. J., et al. 2002. "Crystal Morphology of MV-1 Magnetite." *American Mineralogist* 87: 1727–30.
Clerke, Agnes. 1896. "New Views about Mars." *Edinburgh Review* 184: 368–84.
——. 1902. *A Popular History of Astronomy During the Nineteenth Century.* 4th ed. London: Adam and Charles Black.

Clute, John, and Peter Nicholls. 1995. *The Encyclopedia of Science Fiction.* Rev. ed. New York: St. Martin's Griffin.

Coblentz, W. W. 1925a. "Climactic Conditions on Mars." *Popular Astronomy* 33: 310–16, 363–82.

——. 1925b. "Radiometric Measurements of Stellar and Planetary Temperatures." *Nature* 116: 439–41.

Cockell, Charles S. 2003. *Impossible Extinction: Natural Catastrophe and the Supremacy of the Microbial World.* New York: Cambridge University Press.

Collins, Michael. 1990. *Mission to Mars.* New York: Grove, Weidenfield.

"Coming to Terms with the Human Factor." 1999. *Nature* 402: 217.

Colthup, N. B. 1961. "Identification of Aldehyde in Mars Vegetation Regions." *Science* 134, 529.

Compton, D. G. 1971. *Farewell Earth's Bliss.* Rev. ed. New York: Ace Books.

Cooper, Henry S. F. Jr. 1980. *The Search for Life on Mars: Evolution of an Idea.* New York: Holt, Rinehart, and Winston.

Costa, Richard Hauer. 1985. *H. G. Wells.* Rev. ed. Boston: Twayne.

Cottegnies, Line. 2003. "The Translator as Critic: Aphra Behn's Translation of Fontenelle's *Discovery of New Worlds* (1688)." *Restoration: Studies in English Literary Culture, 1660–1700* 27: 23–38.

Cowan, James. 1896. *Daybreak: The Story of an Old World.* New York: G. H. Richmond.

Craddock, Robert, Ted A. Maxwell, and Alan D. Howard. 1997. "Crater Morphology and Modification in the Sinus Sabaeus and Margaritifer Sinus Regions of Mars." *Journal of Geophysical Research* 102: 321–40.

Creedon, Jeremiah. 1994. "Mars on a Billion Dollars a Day: Can We Hatch Life on the Red Planet?" *Utne Reader* 64 (July-August): 34–36.

Cromie, Robert. 1890. *A Plunge into Space.* New York: Cosmopolitan.

Crosby, Alfred W. 1986. *Ecological Imperialism: The Biological Expansion of Europe, 900–1900.* Cambridge: Cambridge University Press.

Crossley, Robert. 2000. "Percival Lowell and the History of Mars." *Massachusetts Review* 41: 297–318.

Crowe, Michael. 1986. *The Extraterrestrial Life Debate 1750–1900: The Idea of a Plurality of Worlds from Kant to Lowell.* Cambridge: Cambridge University Press.

Crowther, Peter, ed. 2002. *Mars Probes.* New York: DAW Books.

Crumley, Carole, ed. 1994. *Historical Ecology: Cultural Knowledge and Changing Landscapes.* Santa Fe: School of American Research Press.

Csicsery-Ronay, Istvan. 1991. "The SF of Theory: Baudrillard and Haraway." *Science-Fiction Studies* 18: 387–404.

David, Leonard. 1999. "A Master Plan for Mars." *Sky and Telescope* 94 (4 April): 34–40.

——. 2004a. "Life on Mars Likely, Scientist Claims." *Science Tuesday, Space.com.* http://space.com/scienceastronomy/mars_microorganisms_040803.html.

——. 2004b. "NASA Scientist Sees Possible Mat of Martian Microbes." *Space.com.* http://space.com/scienceastronomy/mystery_monday_040809.html.
——. 2004c. "Scientists Seek Scent of Life in Methane at Mars." *Space.com.* http://space.com/scienceastronomy/mars_methane_040824.html.
Davidson, Keay. 1999. *Carl Sagan: A Life.* New York: Wiley.
Davies, Paul. 1995. *Are We Alone? Philosophical Implications of the Discovery of Extraterrestrial Life.* New York: Basic Books.
Davis, Mike. 1998. *Ecology of Fear: Los Angeles and the Imagination of Disaster.* New York: Metropolitan Books.
——. 2001. *Late Victorian Holocausts: El Niño Famines and the Making of the Third World.* London: Verso.
Dawes, W. R. 1865. "On the Planet Mars." *Monthly Notices of the Royal Astronomical Society* 25: 225–68.
Defoe, Daniel. 1903. *Serious Reflections during the Life and Surprising Adventures of Robinson Crusoe.* Ed. G. H. Maynadier. Boston: Old Corner Bookstore/Harvard University Press.
Delany, Samuel. 1994. *Silent Interviews: On Language, Race, Sex, Science Fiction, and Some Comics.* Hanover, N.H.: University Press of New England.
Del Ray, Lester. 1962. *Marooned on Mars.* New York: Holt, Rinehart, and Winston.
——. 1975. *Police Your Planet.* New York: Ballantine. (Orig. pub. 1956.)
Denning, Michael. 1987. *Mechanic Accents: Dime Novels and Working-Class Culture in America.* London: Verso.
Derrida, Jacques. 1994. *Spectres of Marx: The State of the Debt, the Work of Mourning, and the New International.* New York: Routledge.
DeVorkin, David. 1977. "W. W. Campbell's Spectroscopic Study of the Martian Atmosphere." *Quarterly Journal of the Royal Astronomical Society* 18: 37–53.
——. 2000. *Henry Norris Russell: Dean of American Astronomers.* Princeton: Princeton University Press.
Diamond, Jared. 1997. "Kinship with the Stars." *Discover* 18: 44–49.
Dick, Philip K. 1987. *The Collected Stories of Philip K. Dick.* 5 vols. New York: Carol.
——. 1991. *The Three Stigmata of Palmer Eldritch.* New York: Vintage. (Orig. pub. 1965.)
——. 1995. *Martian Time-Slip.* New York: Vintage. (Orig. pub. 1964.)
Dick, Steven J. 1982. *Plurality of Worlds: The Origins of the Extraterrestrial Life Debate from Democritus to Kant.* Cambridge: Cambridge University Press.
——. 1996. *The Biological Universe: The Twentieth-Century Extraterrestrial Life Debate and the Limits of Science.* Cambridge: Cambridge University Press.
Dickson, Gordon. 1978. *The Far Call.* New York: TOR.
DiGregorio, Barry, with Gilbert Levin and Patricia Ann Straat. 1997. *Mars: The Living Planet.* Berkeley: Frog.
Doel, Ronald E. 1996. *Solar System Astronomy in America: Communities, Patronage, and Interdisciplinary Science, 1920–1960.* Cambridge: Cambridge University Press.

Doran, Peter, et al. 1998. "Antarctic paleolake sediments and the search for extinct life on Mars." *Journal of Geophysical Research* 103: 28–36.

Douglas, Ian. 1998. *Semper Mars.* New York: Avon Eos.

——. 1999. *Luna Marine.* New York: Avon Eos.

——. 2000. *Europa Strike.* New York: Avon Eos.

Downing, George R. 1905. "Photographing the Canals of Mars," *New York Herald,* Sunday Magazine, July 30, p. 6.

Drake, Frank. 1997. "Extraterrestrial Intelligence: The Significance of the Search." In *Carl Sagan's Universe*, ed. Yervant Terzian and Elizabeth Bilson. Cambridge: Cambridge University Press, 87–97.

Dreifus, Claudia. 1999. "A New Frontier aboard the Mars Direct: A Conversation with Robert Zubrin." *New York Times.* 2 November, sec. D, p. 3.

Durham, Scott. 1988. "P. K. Dick: From the Death of the Subject to a Theology of Late Capitalism." *Science-Fiction Studies* 15: 173–86.

Dwornik, S. E. 1974. "Similarities: Mars, Earth, and Moon." In *Mars as Viewed by Mariner 9: A Pictorial Presentation by the Mariner 9 Television Team and the Planetary Program Principal Investigators*, NASA SP-329, ed. Harold Masursky et al. Washington, D.C.: NASA Science and Technical Information Office.

Ezell, Edward Clifton, and Linda Newman Ezell. 1984. *On Mars: Exploration of the Red Planet 1958–1978.* NASA SP-4212. Washington, D.C.: NASA.

Farmer, Philip Jose. 1979. *Jesus on Mars.* Los Angeles: Pinnacle Books.

Farren, Mick. 1990. *Mars—The Red Planet.* New York: Ballantine Books.

Fenigstein, Allan, and Peter A. Vanable. 1992. "Paranoia and Self-Consciousness." *Journal of Personality and Social Psychology* 62: 129–38.

Fitting, Peter. 1983. "Reality as Ideological Construct: A Reading of Five Novels by Philip K. Dick." *Science-Fiction Studies* 10: 219–36.

——. 2001. "Estranged Invaders: *The War of the Worlds.*" In *Learning from Other Worlds: Estrangement, Cognition, and the Politics of Science Fiction and Utopia*, ed. Patrick Parrinder. Durham: Duke University Press, 127–45.

Fischer, William B. 1984. *The Empire Strikes Out: Kurd Lasswitz, Hans Dominik, and the Development of German Science Fiction.* Bowling Green, Ohio: Popular Press.

Flammarion, Camille. 1894. *Omega: The Last Days of the World.* New York: Cosmopolitan.

——. 1889. *Urania: A Romance.* London: Chatto and Windus, 1891.

——. 1981.The Planet Mars. Trans. Patrick Moore; typescript in Lowell Observatory Archives. (Trans. of *La planete mars et ses conditions d'habitabilite,* 1892–1909. 2 vols. Paris: Gauthier-Vilar et Fils.)

Flautz, John. 1967. "An American Demagogue in Barsoom." *Journal of Popular Culture* 1: 263–75.

Fogg, Martyn J. 1995. *Terraforming: Engineering Planetary Environments.* Warrendale, Pa.: Society of Automotive Engineers.

Folk, R. L. 1993. "SEM Imaging of Bacteria and Nanobacteria in Carbonate Sediments and Rocks." *Journal of Sedimentary Petrology* 63: 990–99.

Folk, R. L., and R. L. Lynch. 1997. "The Possible Role of Nanobacteria (Dwarf Bacteria) in Clay-Mineral Diagenesis and the Importance of Careful Sample Preparation in High Magnification SEM Study." *Journal of Sedimentary Research* 67: 583–89.

Fontenelle, Bernard le Bovier de. 1990. *Conversations on the Plurality of Worlds.* Trans. H. A. Hargreaves; intro. by Nina Rattner Gelbart. Berkeley: University of California Press.

Forget, François, and Raymond T. Pierrehumbert. 1997. "Warming Early Mars with Carbon Dioxide Clouds that Scatter Infrared Radiation." *Science* 278: 1273–76.

Formisano, Vittorio, Sushil Atreya, Thérèse Encrenaz, Nikolai Ignatiev, Marco Giuranna. 2004. "Detection of Methane in the Atmosphere of Mars." *Science* 306, 1758–61.

Fortier, Edmund A. 1995. "The Mars That Never Was." *Astronomy* 23 (12 December): 36–43.

Forward, Robert L. 1991. *Martian Rainbow.* New York: Del Ray.

Fowler, A. 1916. "Prof. Percival Lowell." *Nature* 98: 231.

Franklin, H. Bruce. 1966. *Future Perfect: American Science Fiction of the Nineteenth Century.* New York: Oxford University Press.

——. 1980. *Robert A. Heinlein: America as Science Fiction.* New York: Oxford University Press.

Franko, Carol. 1994. "Working the 'In-between': Kim Stanley Robinson's Utopian Fiction." *Science-Fiction Studies* 21: 191–211.

Freedman, Carl. 1984. "Towards a Theory of Paranoia: The Science Fiction of Philip K. Dick." *Science-Fiction Studies* 11: 15–24.

——. 1988. "Editorial Introduction: Philip K. Dick and Criticism." *Science-Fiction Studies* 15: 121–30.

——. 2000. *Critical Theory and Science Fiction.* Hanover, N.H.: University Press of New England.

Freud, Sigmund. 1953–74. *The Standard Edition of the Complete Psychological Works of Sigmund Freud.* 24 vols. Ed. James Strachey et al. London: Hogarth Press.

Friedman, E. Imre, Jacek Wierzchos, Carmen Ascasao, and Michael Winklhofer. 2001. "Chains of Magnetite Crystals in the Meteorite ALH84001: Evidence of Biological Origin." *Proceedings of the National Academy of Sciences* 98: 2176–81.

Gantz, Kenneth. 1959. *Not in Solitude.* New York: Doubleday.

Garbedian, H. Gordon. 1928. "Mars Poses Its Riddle of Life." *New York Times Magazine,* December 9, pp. 1–2, 22.

——. 1933. *Major Mysteries of Science.* Garden City, N.Y.: Garden City Publishing.

Gaslin, Glenn. 1998. "Red Tarzana." *New Times* 3, no. 39 (September 24–30): 11–18.

Gauch, Hugh G. Jr. 1993. "Prediction, Parsimony, and Noise." *American Scientist* 81: 468–78.

Gerstell, M. F., J. S. Francisco, Y. L. Lung, and E. T. Aaltonee. 2001. "Keeping Mars Warm with New Super Greenhouse Gases." *Proceedings of the National Academy of Sciences* 98: 2154–57.

Gibson, E. K. Jr., D. S. McKay, K. L. Thomas-Keprta, S. J. Wentworth, F. Westfall, A. Steele, C. S. Romanek, M. S. Bell, and J. Toporski. 2001. "Life on Mars: Evaluation of the Evidence within Martian Meteorites ALH84001, Nakhla, and Shergotty." *Precambrian Research* 106: 15–34.

Gifford, F. A. Jr. 1964. "The Martian Canals According to a Purely Aeolian Hypothesis," *Icarus* 3: 130–35.

Glass, Fred. 1990. "Totally Recalling Arnold: Sex and Violence in the New Bad Future." *Film Quarterly* 44, no. 1 (fall).

Glasstone, Samuel. 1968. *The Book of Mars.* Washington, D.C.: NASA, Scientific and Technical Information Division. Glavin, D. P., M. Schubert, O. Botta, G. Kminek, and J. L. Bada. 2001. "Detecting Pyrolysis Products from Bacteria on Mars." *Earth and Planetary Science Letters* 185: 1–5.

Godwin, Frances. 1638. *The Man in the Moone.* London.

Godwin, Robert, comp. and ed. 2000. *Mars: The* NASA *Mission Reports.* Burling, Ontario: Apogee Books.

——. 2004. *Mars: The* NASA *Mission Reports.* Vol. 2. Burling, Ontario: Apogee Books.

Gold, Barri. 2002. "The Consolation of Physics: Tennyson's Thermodynamic Solution." PMLA 117: 449–64.

Golden, D. C., D. W. Ming, C. S. Schwandt, H. V. Lauer Jr., R. V. Socki, R. A. Morris, G. E. Lofgren, and G. A. McKay. 2001. "A Simple Inorganic Process for Formation of Carbonates, Magnetites and Sulfides in Martian Meteorite ALH84001." *American Mineralologist* 86: 370–75.

Golden, D. C., D. W. Ming, R. V. Morris, A. J. Brearley, H. V. Lauer Jr., A. H. Treiman, M. E. Zolensky, C. S. Schwandt, G. E. Lofgren, and G. A. McKay. 2004. "Evidence for Exclusively Inorganic Formation of Magnetite in Martian Meteorite ALH84001." *American Mineralogist* 89: 681–95.

Goldsmith, Donald. 1997. *The Hunt for Life on Mars.* New York: Dutton.

Goldwert, Marvin. 1993. "Teleology and Paranoia: The Search for Meaning." *Psychological Reports* 72 (February): 326.

Golombek, M. P., et al. 1997. "Overview of the Mars Pathfinder Mission and Assessment of Landing Site Predictions." *Science* 278: 1743–48.

Golumbia, David. 1996. "Resisting 'The World': Philip K. Dick, Cultural Studies, and Metaphysical Realism." *Science-Fiction Studies* 23: 83–102.

Gould, Stephen J. 1994. "A Plea and a Hope for Martian Paleontology." In *Where Next Columbus? The Future of Space Exploration,* ed. Valerie Neal. New York: Oxford University Press, 107–28.

Graham, Elizabeth. 1998. "Metaphors and Metamorphism: Some Thoughts on Environmental Metahistory." In *Advances in Historical Ecology*, ed. William Balée. New York: Columbia University Press, 119–37.

Graham, Loren. 1984. "Bogdanov's Inner Message." *Red Star,* by Alexander Bogdanov. Ed. Loren Graham and Richard Stites. Bloomington: Indiana University Press, 241–52.

Grant, Edward. 1994. *Planets, Stars, and Orbs: The Medieval Cosmos, 1200–1687*. Cambridge: Cambridge University Press.

Gratacap, Louis. 1903. *The Certainty of a Future Life on Mars*. New York: Bretano.

Green, Nathaniel. 1880. "Observations of Mars, at Madeira in August and September 1877." *Memoirs of the Royal Astronomical Society* 40: 123–40.

Greg, Percy. 1978. *Across the Zodiac: The Story of a Wrecked Record*. Abridged by Benjamin Appel. New York: Popular Library. (Orig. pub. 1880.)

Grinspoon, David Harry. 1997. *Venus Revealed: A New Look below the Clouds of Our Mysterious Twin Planet*. New York: Addison Wesley.

——. 2003. *Lonely Planets: The Natural Philosophy of Alien Life*. New York: Ecco/HarperCollins.

Groopman, Jerome. 2000. "Medicine on Mars." *New Yorker*. February 14, pp. 36–41.

Grove, Richard H. 1995. *Green Imperialism: Colonial Expansion, Tropical Island Edens, and the Origins of Environmentalism 1600–1860*. Cambridge: Cambridge University Press.

Guthke, Karl. 1990. *The Last Frontier: Imagining Other Worlds from the Copernican Revolution to Modern Science Fiction*. Trans. Helen Atkins. Ithaca: Cornell University Press.

"The Habitability of Mars." 1909. *Nature* 80: 212.

Haeckel, Ernst. 1906. *Last Words on Evolution: A Popular Retrospect and Summary*. Trans. Joseph McCabe. London: Owen.

Haila, Yrjo, and Richard Levins. 1992. *Humanity and Nature: Ecology, Science and Society*. London: Pluto Press.

Hancock, Graham. 1998. *The Mars Mystery: The Secret Connections between Earth and the Red Planet*. New York: Three Rivers Press.

Hanlon, Michael. 2004. *The Real Mars: Spirit, Opportunity, Mars Express, and the Quest to Explore the Red Planet*. New York: Carroll and Graf.

Hansson, Anders. 1997. *Mars and the Development of Life*. 2nd ed. Chichester, U.K.: Praxis.

Haraway, Donna. 1991. *Simians, Cyborgs, and Women: The Reinvention of Nature*. New York: Routledge.

Harris, Marvin. 1977. *Cannibals and Kings: The Origins of Cultures*. New York: Random House.

Hartmann, William K. 1997. *Mars Underground*. New York: Tom Doherty Associates.

——. 2003. *A Traveler's Guide to Mars: The Mysterious Landscapes of the Red Planet*. New York: Workman Publishing.

Hartmann, William K., and Gerhard Neukum. 2001. "Cratering Chronology and Evolution of Mars." In *Chronology and Evolution of Mars*, ed. R. Kallenbach, J. Geiss, and W. K. Hartmann. Bern: International Space Science Institute, 165–94.

Harvey, R. P., and H. Y. McSween Jr. 1996. "A Possible High-Temperature Origin for the Carbonates in the Martian Meteorite ALH84001." *Nature* 382: 49–51.

Hawken, Paul. 1993. *The Ecology of Commerce: A Declaration of Sustainability.* New York: HarperCollins.

Hayles, N. Katherine. 1991. "Constrained Constructivism: Locating Scientific Inquiry in the Theater of Representation." *New Orleans Review* 18: 76–85.

——. 1999. *How We Became Posthuman: Virtual Bodies in Cybernetics, Literature, and Informatics.* Chicago: University of Chicago Press.

Haynes, Robert. 1990. "Ecce Ecopoeisis: Playing God on Mars." In *Moral Expertise*, ed. D. MacNiven. New York: Routledge, 161–83.

Haynes, Roslynn D. 1980. *H. G. Wells: Discoverer of the Future.* New York: New York University Press.

Head, James W., John F. Mustard, Mikhail A. Kresllavsky, Ralph E. Milliken, and David R. Marchant. 2003. "Recent Ice Ages on Mars." *Nature* 426: 797–802.

Heffernan, William C. 1978. "The Singularity of Our Inhabited World: William Whewell and A. R. Wallace in Dissent." *Journal of the History of Ideas* 39: 81–100.

——. 1981. "Percival Lowell and the Debate over Extraterrestrial Life." *Journal of the History of Ideas* 42: 527–30.

Heinlein, Robert. 1959. "Science Fiction: Its Nature, Faults and Virtues." In *The Science Fiction Novel: Imagination and Social Criticism*, ed. Basil Davenport. Chicago: Advent, 14–48.

——. 1961. *Stranger in a Strange Land.* New York: G. P. Putnam's Sons.

——. 1986. *Double Star*. New York: Ballantine Books. (Orig. pub. 1956.)

Herkenhoff, K. E., et al. 2004. "Evidence from Opportunity's Microscopic Imager for Water on Meridiani Planum." *Science* 306, 1727–30.

Herschel, Sir William. 1912. *The Scientific Papers of Sir William Herschel.* 2 vols. London: The Royal Society and the Royal Astronomical Society.

Hetherington, Norriss S. 1976. "Amateur versus Professional: The British Astronomical Association and the Controversy over Canals on Mars." *Journal of the British Astronomical Association* 86: 303–8.

——. 1981. "Percival Lowell: Professional Scientist or Interloper?" *Journal of the History of Ideas* 42: 159–61.

——. 1988. *Science and Objectivity: Episodes in the History of Astronomy.* Ames: Iowa State University Press.

Heward, E. Vincent. 1907. "Mars, It Is a Habitable World?" *Fortnightly Review* (August): 215–28.

Hillegas, Mark R. 1970. "Martians and Mythmakers: 1877–1938." In *Challenges in American Culture*, ed. Ray B. Browne, Larry N. Landrum, and William K. Bottoroff. Bowling Green, Ohio: Popular Press, 150–77.

——. 1975. "Victorian 'Extraterrestrials.'" *Harvard Studies in English* 6: 391–414.

Hipolito, Jane, and Willis McNelly, eds. 1971. *The Book of Mars.* London: Orbit.

Hoagland, Richard. 1996. *The Monuments of Mars: A City on the Edge of Forever.* 4th ed. Berkeley: Frog.

Hoberek, Andrew P. 1997. "The 'Work' of Science Fiction: Philip K. Dick and

Occupational Masculinity in the Post-World War II United States" *Modern Fiction Studies* 43: 374–404.

Hoffman, Nick. 2000. "White Mars: A New Model for Mars' Surface and Atmosphere Based on CO_2." *Icarus* 146: 326–42.

Hofstadter, Richard. 1965. *The Paranoid Style in American Politics and Other Essays.* New York: Knopf.

Holden, Edward S. 1894. "The Lowell Observatory in Arizona." *Publications of the Astronomical Society of the Pacific* 35: 160–69.

Holmsten, Brian, and Alex Lubertozzi, eds. 2001. *The Complete War of the Worlds: Mars' Invasion of Earth from H. G. Wells to Orson Welles.* Napierville, Ill.: Sourcebooks.

Holtsmark, Erling B. 1986. *Edgar Rice Burroughs.* Boston: Twayne.

Horowitz, Norman. 1986. *To Utopia and Back: The Search for Life in the Solar System.* New York: W. H. Freeman.

Horowitz, Norman, Roy Cameron, and Jerry Hubbard. 1972. "Microbiology of the Dry Valleys of Antarctica." *Science* 176: 242–45.

Horowitz, N. H., G. L. Hobby, and Jerry S. Hubbard. 1977. "Viking on Mars: The Carbon Assimilation Experiments." *Journal of Geophysical Research* 82: 4659–62.

Housden, C. E. 1914. *The Riddle of the Planet Mars.* London: Longmans, Green.

Hoyt, William Graves. 1976. *Lowell and Mars.* Tucson: University of Arizona Press.

Hughes, David Y. 1977. "The Garden in Wells's Early Science Fiction." In *H. G. Wells and Modern Science Fiction*, ed. Darko Suvin and Robert Philmus. Lewisburg, PA.: Bucknell University Press.

Hume, Kathryn. 1983. "The Hidden Dynamics of *The War of the Worlds.*" *Philological Quarterly* 62: 279–92.

Hunt, Charles D., and Michael O. van Pelt. 2004. "Comparing NASA and ESA Cost Estimating Methods for Human Missions to Mars." Mars Society, http://www.marssociety.org.

Hunt, H. F. 1909. "The Functions of the Martian Canals." *Nature* 82, 69.

Huntington, John. 1982. *The Logic of Fantasy: H. G. Wells and Science Fiction.* New York: Columbia University Press.

Huygens, Christiaan. 1698. *The Celestial Worlds Discovered.* London.

Ihde, Don. 1990. *Technology and the Lifeworld: From Garden to Earth.* Bloomington: Indiana University Press.

Ingerson, Alice E. 1994. "Tracking and Testing the Nature-Culture Divide." In *Historical Ecology: Cultural Knowledge and Changing Landscapes*, ed. Carole Crumley. Santa Fe: School of American Research Press, 43–66.

Jakosky, Bruce. 1998. *The Search for Life on Other Planets.* Cambridge: Cambridge University Press.

Jakosky, Bruce, and Roger Phillips. 2001. "Mars' Volatiles and Climate History." *Nature* 412: 237–44.

Jakosky, Bruce, K. Nealson, C. Bakermans, R. Ley, and M. T. Mellon. 2003. "Subfreezing Activity of Microorganisms and the Potential Habitability of Mars' Polar Regions." *Astrobiology* 3: 343–50.

James, Edward. 1994. *Science Fiction in the Twentieth Century*. New York: Oxford University Press.

Jameson, Fredric. 1982. "Progress vs. Utopia: Can We Imagine the Future?" *Science-Fiction Studies* 9: 147–58.

——. 1987. "Science Fiction as a Spatial Genre: Generic Discontinuities and the Problem of Figuration in Vonda McIntyre's *The Exile Waiting*." *Science-Fiction Studies* 14: 44–59.

——. 1989. "Nostalgia for the Present." *South Atlantic Quarterly* 88: 517–37.

——. 2001. " 'If I Find One Good City I Will Spare the Man': Realism and Utopia in Kim Stanley Robinson's *Mars* Trilogy." In *Learning from Other Worlds: Estrangement, Cognition, and the Politics of Science Fiction and Utopia*, ed. Patrick Parrinder. Durham: Duke University Press, 208–32.

Joels, Kerry Mark. 1985. *The Mars One Crew Manual*. New York: Ballantine.

Johnson, William B., and Thomas D. Clareson. 1963–64. "The Interplay of Science and Fiction: The Canals of Mars." *Extrapolation* 5: 37–48.

Jones, Alice Ilgenfritz, and Ella Merchant. 1991. *Unveiling a Parallel: A Romance*. Ed. Carol A. Kolmerton. Syracuse: Syracuse University Press. (Orig. pub. 1893.)

Judson, Olivia. 2004. "Some Things Are Better Left on Mars." *New York Times*, April 19.

Kaplan, Lewis D., Guido Munch, and Hyron Spinrad. 1964. "An Analysis of the Spectrum of Mars." *Astrophysical Journal* 139: 1–15.

Kargel, Jeffrey S. 2004. *Mars: A Warmer, Wetter Planet*. New York: Springer-Praxis.

Karl, D. M., F. Bird, K. Björkman, T. Houlihan, R. Shackelford, L. Tupas. 1999. "Microorganisms in the Accreted Ice of Lake Vostok, Antarctica." *Science* 286: 2144–46.

Kasting, James. 1997. "The Early Mars Climate Question Heats Up." *Science* 278: 1245.

Kemp, Peter. 1982. *H. G. Wells and the Culminating Ape: Biological Themes and Imaginative Obsessions*. London: Macmillan.

Kerr, Richard A. 2003. "Eons of a Cold, Dry, Dusty Mars." *Science* 301: 1037–38.

Ketterer, David. 1984. "James Blish's *Welcome to Mars* and the Haertel Complex." *Science-Fiction Studies* 11: 284–90.

Kieffer, Hugh H., Bruce M. Jakosky, Conway Snyder, and Mildred S. Matthews, ed. 1991. *Mars*. Tucson: University of Arizona Press.

Kingsland, Sharon. 1985. *Modeling Nature: Episodes in the History of Population Ecology*. Chicago: University of Chicago Press.

Kirschvink J. L., and H. Vali. 1999. "Criteria for the Identification of Bacterial Magnetofossils on Earth or Mars." *Lunar and Planetary Science* 30, abstract 1681.

Klein, Harold. 1977. "The Viking Biological Investigation: General Aspects." *Journal of Geophysical Research* 82: 4677–80.

——. 1979. "The Search for Life on Mars." *Review of Geophysics and Space Physics* 17: 1655–62.

Kline, Otis Adelbert. 1961. *The Outlaws of Mars.* New York: Ace Books. (Orig. pub. 1933.)
Klingelhöfer, G., et al. 2004. "Jarosite and Hematite at Meridiani Planum from Opportunity's Mössbauer Spectrometer." *Science* 306, 1740-1745.
Kluger, Jeffrey. 1992. "Mars, in the Earth's Image." *Discover* 13, no. 9 (September): 72–76.
——. 2000. "Will We Live on Mars?" *Time,* 10 April, pp. 61–63.
Koch, Howard. 1970. *The Panic Broadcast: Portrait of an Event.* New York: Little, Brown.
Kornbluth, C. M. 1959. "The Failure of the Science Fiction Novel as Social Criticism." In *The Science Fiction Novel: Imagination and Social Criticism,* ed. Basil Davenport. Chicago: Advent, 49–76.
Kornbluth, C. M., and Judith Merril. [Cyril Judd]. 1952. *Gunner Cade.* New York: Ace Books.
——. 1953. *Outpost Mars.* New York: Ace Books.
Koyré, Alexandre. 1957. *From the Closed World to the Infinite Universe.* Baltimore: Johns Hopkins University Press.
Krasnopolsky, V., G. L. Bjoraker, M. J. Mumma, and D. E. Jennings. 1997. "High Resolution Spectroscopy of Mars at 3.7 and 8 μm: A Sensitive Search for H_2O_2, H_2CO, HCl, and CH_4, and Detection of HDO." *Journal of Geophysical Research* 102: 6525–34.
Kuiper, Gerard, and Barbara Middlehurst, eds. 1961. *Planets and Satellites.* Chicago: University of Chicago Press.
Kyle, Richard. 1970. "Out of Time's Abyss: The Martian Stories of Edgar Rice Burroughs." *Riverside Quarterly* 4: 110–22.
Lakoff, George, and Mark Johnson. 1980. *Metaphors We Live By.* Chicago: University of Chicago Press.
Landis, Geoffrey. 2000. *Mars Crossing.* New York: TOR.
Lanier, Sterling. 1983. *Menace under Marswood.* New York: Ballantine Books.
Lankford, John. 1981a. "Amateurs versus Professionals: The Controversy over Telescope Size in Later Victorian Science." *Isis* 72: 11–28.
——. 1981b. "Amateurs and Astrophysics: A Neglected Aspect in the Development of a Scientific Specialty." *Social Studies of Science* 11: 275–303.
——. 1997. *American Astronomy: Community, Careers, and Power, 1859–1940.* Chicago: University of Chicago Press.
Laplace, Pierre Simon. 1796. *Exposition du système du monde.* Paris.
Lasswitz, Kurd. 1971. *Two Planets.* Trans. Hans H. Rudnick; abridged by Erich Lasswitz. Carbondale: Southern Illinois University Press. (Orig. pub. 1897.)
Latour, Bruno. 1987. *Science in Action: How to Follow Scientists and Engineers through Society.* Cambridge: Harvard University Press.
——. 1993. *We Have Never Been Modern.* Trans. Catherine Porter. Cambridge: Harvard University Press.
Lawler, Andrew, and Richard Kerr. 1999. "Changes to Missions Could Delay Science." *Science* 286: 2248.

Lears, T. Jackson. 1981. *No Place of Grace: Antimodernism and the Transformation of American Culture.* New York: Pantheon.

Lederberg, Joshua. 1966. "Exobiology: Approaches to Life Beyond the Earth." In *Extraterrestrial Life: An Anthology and Bibliography,* ed. Colin Pittendrigh. Washington, D.C.: National Academy of Sciences Research Council, 124–37. (Orig. pub. 1960.)

Leib, Michael. 1998. *Children of Ezekiel: Aliens, UFOs, the Crisis of Race, and the Advent of End Time.* Durham: Duke University Press.

"Lessons Mars Has for Mother Earth." 1909. *Philadelphia Record.* May 5.

Levin, Gilbert. 1972. "Detection of Metabolically Produced Labeled Gas: The Viking Mars Lander." *Icarus* 16: 153–66.

Levin, Gilbert V. 1997. "The Viking Labeled Release Experiment and Life on Mars." *Instruments, Methods, and Missions for the Investigation of Extraterrestrial Microorganisms, SPIE Proceedings*, 3111, 146–61.

——. 2001. "The Oxides of Mars." *Instruments, Methods, and Missions for Astrobiology, SPIE Proceedings* 4495, 131–35.

——. 2002. "Iron (VI) Seems an Unlikely Explanation for Viking Labeled Release Results." *Icarus* 159: 266–67.

——. 2004. "Interpretation of New Results from Mars with Respect to Life." *Instruments, Methods, and Missions for Astrobiology, SPIE Proceedings*, 5555, 126–38.

Levin, Gilbert, and Patricia Ann Straat. 1976. "Labeled Release—An Experiment in Radiorespirometry." *Origins of Life* 7: 293–311.

——. 1977. "Recent Results from the Viking Labeled Release Experiment on Mars." *Journal of Geophysical Research* 82: 4663–67.

——. 1981. "A Search for a Nonbiological Explanation of the Viking Labeled Release Life Detection Experiment." *Icarus* 45: 494–516.

Levin, Gilbert, and Ron L. Levin. 1998. "Liquid Water and Life on Mars." *Instruments, Methods, and Missions for Astrobiology, SPIE Proceedings*, 3441, 30–41.

Levin, Gilbert, Lawrence Kuznetz, and Arthur Lafleur. 2000. "Approaches to Resolving the Question of Life on Mars." *Instruments, Methods, and Missions for Astrobiology, SPIE Proceedings* 4137, 48–62.

Levin, G. V., J.D. Miller, P. A. Straat, and R. E. Hoover. 2002. "A Sterile Robotic Mars Soil Analyzer." *Instruments, Methods, and Missions for Astrobiology, SPIE Proceedings*, 4859, 78–86.

Levine, George. 2002. *Dying to Know: Scientific Epistemology and Narrative in Victorian England.* Chicago: University of Chicago Press.

Levitt, I. M. 1956. *A Space Traveler's Guide to Mars.* New York: Henry Holt.

Levoy, Conway. 2001. "Weather and Climate on Mars." *Nature* 412: 245–49.

Lewis, C. S. 1997. *Out of the Silent Planet.* New York: Quality Paperback Book Club. (Orig. pub. 1938.)

Lewis, John. S. 1996. *Mining the Sky: Untold Riches from the Asteroids, Comets, and Planets*. Reading, MA: Addison-Wesley.

Lewis, Richard S. 1978. *From Vinland to Mars: A Thousand Years of Exploration.* New York: Quadrangle.

Lewontin, Richard. 1991. "Facts and the Factitious in the Natural Sciences." *Critical Inquiry* 18: 140–53.

Lewontin, Richard, and Richard Levins. 1985. *The Dialectical Biologist.* Cambridge, Mass.: Harvard University Press.

Ley, Willy. 1966. *Mariner IV to Mars.* New York: New American Library.

Ley, Willy, and Werner von Braun. 1956. *The Exploration of Mars.* New York: Viking.

Lobitz, Brad, Byron L. Wood, Maurice M. Averner, and Christopher P. McKay. 2001. "Use of Spacecraft Data to Derive Regions on Mars Where Liquid Water Would be Stable." *Proceedings of the National Academy of Sciences* 98: 2132–37.

Locke, George. 1975. *Voyages in Space: A Bibliography of Interplanetary Fiction, 1801–1914.* London: Ferret Fantasy.

Locke, John. 1960. *Two Treatises of Government.* Ed. Peter Laslett. Cambridge: Cambridge University Press.

Lockyer, William J. S. 1896. "Mars as He Now Appears." *Nature* 50: 476–78.

——. 1906. "Lowell's Observations of the Planet Mars." *Nature* 74: 587–89.

Lovelock, James, and Michael Allaby. 1984. *The Greening of Mars.* New York: St. Martin's Press.

Lowell, Percival. 1895. *Mars.* New York: Houghton Mifflin.

——. 1905a. "Means, Methods and Mistakes in the Study of Planetary Evolution." Lowell Observatory Archives. Manuscript dated April 13.

——. 1905b. "Photographing the Canals of Mars." *New York Herald,* Sunday magazine. July 30, p. 6.

——. 1906a. "First Photographs of the Canals of Mars." *Proceedings of the Royal Society* 77: 132–36.

——. 1906b. *Mars and Its Canals.* London: Macmillan.

——. 1907. "The Canals of Mars, Optically and Psychologically Considered—A Reply to Professor Newcomb." *The Astrophysical Journal* 26: 131–40.

——. 1908. "Planetary Photography." *Nature* 77: 402–4.

——. 1909. *Mars as the Abode of Life.* New York: Macmillan.

——. 1910a. "Recent Discoveries about Mars and the Martians." Lowell Observatory Archives. Typescript of lecture delivered at the Museum of Natural History, February-March.

——. 1910b. "The Portent of Socialism." Lowell Observatory Archives. Address to the Victorian Club, December 8.

——. 1916a. *Immigration versus the United States. An Address Delivered at Phoenix, Arizona February 17, 1916.* Lynn, Mass.: Thomas Nichols and Son.

——. 1916b. "Mars and the Earth." Lowell Observatory Archives. Manuscript dated August.

Lupoff, Richard. 1976. *Barsoom: Edgar Rice Burroughs and the Martian Vision.* Baltimore: Mirage.

Mackey, Douglas A. 1988. *Philip K. Dick.* Boston: Twayne.
Maclaurin, Colin. 1748. *An Account of Sir Isaac Newton's Philosophical Discoveries.* London.
Macpherson, Hector Jr. 1905. "Mars and Its Canals," *Scottish Review,* November 23, pp. 464–65.
Malin, Michael C. 1999. "Visions of Mars: A Tour of the Red Planet Turns Up Geologic Vistas that Sometimes are Hauntingly Earth-like." *Sky and Telescope* 94, no. 4 (April): 42–49.
Malin, Michael C., and Kenneth S. Edgett. 2000a. "Evidence for Recent Groundwater Seepage and Surface Runoff on Mars." *Science* 288: 2330–35.
——. 2000b. "Sedimentary Rocks of Early Mars." *Science* 290: 1927–37.
Malmgren, Carl D. 1991. *Worlds Apart: Narratology of Science Fiction.* Bloomington: Indiana University Press.
Markley, Robert. 1993. *Fallen Languages: Crises of Representation in Newtonian England, 1660–1740.* Ithaca: Cornell University Press.
——. 1996. "Boundaries: Mathematics, Alienation, and the Metaphysics of Cyberspace." In *Virtual Realities and Their Discontents,* ed. Robert Markley. Baltimore: Johns Hopkins University Press, 55–77.
——. 1997. "Alien Assassinations: *The X-Files* and the Paranoid Structure of History." *Camera Obscura* 40–41: 77–103.
——. 1999a. "After the Science Wars: From Old Battles to New Directions in the Cultural Study of Science." *After the Disciplines: The Emergence of Cultural Studies,* ed. Michael Peters. London: Bergin and Garvey, 47–70.
——. 1999b. " 'Land Enough in the World': Locke's Golden Age and the Infinite Extensions of 'Use.' " *South Atlantic Quarterly* 98: 817–37.
——. 1999c. "Foucault, Modernity, and the Cultural Study of Science." *Configurations: A Journal of Literature, Science, and Technology* 7: 153–73.
Markley, Robert, and Harrison Higgs, Michelle Kendrick, and Helen Burgess. 2001. *Red Planet: Scientific and Cultural Encounters with Mars.* DVD-ROM. Philadelphia: University of Pennsylvania Press.
Marsak, Leonard. 1959. *Bernard de Fontenelle: The Idea of Science in the French Enlightenment. Transactions of the American Philosophical Society,* new series 49, part 7.
"Martian Canal Studies[.] Results of Percival Lowell's Observations[.] His Drawings Confirmed by Recent Photographs." 1905. *Boston Transcript,* October 17.
Masurksy, Harold, et al. 1974. *Mars as Viewed by Mariner 9: A Pictorial Presentation by the Mariner 9 Television Team and the Planetary Program Principal Investigators.* NASA SP-329. Washington, D.C.: NASA Science and Technical Information Office.
Maunder, E. Walter. 1894. "The Canals of Mars." *Knowledge.* November 1, pp. 249–52.
Maunder, E. Walter, and J. E. Evans. 1903. "Experiments as to the Actuality of the 'Canals' of Mars." *Monthly Notices of the Royal Astronomical Society* 58: 488–99.
McAuley, Paul. 2001. *The Secret of Life.* New York: TOR Books.

McConnell, Frank. 1981. *The Science Fiction of H. G. Wells.* New York: Oxford University Press.

McCurdy, Howard E. 1997. *Space and the American Imagination.* Washington, D.C.: Smithsonian Institution Press.

McDonald, Ian. 1988. *Desolation Road.* New York: Bantam.

McElroy, Bernard. 1985. "The Art of Projective Thinking: Franz Kafka and the Paranoid Vision." *Modern Fiction Studies* 31: 217–32.

McKay, Christopher P., ed. 1985. *The Case for Mars II.* American Astronomical Society Science and Technology Series, vol. 62. San Diego: Univelt.

——. 1997. "The Search for Life on Mars." *Origins of Life and Evolution of the Biosphere* 27: 263–89.

McKay, C. P., and C. R. Stoker. 1989. "The Early Environment and Its Evolution on Mars." *Review of Geophysics* 27: 189.

McKay, C. P., R. L. Mancinelli, C. R. Stoker, and R. A. Wharton Jr. 1992. "The Possibility of Life on Mars During a Water-Rich Past." In *Mars,* ed. H. H. Kieffer, B. M. Jakosky, C. W. Snyder, M. S. Mathews. Tucson: University of Arizona Press, 1234–45.

McKay, Christopher P., Owen B. Toon, and James F. Kasting. 1991. "Making Mars Habitable." *Nature* 352, 489–95.

McKay, David S., Everett K. Gibson Jr., Kathie Thomas-Keprta, Hojatollah Vali, Christopher Romanek, Simon J. Clemett, Xavier D. F. Chillier, Claude R. Maechling, and Richard N. Zare. 1996. "Search for Past Life on Mars: Possible Relic Biogenic Activity in Martian Meteorite ALH84001." *Science* 273: 924–30.

McKim, Richard. 1993. "The Life and Times of E. M. Antoniadi, 1870–1944." *Journal of the British Astronomical Association* 103: 164–70, 219–27.

McNeill, J. R. 2000. *Something New under the Sun: An Environmental History of the Twentieth-Century World.* New York: Norton.

McNichol, Tom. 2001. "The New Red Menace." *Wired* 9, no.7: 140–47.

McSween, H. Y Jr., and R. P. Harvey. 1998. "An Evaporation Model for Formation of Carbonates in the ALH84001 Martian Meteorite." *International Geophysical Review* 40: 774–83.

Menuck, Morton. 1992. "Differentiating Paranoia and Legitimate Fears." *American Journal of Psychiatry* 149 (January): 140–41.

Merril, Judith. 1971. "What Do You Mean: Science? Fiction?" In *Science Fiction: The Other Side of Realism,* ed. Thomas D. Clareson. Bowling Green, Ohio: Popular Press, 53–95.

Meyer, Thomas R., and Christopher P. McKay. 1996. "Using the Resources of Mars for Human Settlement." In *Strategies for Mars: A Guide to Human Exploration*, ed. Carol Stoker and Carter Emmart. San Diego: Univelt, for the American Astronautical Society, 393–442.

Miklitsch, Robert. 1995. "*Total Recall:* Production, Revolution, Simulation-Alienation Effect." *Camera Obscura* 32: 4–39.

Miller, Joseph D., Patricia A. Straat, and Gilbert V. Levin. 2001. "Periodic Analysis of the Viking Lander Labeled Release Experiment." In *Instruments, Methods, and Missions for Astrobiology, SPIE Proceedings*, 4495: 96–107.

Miller, Schuyler P[eter]. 1954. *The Titan and Other Stories.* London: Weidenfeld and Nicholson.

Miller, Walter M. Jr. "Crucifixus Etiam." 1980. In *The Best of Walter M. Miller, Jr.* New York: Pocket Books.

Mirowski, Philip. 1989. *More Heat than Light: Economics as Social Physics, Physics as Nature's Economics.* Cambridge: Cambridge University Press.

Mirowsky, John, and Catherine E. Ross. 1983. "Paranoia and the Structure of Powerlessness." *American Sociological Review* 48 (April): 228–39.

Mishkin, Andrew. 2003. *Sojourner: An Insider's View of the Mars Pathfinder Mission.* New York: Berkley.

Mitman, Gregg. 1992. *The State of Nature: Ecology, Community, and American Social Thought, 1900–1950.* Chicago: University of Chicago Press.

Mizejewski, Linda. 1993. "Total Recoil: The Schwarzenegger Body on Postmodern Mars." *Postscript* 12, no. 3 (summer).

Moffitt, Donald. 1989. *Crescent in the Sky.* New York: Del Ray.

Moorcock, Michael. 1981. *Warrior of Mars.* Omnibus ed. London: New English Library.

Moore, James R. 1979. *The Post-Darwinian Controversies: A Study of the Protestant Struggle to Come to Terms with Darwin in Great Britain and America, 1870–1900.* Cambridge: Cambridge University Press.

Moore, Jim. 1986. "Socializing Darwinism: Historiography and the Fortunes of a Phrase." In *Science as Politics*, ed. Les Levidow. London: Free Association Books, 38–80.

Moore, Patrick. 1977. *A Guide to Mars.* New York: Norton.

——. 1999. *Patrick Moore on Mars.* London: Cassell.

Morse, Edward S. 1907. *Mars and Its Mystery.* Boston: Little, Brown.

Morton, Oliver. 2002. *Mapping Mars: Science, Imagination, and the Birth of a World.* London: Fourth Estate.

Mullen, Richard D. 1971. "The Undisciplined Imagination: Edgar Rice Burroughs and Lowellian Mars." In *The Other Side of Realism: Essays on Modern Fantasy and Science Fiction*, ed. Thomas D. Clareson. Bowling Green, Ohio: Popular Press, 229–47.

——. 1995. "From Standard Magazines to Pulps and Big Slicks: A Note on the History of U.S. General and Fiction Magazines." *Science-Fiction Studies* 22: 144–56.

Murray, Bruce. 1989. *Journey into Space: The First Three Decades of Space Exploration.* New York: Norton.

——. 1997. "From the Eyepiece to the Footpad: The Search for Life on Mars." In *Carl Sagan's Universe*, ed. Yervant Terzian and Elizabeth Bilson. Cambridge: Cambridge University Press, 35–48.

Murray, Bruce, Michael C. Malin, and Ronald Greeley. 1981. *Earthlike Planets: Surfaces of Mercury, Venus, Earth, Moon, Mars.* San Francisco: W. H. Freeman.

Mustard, John F., Christopher D. Cooper, and Moses K. Rifkin. 2001. "Evidence for Recent Climate Change from the Identification of Youthful Near-Surface Ground Ice." *Nature* 412: 411–14.

Mutch, Thomas, R. E. Arvidson, J. W. Head, K. L. Jones, and R. S. Saunders. 1976. *The Geology of Mars.* Princeton: Princeton University Press.

Navarro-Gonzalez, Rafael, et al. 2003. "Mars-Like Soils in the Atacama Desert, Chile and the Dry Limit of Microbial Life." *Science* 302: 1018–21.

Neukum, G., et al. 2004. "Recent and Episodic Volcanic and Glacial Activity on Mars Revealed by the High Resolution Stereo Camera." *Nature*, 432, 971–979.

Newcomb, Simon. 1907. "The Optical and Psychological Principles Involved in the Interpretation of the So-called Canals of Mars." *Astrophysical Journal* 26: 1–17.

Nicholson, Marjorie Hope. 1935. "The Telescope and the Imagination." *Modern Philology* 32: 428–62.

——. 1940. "Cosmic Voyages." *ELH* 7: 83–107.

Niderst, Alain. 1972. *Fontenelle a la recherce de lui-meme 1657–1702.* Paris: A.-G. Nizet.

Nisbet, E. G., and N. H. Sleep. 2001. "The Habitat and Nature of Early Life." *Nature* 409: 1083–91.

Niven, Larry. *Rainbow Mars.* 1999. London: Orbit.

Noble, Edmund. 1908. "Does the Universe Exist for Man Alone?" *Boston Herald,* June 28.

Numbers, Ronald. 1977. *Creation by Natural Law: Laplace's Nebular Hypothesis in American Thought.* Seattle: University of Washington Press.

Oberg, James Edward. 1982. *New Earths: Transforming Other Planets for Humanity.* Harrisburg, PA: Stackpole Books.

O'Connor, James. 1998. *Natural Causes: Essays in Ecological Marxism.* New York: Guilford.

O'Neill, Gerard K. 2000. *The High Frontier: Human Colonies in Space.* 3rd ed. New York: Apogee Books.

Ordway Frederick I., III. 1996. "Mars Mission Concepts: The Von Braun Era." In *Strategies for Mars: A Guide to Human Exploration,* American Astronautical Science and Technology Series, vol. 86, ed. Carol Stoker and Carter Emmart. San Diego: Univelt, 69–95.

Oyama, Vince, and Bonnie Berdahl. 1977. "The Viking Gas Exchange Experiment Results from Chryse and Utopia Surface Samples." *Journal of Geophysical Research* 82: 4669–76.

——. 1979. "A Model of Martian Surface Chemistry." *Journal of Molecular Evolution* 14: 199.

Pace, Norman R. 1997. "A Molecular View of Microbial Diversity and the Biosphere." *Science* 276: 734–40.

Palmer, Raymond A. 1952. "Editorial." *Other Worlds: Science Stories,* August, pp. 2–4.

Park, Robert. 2000. *Voodoo Science: The Road from Foolishness to Fraud.* New York: Oxford University Press.

Parrinder, Patrick. 1995. *Shadows of the Future: H. G. Wells, Science Fiction, and Prophecy.* Liverpool: Liverpool University Press.

Paxson, James J. 1999. "Kepler's Allegory of Containment, the Making of Modern Astronomy, and the Semiotics of Mathematical Thought." *Intertexts* 3: 105–23.

Peirce, Benjamin. 1881. *Ideality in the Physical Sciences.* Boston: Little, Brown.

Penley, Constance. 1997. *NASA/TREK: Popular Science and Sex in America.* London: Verso.

Pesek, Ludek. 1975. *The Earth Is Near.* Trans. Anthea Bell. New York: Dell. (Orig. pub. 1970.)

Philmus, Robert M. 1970. *Into the Unknown: The Evolution of Science Fiction from Francis Godwin to H. G. Wells.* Berkeley: University of California Press.

Philmus, Robert M., and David Y. Hughes, eds. 1975. *H. G. Wells: Early Writings in Science and Science Fiction.* Berkeley: University of California Press.

Pickering, Andrew. 1995. *The Mangle of Practice: Time, Agency, and Science.* Chicago: University of Chicago Press.

Pickering, William H. 1921. *Mars.* Boston: Gorham Press.

Pierce, John J. 1987. *Foundations of Science Fiction.* Westport, Conn.: Greenwood.

——. 1989. *When World Views Collide: A Study in Imagination and Evolution.* Westport, Conn.: Greenwood.

Pittendrigh, Colin, Wolf Vishniac, and J. P. T. Pearman, eds. 1966. *Biology and the Exploration of Mars.* NASA publication 1296. Washington, D.C.: NASA.

Plotkin, Howard. 1993. "William H. Pickering in Jamaica: The Founding of Woodlawn and Studies of Mars." *Journal for the History of Astronomy* 24: 101–22.

Plotnitsky, Arkady. 2002. *The Knowable and the Unknowable: Modern Science, Neoclassical Thought, and the "Two Cultures."* Ann Arbor: University of Michigan Press.

Pohl, Frederik. 1976. *Man Plus.* New York: Baen Books.

——. 1988. *The Day the Martians Came.* New York: St. Martin's Press.

Pohl, Frederik, and Thomas T. Thomas. 1995. *Mars Plus.* New York: Baen Books.

Polan, Dana B. 1986. *Power and Paranoia: History, Narrative, and the American Cinema.* New York: Columbia University Press.

Politics and Life on Mars: A Story of a Neighbouring Planet. 1883. London: Low, Marston.

Pollack, James. 1979. "Climactic Changes on the Terrestrial Planets." *Icarus* 37: 479–553.

Pollack, J. B., and Carl Sagan. 1993. "Planetary Engineering." *Resources of Near-earth Space.* Ed. J. Lewis, M. S. Matthews, and M. L. Guerrieri. Tucson: University of Arizona Press, 921–50.

Pope, Gustavus. 1894. *Journey to Mars.* New York: G. W. Dillingham.

Porges, Irwin. 1975. *Edgar Rice Burroughs: The Man Who Created Tarzan.* Provo, Utah: Brigham Young University Press.

Powers, Robert. 1986. *Mars: Our Future on the Red Planet.* New York: Houghton Mifflin.

Priscu, John C., et al. 1999. "Geomicrobiology of Subglacial Ice above Lake Vostok, Antarctica." *Science* 286: 2141–44.
Proctor, Mary. 1929. *Romance of the Planets.* New York: Harper Brothers.
"Prof. Lowell Attacked." 1907. *Boston Advertiser,* November.
Rabkin, Eric S. 1988. "Irrational Expectations; or How Economics and the Post-Industrial World Failed Philip K. Dick." *Science-Fiction Studies* 15: 161–72.
Raby, Peter. 2001. *Alfred Russel Wallace: A Life.* Princeton: Princeton University Press.
Raeburn, Paul. 1998. *Uncovering the Secrets of the Red Planet: Mars.* Washington, D.C.: National Geographic Society.
Randles, W. G. L. 1999. *The Unmaking of the Medieval Christian Cosmos, 1500–1760.* Aldershot, U.K.: Ashgate.
Rea, D. G., T. Belsky, and M. Calvin. 1963. "Interpretation of the 3- to 4-Micron Infrared Spectrum of Mars." *Science* 141: 923–27.
Reichhardt, Tony. 1999. "Concern over Mars Lander as Inquiry Reports on Orbiter Loss." *Nature* 402: 221–22.
Rieder, R., et al., 2004. "Chemistry of Rocks and Soils at Meridiani Planum from the Alpha Particle X-ray Spectrometer." *Science* 306, 1746–49.
Rendall, Steven. 1971. "Fontenelle and His Public." *MLN* 86: 496–508.
Richardson, Robert S. 1964. *Mars.* New York: Harcourt Brace and World.
Ritchie, John Jr. 1909. "Percival Lowell's Mars: A Generous Reception by European Experts." *Boston Transcript,* March 31.
Robinson, Kim Stanley. 1984. *Icehenge.* London: HarperCollins.
——. 1993. *Red Mars.* New York: Bantam.
——. ed. 1994a. *Future Primitive: The New Ecotopias.* New York: TOR.
——. 1994b. *Green Mars.* London: HarperCollins.
——. 1996. *Blue Mars.* New York: Bantam.
——. 1997. *Antarctica.* London: HarperCollins.
——. 1999. *The Martians.* London: HarperCollins.
——. 2002. *Vinland the Dream and Other Stories.* London: HarperCollins.
Roe, Emery M. 1992. "Applied Narrative Analysis: The Tangency of Literary Criticism, Social Science and Policy Analysis." *New Literary History* 23: 555–81.
Rolston, William E. 1908. "Water Vapour in the Martian Atmosphere." *Nature* 77, 442.
Rose, Mark. 1981. *Alien Encounters: Anatomy of Science Fiction.* Cambridge: Harvard University Press.
Rothenberg, Marc. 1981. "Organization and Control: Professionals and Amateurs in American Astronomy, 1899–1918." *Social Studies of Science* 11: 305–25.
Rothschild, Lynn. 1990. "Earth Analogs for Martian Life: Microbes in Evaporites, a New Model System for Life on Mars." *Icarus* 88: 246–60.
Rothschild, Lynn, and Rocco L. Mancinelli. 2001. "Life in Extreme Environments." *Nature* 409: 1092–101.
Rotman, Brian. 1993. *Ad Infinitum: The Ghost in Turing's Machine. Taking God Out of Mathematics and Putting the Body Back In.* Stanford: Stanford University Press.
Rottensteiner, Franz. 1971. "Kurd Lasswitz: A German Pioneer of Science Fiction." In

The Other Side of Realism, ed. Thomas Clareson. Bowling Green, Ohio: Popular Press, 289–306.

Rouse, Joseph. 1991. "Philosophy of Science and the Persistent Narratives of Modernity." *Studies in the History and Philosophy of Science* 22: 141–69.

Rovin, Jeff. 1978. *Mars!* Los Angeles: Corwin Books.

Roy, John Flint. 1976. *A Guide to Barsoom: The Mars of Edgar Rice Burroughs.* New York: Ballantine.

Russell, Henry Norris. 1934. "Fading Belief in Life on Other Planets." *Scientific American* 150: 296–97.

Russell, W. M. S. 1990. "H. G. Wells and Ecology." In *H. G. Wells under Revision*, ed. Patrick Parrinder and Christopher Rolfe. London: Associated University Presses, 145–52.

Sagan, Carl. 1971. "The Long Winter Model of Martian Biology: A Speculation." *Icarus* 15: 511–14.

——. 1973. "Planetary Engineering on Mars." *Icarus* 20: 513–14.

——. 1994. *Pale Blue Dot: A Vision of the Human Future in Space.* New York: Random House.

Sagan, Carl, and James B. Pollock. 1966. "On the Nature of the Canals of Mars." *Nature* 212: 117–21.

Sagan, Carl, Owen B. Toon, and P. J. Gierasch. 1973. "Climactic Change on Mars." *Science* 181: 1045–49.

Salisbury, F. B. 1962. "Martian Biology." *Science* 136: 17–26.

Sanborn, Geoffrey. 1998. *The Sign of the Cannibal: Melville and the Making of a Postcolonial Reader.* Durham: Duke University Press.

Sawyer, Kathy. 2001. "A Mars Never Dreamed Of." *National Geographic* 199, no. 2: 30–51.

Saxton, Alexander. 1991. *The Rise and Fall of the White Republic: Class Politics and Mass Culture in Nineteenth-Century America.* London: Verso.

Schiaparelli, Giovanni Virginio. 1894. "Schiaparelli on Mars." Trans. W. H. Pickering. *Nature* 51: 87–90.

——. 1895. "La vita sol pianeta marte." *Natura ad arte* 4: 83–95.

——. 1996. *Astronomical and Physical Observations of the Axis of Rotation and the Topography of the Planet Mars. First Memoir, 1877–1878.* Trans. William Sheehan. Association of Lunar and Planetary Observers, monograph no. 5. October.

Schindler, Kevin. 1998. *100 Years of Good Seeing: The History of the 24-Inch Clark Telescope.* Rev. ed. Flagstaff, Ariz.: Lowell Observatory.

Schleifer, Ronald. 2000. *Modernism and Time: The Logic of Abundance in Literature, Science, and Culture, 1880–1930.* Cambridge: Cambridge University Press.

Schmerz, Joanna. 1993. "On Reading the Politics of *Total Recall.*" *Postscript* 12, no. 3 (summer).

Schmidt, Stanley, and Robert Zubrin, eds. 1996. *Islands in the Sky: Bold New Ideas for Colonizing Space.* New York: Wiley.

Schnell, Lisa. 1992. "Parenthetical Disturbances: Aphra Behn and the Rhetoric of Relativity." *Recherces Sémiotiques/Semiotic Inquiry* 12: 95–113.
Schofield, J. T., J. R. Barnes, D. Crisp, R. M. Haberle, S. Larsen, A. J. Magalhães, J. R. Murphy, A. Sieff, and G. Wilson. 1997. "The Mars Pathfinder Atmospheric Structure Investigation/Meteorology (ASI/MET) Experiment." *Science* 28: 1752–58.
Schorn, Ronald A. 1971. "The Spectroscopic Search for Water on Mars: A History." In *Planetary Atmospheres*, ed. Carl Sagan, Tobias C. Owen, and Harlan Smith. New York: Springer-Verlag, 223–36.
Sears, Paul B. 1935. *Deserts on the March.* Norman: University of Oklahoma Press.
Segura, Teresa L., Owen B. Toon, Anthony Colaprete, and Kevin Zahnle. 2002. "Environmental Effects of Large Impacts on Mars." *Science* 298: 1977–80.
Serres, Michel. 1982. *The Parasite.* Trans. Lawrence Scher. Baltimore: Johns Hopkins University Press.
Service, Robert. 2000. *Lenin: A Biography.* Cambridge: Harvard University Press.
Serviss, Garrett P. 1896. "If We Could Move to Mars." *Harpers,* February.
——. 1904. "A World's Battle for Life." *New York Journal.* May.
——. 1905. "Photographs of Mars." Unidentified newspaper in Lowell Observatory Archives.
——. 1908. *Astronomy with the Naked Eye.* New York.
——. 1947. *Edison's Invasion of Mars.* New York. (Orig. pub. 1898.)
Shapin, Steven. 1994. *A Social History of Truth: Civility and Science in Seventeenth-Century England.* Chicago: University of Chicago Press.
——. 1999. "Rarely Pure and Never Simple." *Configurations* 8: 1–17.
Shapiro, Robert. 1999. *Planetary Dreams: The Quest to Discover Life Beyond Earth.* New York: Wiley.
Shaviro, Steven. 1996. *Doom Patrols.* London: Serpent's Tail.
Sheehan, William. 1988. *Planets and Perceptions: Telescopic Views and Interpretations, 1609–1909.* Tucson: University of Arizona Press.
——. 1995. *The Immortal Fire Within: The Life and Work of Edward Emerson Barnard.* Cambridge: Cambridge University Press.
——. 1996. *The Planet Mars.* Tucson: University of Arizona Press.
Sheehan, William, and James O'Meara. 2001. *Mars: The Lure of the Red Planet.* Amherst, N.Y.: Prometheus Books.
Shiner, Lewis. 1984. *Frontera.* New York: Baen Books.
Shirley, Donna, with Danelle Morton. 1999. *Managing Martians.* New York: Broadway Books.
——, moderator. 2004. "Terraforming Mars: Experts Debate How, Why, and Whether." Science Tuesday. http://www.space.com/scienceastronomy/terraform_debate_040727-7.html.
Shklovskii, I. S., and Carl Sagan. 1966. *Intelligent Life in the Universe.* Trans., in part, by Paula Fern. San Francisco: Holden-Day.

Sinton, W. M. 1957. "Spectroscopic Evidence of Vegetation on Mars." *Astrophysical Journal* 126: 231–37.

——. 1959. "Further Evidence of Vegetation on Mars." *Science* 130: 1234–37.

——. 1961. "Identification of Aldehydes in Vegetation Regions of Mars." *Science* 134: 529.

Slipher, Earl C. 1918. Working Papers, Mars Drawings, Box 2, Folder 1A. Lowell Observatory Archives.

Slipher, Earl C., et al. 1952. "The Study of Planetary Atmospheres." United States Air Force Contract 19(122)-162. Flagstaff, Ariz.: Lowell Observatory.

——. 1962. *The Photographic Story of Mars.* Flagstaff, Ariz.: Northland Press.

Slusser, George. 1988. "History, Historicity, Story." *Science-Fiction Studies* 15: 187–213.

Slusser, George, and Eric. S. Rabkin, eds. 1985. *Shadows of the Magic Lamp: Fantasy and Science Fiction in Film.* Carbondale: Southern Illinois University Press.

Smith, Barbara Herrnstein. 1997. "Microdynamics of Incommensurability: Philosophy of Science Meets Science Studies." In *Mathematics, Science, and Postclassical Theory,* ed. Barbara Herrnstein Smith and Arkady Plotnitsky. Durham: Duke University Press, 243–66.

Smith, Robert W. 1997. "Engines of Discovery: Scientific Instruments and the History of Astronomy and Planetary Science in the United States in the Twentieth Century." *Journal of the History of Astronomy* 28: 49–77.

Sobchack, Vivian. 1987. *Screening Space: The American Science Fiction Film.* New York: Ungar.

——. 1996. "The Fantastic." In *The Oxford History of World Cinema,* ed. Geoffrey Nowell-Smith. New York: Oxford University Press, 312–21.

Soderblom, L. A., et al. "Soils of Eagle Crater and Meridiani Planum at the Opportunity Rover Landing Site." *Science* 306, 1723–26.

Spencer-Jones, H. 1940. *Life on Other Worlds.* New York: Macmillan.

Spinrad, Norman. 1970. *Science Fiction in the Real World.* Carbondale: Southern Illinois University Press.

Squyres, Steven, and James F. Kasting. 1994. "Early Mars: How Warm and How Wet?" *Science* 265: 744–49.

Squyres, S. W., et al. 2004a. "The Opportunity Rover's Athena Science Investigation at Meridiani Planum, Mars." *Science* 306, 1698–1703.

Squyres, S. W., et al. 2004b. "In Situ Evidence for an Ancient Aqueous Environment at Meridiani Planum, Mars." *Science* 306, 1709–14.

Stabenow, Dana. 1995. *Red Planet Run.* New York: Ace Books.

"A Steam-Heated Planet." 1908. *New York Times.* March 1, part 2, p. 8.

Stecopoulos, Harry. 1997. "The World According to Normal Bean: Edgar Rice Burroughs's Popular Culture." In *Race and the Subject of Masculinities,* ed. Harry Stecopoulos and Michael Uebel. Durham: Duke University Press, 170–91.

Steele, Allen. 1992. *Labyrinth of Night.* New York: Ace Books.

Stepan, Nancy Leys. 2001. *Picturing Tropical Nature.* Ithaca: Cornell University Press.

Stevenson, Donald. 2001. "Mars' Core and Magnetism." *Nature* 412: 214–19.

Stites, Richard. 1984. "Fantasy and Revolution: Alexander Bogdanov and the Origins of Bolshevik Science Fiction." *Red Star,* by Alexander Bogdanov. Ed. Loren Graham and Richard Stites. Bloomington: Indiana University Press, 1–16.

——. 1989. *Revolutionary Dreams: Utopian Vision and Experimental Life in the Russian Revolution.* New York: Oxford University Press.

Strauss, David. 1993. " 'Fireflies Flashing in Unison': Percival Lowell, Edward Morse and the Birth of Planetology." *Journal for the History of Astronomy* 24: 157–69.

——. 1994. "Percival Lowell, W. H. Pickering and the Founding of the Lowell Observatory." *Annals of Science* 51: 37–58.

——. 1998. "Reflections on *Lowell and Mars.*" *Annals of Science* 55: 95–103.

——. 2001. *Percival Lowell: The Culture and Science of a Boston Brahmin.* Cambridge: Harvard University Press.

Stoker, Carol, and Carter Emmart, eds. 1996. *Strategies for Mars: A Guide to Human Exploration.* San Diego: Univelt.

Strughold, Hubertus. 1953. *The Green and Red Planet: A Physiological Study of the Possibility of Life on Mars.* Albuquerque: University of New Mexico Press.

Struve, Otto, and Velta Zebergs. 1962. *Astronomy of the Twentieth Century.* New York: Macmillan.

Suvin, Darko. 1979. *Metamorphoses of Science Fiction: On the Poetics and History of a Literary Genre.* New Haven: Yale University Press.

——. 1983. *Victorian Science Fiction in the UK: The Discourses of Knowledge and Power.* Boston: G.K. Hall.

——. 1988. *Position and Presuppositions in Science Fiction.* Kent, Ohio: Kent State University Press.

Swanson, David, Philip Bohnert, and Jackson Smith. 1970. *The Paranoid.* Boston: Little, Brown.

Taliaferro, John. 1999. *Tarzan Forever: The Life of Edgar Rice Burroughs, Creator of Tarzan.* New York: Scribner.

Tatarewicz, Joseph N. 1990. *Space Technology and Planetary Astronomy.* Bloomington: Indiana University Press.

Taylor, A. P., and J. C. Barry. 2004. "Magnetosomal Matrix: Ultrafine Structure may Template Biomineralization of Magnetosomes." *Journal of Microscopy* 213 (2 February): 180–97.

Taylor, Michael Ray. 1999. *Dark Life: Martian Nanobacteria, Rock-Eating Cave Bugs, and Other Extreme Organisms of Inner Earth and Outer Space.* New York: Scribner.

Telotte, J. P. 2001. *Science Fiction Film.* New York: Cambridge University Press.

Terzian, Yervant, and Elizabeth Bilson, eds. 1997. *Carl Sagan's Universe.* Cambridge: Cambridge University Press.

"Theory of Martian Life Corroborated." 1907. *New York Times.* November 5.

Thomas-Keprta, Kathie L., Simon J. Clemett, Dennis A. Bazylinski, Joseph L. Kirschvink, David S. McKay, Susan J. Wentworth, Hojatollah Vali, Everett K. Gibson Jr., Mary Fae McKay, and Christopher S. Romanek. 2001. "Truncated Hexa-

octahedral Magnetite Crystals in ALH84001: Presumptive Biosignatures." *Proceedings of the National Academy of Sciences* 98: 2164–69.
Thomas-Keprta, Kathie L., Simon J. Clemett, Dennis A. Bazylinski, Joseph L. Kirschvink, David S. McKay, Susan J. Wentworth, Hojatollah Vali, Everett K. Gibson Jr., and Christopher S. Romanek. 2002. "Magnetofossils from Ancient Mars: A Robust Biosignature in the Martian Meteorite ALH84001." *Applied and Environmental Microbiology* 68: 3663–72.
Tichi, Cecelia. 1979. *New World, New Earth: Environmental Reform in American Literature from the Puritans through Whitman.* New Haven: Yale University Press.
Tolstoy, Alexei N. 1981. *Aelita.* Trans. Antonia W. Bouis. New York: Macmillan.
Toon, Owen B. 1997. "Environments of Earth and Other Worlds." In *Carl Sagan's Universe,* ed. Yervant Terzian and Elizabeth Bilson. Cambridge: Cambridge University Press, 51–63.
"Tragic Struggle of the Dying Martians." 1907. *London Daily News.* July 6.
Treiman, Allan. 1999. "Microbes in a Martian Meteorite? An Update on the Controversy." *Sky and Telescope* 94, no. 4 (April): 52–58.
——. 2003. "Submicron Magnetite Grains and Carbon Compounds in Martian Meteorite ALH84001: Inorganic Abiotic Formation by Shock and Thermal Metamorphism." *Astrobiology* 3: 369–92.
Tsapin, A. I., M. G. Goldfeld, G. D. McDonald, and K. H. Nealson. 2000. "Iron (VI): Hypothetical Candidate for the Martian Oxidant." *Icarus* 147: 68–78.
Tsapin, A., M. Goldfeld, and K. Nealson. 2002. "Viking Experiments and Hypothesis that Fe(VI) is a Possible Candidate as a Martian Oxidant." *Icarus* 159: 268.
Turner, Frederick. 1988. *Genesis* . Dallas: Saybrook.
——. 1989. "Life on Mars: Cultivating a Planet—and Ourselves." *Harper's,* August, pp. 33–34.
Turner, Martin J. L. 2004. *Expedition Mars.* London: Springer/Praxis.
Valley, John W., John M. Eiler, Colin M. Graham, Everett K. Gibson, Christopher S. Romanek, and Edward M. Stolper. 1997. "Low Temperature Carbonate Concretions in the Martian Meteorite ALH84001: Evidence from Stable Isotopes and Mineralogy." *Science* 275: 1633–37.
Vaucouleurs, Gerard de. 1954. *Physics of the Planet Mars: An Introduction to Areophysics.* London: Faber and Faber.
von Braun, Werner. 1992. *The Mars Project.* Champaign-Urbana: University of Illinois Press. (Orig. pub. 1953.)
von Braun, Werner, and Frederick I. Ordway III. 1979. *New Worlds: Discoveries from Our Solar System.* New York: Anchor Doubleday.
Wachhorst, Wyn. 2000. *The Dream of Spaceflight: Essays on the Near Edge of Infinity.* New York: Basic Books.
Wagar, W. Warren. 1983. "The Rebellion of Nature." In *The Ends of the World,* ed. Eric S. Rabkin, Martin H. Greenberg, and Joseph Olander. Carbondale: Southern Illinois University Press, 139–72.
Walker, Gabrielle. 1999. "Mars on Earth." *New Scientist* 162, no. 2182 (17 April): 48–53.

Wallace, Alfred Russel. 1907. *Is Mars Habitable? A Critical Examination of Professor Percival Lowell's Book "Mars and Its Canals," with an Alternative Explanation.* London: Macmillan.

Walsh, John. 1964. "Space: National Academy Proposal Recommends Exploration of Mars as Major Goal in 1971–85 Period." *Science* 146: 1025–27.

Walter, Malcolm. 1999. *The Search for Life on Mars.* Cambridge, Mass.: Perseus.

Ward, Lester. 1968. *Dynamic Sociology, or, Applied Social Science.* 2 vols. New York: Greenwood Press. (Orig. pub. 1883.)

Watson, Ian. 1977. *The Martian Inca.* New York: Ace Books.

Weinberg, Steven. 1996. "Sokal's Hoax." *New York Review of Books* 43, no. 13 (August 8): 11–15.

Wells. H. G. 1908. "The Things That Live on Mars." *Cosmopolitan Magazine* 54: 335–42.

——. 1920. *Russia in the Shadows.* London: Doran.

——. 1934. *Experiment in Autobiography: Discoveries and Conclusions of a Very Ordinary Brain (since 1866).* New York: Macmillan.

——. 1975. "Intelligence on Mars." In *H. G. Wells: Early Writings in Science and Science Fiction,* ed. Robert Philmus and David Y. Hughes. Berkeley: University of California Press. (Orig. pub. 1896.)

——. 1993. *A Critical Edition of The War of the Worlds: H. G. Wells's Scientific Romance.* Ed. David Y. Hughes and Harry M. Geduld. Bloomington: Indiana University Press.

Wharton, David. 2002. *Life at the Limits: Organisms in Extreme Environments.* Cambridge: Cambridge University Press.

Whewell, William. 1833. *Astronomy and General Physics Considered with Reference to Natural Theology.* London.

White, Eric. 1990. "Contemporary Cosmology and Narrative Theory." In *Literature and Science: Theory and Practice,* ed. Stuart Peterfreund. Boston: Northeastern University Press, 91–112.

——. 1995. " 'Once They Were Men, Now They're Land Crabs': Monstrous Becomings in Evolutionist Cinema." In *Posthuman Bodies,* ed. Judith Halberstam and Ira Livingston. Bloomington: Indiana University Press, 244–65.

Wicks, Mark. 1911. *To Mars via the Moon.* London: Seelye.

Wilford, John Noble. 1990. *Mars Beckons: The Mysteries, the Challenges, the Expectations of our Next Great Adventure in Space.* New York: Knopf.

Wilhelms, Don E. 1993. *To a Rocky Moon: A Geologist's History of Lunar Exploration.* Tucson: University of Arizona Press.

Wilkins, John. 1640. *Discovery of a New World in the Moone.* 2nd ed. London.

Williams, Michael Lindsay. 1986. *Martian Spring.* New York: Avon Books.

Williams, Paul. 1986. *Only Apparently Real: The World of Philip K. Dick.* New York: Arbor House.

Williamson, Jack. 1992. *Beachhead.* New York: TOR.

Wilson, Christopher. 1992. *White Collar Fictions: Class and Social Representation in U.S. Fiction, 1880–1925.* Athens: University of Georgia Press.

Winterhalder, Bruce. 1994. "Concepts in Historical Ecology: The View from Evolutionary Biology." In *Historical Ecology: Cultural Knowledge and Changing Landscapes,* ed. Carole Crumley. Santa Fe: School of American Research Press, 17–41.

Woese, C. R. 1987. "Bacterial Evolution." *Microbiology Review* 51: 221–71.

——. 2002. "On the Evolution of Cells." *Proceedings of the National Academy of Sciences* 99: 8742–47.

Wollheim, Donald A. 1963. *The Secret of the Martian Moons.* New York: Tempo Books. (Orig. pub. 1955.)

Wood, Denis. 1992. *The Power of Maps.* New York: Guilford.

Woolgar, Steve. 1988. *Science: The Very Idea.* London: Tavistock.

Worster, Donald. 1993. *The Wealth of Nature: Environmental History and the Ecological Imagination.* New York: Oxford University Press.

——. 1994. *Nature's Economy: A History of Ecological Ideas.* 2nd. ed. Cambridge: Cambridge University Press.

Worthington, James H. 1910. "Markings of Mars." *Nature* 85, 40.

Wright, Peter. 1996. "Selling Mars: Burroughs, Barsoom and Expedient Xenography." *Foundation: The Review of Science Fiction* 68 (autumn): 24–26.

Wyndham, John. 1952. "Dumb Martian." *Galaxy* 4, no. 4: 49–74.

Yen, A. S., S. S. Kim, M. H. Hecht, M. S. Frant, and B. Murray. 2000. "Evidence That the Reactivity of the Martian Soil Is Due to Superoxide Ions." *Science* 289: 1909–12.

Young, Robert M. 1985. *Darwin's Metaphor: Nature's Place in Victorian Culture.* Cambridge: Cambridge University Press.

Youngblood, Denise J. 1985. *Soviet Cinema in the Silent Era, 1918–1935.* Ann Arbor: UMI Research Press.

Yue, Gang. 1999. *The Mouth that Begs: Hunger, Cannibalism, and the Politics of Eating in Modern China.* Durham: Duke University Press.

Zahnle, Kevin. 2001. "Decline and Fall of the Martian Empire." *Nature* 412: 209–13.

Zent, A. P., and C. P. McKay. 1994. "The Chemical Reactivity of the Martian Soil and Implications for Future Missions." *Icarus* 108: 146–57.

Žižek, Slavoj. 1989. *The Sublime Object of Ideology.* London: Verso.

Zuber, Maria T. 2001. "The Crust and Mantle of Mars." *Nature* 412: 220–27.

——. 2003. "Mars: The Inside Story." *Sky and Telescope* 106, no. 6: 42–48.

Zubrin, Robert. 1996a. "The Significance of the Martian Frontier." In *Strategies for Mars: A Guide to Human Exploration*, ed. Carol Stoker and Carter Emmart. San Diego: Univelt, for the American Astronautical Society, 13–24.

——. 1996b. *The Case for Mars: The Plan to Settle the Red Planet and Why We Must.* New York: Free Press.

——, ed. 1997. *From Imagination to Reality: Mars Exploration Studies of the Journal of the British Interplanetary Society. Part II: Base Building, Colonization and Terraformation.* San Diego: Univelt, for the American Astronautical Society and the British Planetary Society.

——. 1999. *Entering Space: Creating a Spacefaring Civilization.* New York: Tarcher/ Putnam.
——. 2001. *First Landing.* New York: Ace Books.
——. 2004. "Methane Detection Points to Life on Mars." http://masrsociety.org.
Zubrin, Robert, and Maggie Zubrin, eds. 1999. *Proceedings of the Founding Convention of the Mars Society: August 13–16, 1998.* 3 vols. San Diego: Univelt.

INDEX

Robert Markley is a professor of English
at the University of Illinois.

Library of Congress Cataloging-in-Publication Data

Markley, Robert

Dying planet : Mars in science and the imagination / Robert Markley.

p. cm.

Includes bibliographical references and index.

ISBN 0-8223-3600-6 (cloth : alk. paper)

ISBN 0-8223-3638-3 (pbk. : alk. paper)

1. Science fiction, American—History and criticism.

2. Mars (Planet)—In literature.

3. Mars (Planet)

I. Title.

PS374.S35M37 2005 813.'087620936—dc22

2005006509

•●⬤●•